The First Million Digits of Pi

edited by

David E. McAdams

For more information, see http://www.piday.org.
Editor's website is http://www.demcadams.com.

This book is for educational and entertainment purposes only. The publisher and author are not offering it for mathematical advice.

Other Books by David E. McAdams

Parrot Colors – An introduction to the concept of colors. For preschoolers.
Flower Colors – An introduction to the concept of colors. For preschoolers.
Shapes – An introduction to shapes. For preschoolers.
Numbers – An introduction to the concept of numbers. For grades K-2.
What is Bigger Than Anything? (Infinity) – An introduction to the concept of infinity. For grades 1-3.
Swing sets (Sets) – An introduction to set theory. For grades 2-4.
One Penny, Two – If Sig's penny doubles each day, how long until he can buy a dark green sports car? For grades 3-6.
Learning With Money Activity Kit – Teach large numbers and counting with over $1,000,000 in play money.
My Favorite Fractals (volumes 1, 2) – Picture books of wondrous fractals presented as high resolution images. For all ages.
All Math Words Dictionary – A math dictionary for students of pre-algebra, algebra, geometry, and pre-calculus.
The First Million Digits of Pi – The first million digits of pi. For all ages.
e to One Million Digits – The first million digits of the Euler's constant e. For all ages.
The Square Root of 2 to One Million Digits – The first million digits of the square root of 2. For all ages.
The First Hundred Thousand Prime Numbers – The first hundred thousand prime numbers. For all ages.
Orders of Ten – A book that illustrates orders of ten with dots (1, 10, 100, … dots). For ages 10-15.
Geometric Nets Project Book – 80 geometric nets to copy, cut out, and tape together into 3 dimensional polyhedra. For ages 9 and up.
Geometric Nets Mega Project Book – 253 geometric nets to copy, cut out, and tape together into 3 dimensional polyhedra. For ages 9 and up

For an up to date list, see http://www.DEMcAdams.com.

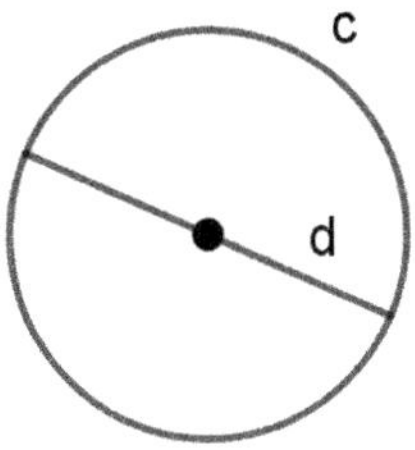

$$\Pi = \frac{c}{d}$$

$$\Pi = \frac{circumference}{diameter}$$

Π ≈ 3.141592653589793238462643383279502884197169399375105820974944
592307816406286208998628034825342117067982148086513282306647093844
609550582231725359408128481117450284102701938521105559644622948954
930381964428810975665933446128475648233786783165271201909145648566
923460348610454326648213393607260249141273724587006606315588174881
520920962829254091715364367892590360011330530548820466521384146951
941511609433057270365759591953092186117381932611793105118548074462
379962749567351885752724891227938183011949129833673362440656643086
021394946395224737190702179860943702770539217176293176752384674818
467669405132000568127145263560827785771342757789609173637178721468
440901224953430146549585371050792279689258923542019956112129021960
864034418159813629774771309960518707211349999998372978049951059731
732816096318595024459455346908302642522308253344685035261931188171
010003137838752886587533208381420617177669147303598253490428755468
731159562863882353787593751957781857780532171226806613001927876611
195909216420198938095257201065485863278865936153381827968230301952
035301852968995773622599413891249721775283479131515574857242454150
695950829533116861727855889075098381754637464939319255060400927701
671139009848824012858361603563707660104710181942955596198946767837
449448255379774726847104047534646208046684259069491293313677028989
152104752162056966024058038150193511253382430035587640247496473263
914199272604269922796782354781636009341721641219924586315030286182
974555706749838505494588586926995690927210797509302955321165344987
202755960236480665499119881834797753566369807426542527862551818417
574672890977772793800081647060016145249192173217214772350141441973
568548161361157352552133475741849468438523323907394143334547762416
862518983569485562099219222184272550254256887671790494601653466804
988627232791786085784383827967976681454100953883786360950680064225
125205117392984896084128488626945604241965285022210661186306744278
622039194945047123713786960956364371917287467764657573962413890865
832645995813390478027590099465764078951269468398352595709825822620
522489407726719478268482601476990902640136394437455305068203496252
451749399651431429809190659250937221696461515709858387410597885959
772975498930161753928468138268683868942774155991855925245953959431
049972524680845987273644695848653836736222626099124608051243884390
451244136549762780797715691435997700129616089441694868555848406353
422072225828488648158456028506016842739452267467678895252138522549
954666727823986456596116354886230577456498035593634568174324112515
076069479451096596094025228879710893145669136867228748940560101503
308617928680920874760917824938589009714909675985261365549781893129
784821682998948722658804857564014270477555132379641451523746234364
542858444795265867821051141354735739523113427166102135969536231442

952484937187110145765403590279934403742007310578539062198387447808
478489683321445713868751943506430218453191048481005370614680674919
278191197939952061419663428754440643745123718192179998391015919561
814675142691239748940907186494231961567945208095146550225231603881
930142093762137855956638937787083039069792077346722182562599661501
421503068038447734549202605414665925201497442850732518666002132434
088190710486331734649651453905796268561005508106658796998163574736
384052571459102897064140110971206280439039759515677157700420337869
936007230558763176359421873125147120532928191826186125867321579198
414848829164470609575270695722091756711672291098169091528017350671
274858322287183520935396572512108357915136988209144421006751033467
110314126711136990865851639831501970165151168517143765761835155650
884909989859982387345528331635507647918535893226185489632132933089
857064204675259070915481416549859461637180270981994309924488957571
282890592323326097299712084433573265489382391193259746366730583604
142813883032038249037589852437441702913276561809377344403070746921
120191302033038019762110110044929321516084244485963766983895228684
783123552658213144957685726243344189303968642624341077322697802807
318915441101044682325271620105265227211166039666557309254711055785
376346682065310989652691862056476931257058635662018558100729360659
876486117910453348850346113657686753249441668039626579787718556084
552965412665408530614344431858676975145661406800700237877659134401
712749470420562230538994561314071127000407854733269939081454664645
880797270826683063432858785698305235808933065757406795457163775254
202114955761581400250126228594130216471550979259230990796547376125
517656751357517829666454779174501129961489030463994713296210734043
751895735961458901938971311179042978285647503203198691514028708085
990480109412147221317947647772622414254854540332157185306142288137
585043063321751829798662237172159160771669254748738986654949450114
654062843366393790039769265672146385306736096571209180763832716641
627488880078692560290228472104031721186082041900042296617119637792
133757511495950156604963186294726547364252308177036751590673502350
728354056704038674351362222477158915049530984448933309634087807693
259939780541934144737744184263129860809988868741326047215695162396
586457302163159819319516735381297416772947867242292465436680098067
692823828068996400482435403701416314965897940924323789690706977942
236250822168895738379862300159377647165122893578601588161755782973
523344604281512627203734314653197777416031990665541876397929334419
521541341899485444734567383162499341913181480927777103863877343177
207545654532207770921201905166096280490926360197598828161332316663
652861932668633606273567630354477628035045077723554710585954870279
081435624014517180624643626794561275318134078330336254232783944975
382437205835311477119926063813346776879695970309833913077109870408
591337464144282277263465947047458784778720192771528073176790770715
721344473060570073349243693113835049316312840425121925651798069411
352801314701304781643788518529092854520116583934196562134914341595
625865865570552690496520985803385072242648293972858478316305777756
068887644624824685792603953527734803048029005876075825104747091643
961362676044925627420420832085661190625454337213153595845068772460
290161876679524061634252257719542916299193064553779914037340432875
262888963995879475729174642635745525407909145135711136941091193932
519107602082520261879853188770584297259167781314969900901921169717
372784768472686084900337702424291651300500516832336435038951702989
392233451722013812806965011784408745196012122859937162313017114448
464090389064495444006198690754851602632750529834918740786680881833
851022833450850486082503930213321971551843063545500766828294930413
776552793975175461395398468339363830474611996653858153842056853386
218672523340283087112328278921250771262946322956398989893582116745

6270102183564622013496715188190973038119800497340723961036854066431
9395097901906996395524530054505806855019567302292191393391856803449
0398205955100226353536192041994745538593810234395544959778377902374
2161727111723643435439478221818528624085140066604433258885698670543
1547069657474585503323233421073015459405165537906866273337995851156
2578432298827372319898757141595781119635833005940873068121602876496
2867446047746491599505497374256269010490377819868359381465741268049
2564879855614537234786733039046883834363465537949864192705638729317
4872332083760112302991136793862708943879936201629515413371424892830
7220126901475466847653576164773794675200490757155527819653621323926
4061601363581559074220202031872776052772190055614842555187925303435
1398442532234157623361064250639049750086562710953591946589751413103
4822769306247435363256916078154781811528436679570611086153315044521
2747392454494542368288606134084148637767009612071512491404302725386
0764823634143346235189757664521641376796903149501910857598442391986
2916421939949072362346468441173940326591840443780513338945257423995
0829659122850855582157250310712570126683024029295252201187267675622
0415420516184163484756516999811614101002996078386909291603028840026
9104140792886215078424516709087000699282120660418371806535567252532
5675328612910424877618258297651579598470356222629348600341587229805
3498965022629174878820273420922224533985626476691490556284250391275
7710284027998066365825488926488025456610172967026640765590429099456
8150652653053718294127033693137851786090407086671149655834343476933
8578171138645587367812301458768712660348913909562009939361031029161
6152881384379099042317473363948045759314931405297634757481193567091
1013775172100803155902485309066920376719220332290943346768514221447
7379393751703443661991040337511173547191855046449026365512816228824
4625759163330391072253837421821408835086573917715096828874782656995
9957449066175834413752239709683408005355984917541738188399944697486
7626551658276584835884531427756879002909517028352971634456212964043
5231176006651012412006597558512761785838292041974844236080071930457
6189323492292796501987518721272675079812554709589045563579212210333
4669749923563025494780249011419521238281530911407907386025152274299
5818072471625916685451333123948049470791191532673430282441860414263
6395480004480026704962482017928964766975831832713142517029692348896
2766844032326092752496035799646925650493681836090032380929345958897
0695365349406034021665443755890045632882250545255640564482465151875
4711962184439658253375438856909411303150952617937800297412076651479
3942590298969594699556576121865619673378623625612521632086286922210
3274889218654364802296780705765615144632046927906821207388377814233
5628236089632080682224680122482611771858963814091839036736722208883
2151375560037279839400415297002878307667094447456013455641725437090
6979396122571429894671543578468788614445812314593571984922528471605
0492212424701412147805734551050080190869960330276347870810817545011
9307141223390866393833952942578690507643100638351983438934159613185
4347546495569781038293097164651438407007073604112373599843452251610
5070270562352660127648483084076118301305279320542746286540360367453
2865105706587488225698157936789766974220575059683440869735020141020
6723585020072452256326513410559240190274216248439140359989535394590
9440704691209140938700126456001623742880210927645793106579229552498
8727584610126483699989225695968815920560010165525637567856672279661
9885782794848855834397518744545512965634434803966420557982936804352
2027709842942325330225763418070394769941597915945300697521482933665
5566156787364005366656416547321704390352132954352916941459904160875
3201868379370234888689479151071637852902345292440773659495630510074
2108714261349745956151384987137570471017879573104229690666702144986
3746459528082436944578977233004876476524133907592043401963403911473
202338071

509522201068256342747164602433544005152126693249341967397704159568
375355516673027390074972973635496453328886984406119649616277344951
827369558822075735517665158985519098666539354948106887320685990754
079234240230092590070173196036225475647894064754834664776041146323
390565134330684495397907090302346046147096169688688501408347040546
074295869913829668246818571031887906528703665083243197440477185567
893482308943106828702722809736248093996270607472645539925399442808
113736943388729406307926159599546262462970706259484556903471197299
640908941805953439325123623550813494900436427852713831591256898929
519642728757394691427253436694153236100453730488198551706594121735
246258954873016760029886592578662856124966552353382942878542534048
308330701653722856355915253478445981831341129001999205981352205117
336585640782648494276441137639386692480311836445369858917544264739
988228462184490087776977631279572267265556259628254276531830013407
092233436577916012809317940171859859993384923549564005709955856113
498025249906698423301735035804408116855265311709957089942732870925
848789443646005041089226691783525870785951298344172953519537885534
573742608590290817651557803905946408735061232261120093731080485485
263572282576820341605048466277504500312620080079980492548534694146
977516493270950493463938243222718851597405470214828971117779237612
257887347718819682546298126868581705074027255026332904497627789442
362167411918626943965067151577958675648239939176042601763387045499
017614364120469218237076488783419689686118155815873606293860381017
121585527266830082383404656475880405138080163363887421637140643549
556186896411228214075330265510042410489678352858829024367090488711
819090949453314421828766181031007354770549815968077200947469613436
092861484941785017180779306810854690009445899527942439813921350558
642219648349151263901280383200109773868066287792397180146134324457
264009737425700735921003154150893679300816998053652027600727749674
584002836240534603726341655425902760183484030681138185510597970566
400750942608788573579603732451414678670368809880609716425849759513
806930944940151542222194329130217391253835591503100333032511174915
696917450271494331515588540392216409722910112903552181576282328318
234254832611191280092825256190205263016391147724733148573910777587
442538761174657867116941477642144111126358355387136101102326798775
641024682403226483464176636980663785768134920453022408197278564719
839630878154322116691224641591177673225326433568614618654522268126
887268445968442416107854016768142080885028005414361314623082102594
173756238994207571362751674573189189456283525704413354375857534269
869947254703165661399199968262824727064133622217892390317608542894
3733935618891651250424440400895271983787386480584726895462438823437
517885201439560057104811949884239060613695734231559079670346149143
447886360410318235073650277859089757827273130504889398900992391350
337325085598265586708924261242947367019390772713070686917092646254
842324074855036608013604668951184009366860954632500214585293095000
090715105823626729326453738210493872499669933942468551648326113414
611068026744663733437534076429402668297386522093570162638464852851
490362932019919968828517183953669134522244470804592396602817156551
565666111359823112250628905854914509715755390024393153519090210711
945730024388017661503527086260253788179751947806101371500448991721
002220133501310601639154158957803711779277522597874289191791552241
718958536168059474123419339842021874564925644346239253195313510331
147639491199507285843065836193536932969928983791494193940608572486
396883690326556436421664425760791471086998431573374964883529276932
822076294728238153740996154559879825989109371712621828302584811238
901196822142945766758071865380650648702613389282299497257453033283
896381843944770779402284359883410035838542389735424395647555684095
224844554139239410001620769363684677641301781965937997155746854194

633489374843912974239143365936041003523437770658886778113949861647
874714079326385873862473288964564359877466763847946650407411182565
837887845485814896296127399841344272608606187245545236064315371011
274680977870446409475828034876975894832824123929296058294861919667
091895808983320121031843034012849511620353428014412761728583024355
983003204202451207287253558119584014918096925339507577840006746552
603144616705082768277222353419110263416315714740612385042584598841
990761128725805911393568960143166828317632356732541707342081733223
046298799280490851409479036887868789493054695570307261900950207643
349335910602454508645362893545686295853131533718386826561786227363
716975774183023986006591481616404944965011732131389574706208847480
236537103115089842799275442685327797431139514357417221975979935968
525228574526379628961269157235798662057340837576687388426640599099
350500081337543245463596750484423528487470144354541957625847356421
619813407346854111766883118654489377697956651727966232671481033864
391375186594673002443450054499539974237232871249483470604406347160
632583064982979551010954183623503030945309733583446283947630477564
501500850757894954893139394489921612552559770143685894358587752637
962559708167764380012543650237141278346792610199558522471722017772
370041780841942394872540680155603599839054898572354674564239058585
021671903139526294455439131663134530893906204678438778505423939052
473136201294769187497519101147231528932677253391814660730008902776
896311481090220972452075916729700785058071718638105496797310016787
085069420709223290807038326345345203802786099055690013413718236837
099194951648960075504934126787643674638490206396401976668559233565
463913836318574569814719621084108096188460545603903845534372914144
651347494078488442377217515433426030669883176833100113310869042193
903108014378433415137092435301367763108491351615642269847507430329
716746964066653152703532546711266752246055119958183196376370761799
191920357958200759560530234626775794393630746305690108011494271410
093913691381072581378135789400559950018354251184172136055727522103
526803735726527922417373605751127887218190844900617801388971077082
293100279766593583875890939568814856026322439372656247277603789081
445883785501970284377936240782505270487581647032458129087839523245
323789602984166922548964971560698119218658492677040395648127810217
991321741630581055459880130048456299765112124153637451500563507012
781592671424134210330156616535602473380784302865525722275304999883
701534879300806260180962381516136690334111138653851091936739383522
934588832255088706450753947395204396807906708680644509698654880168
287434378612645381583428075306184548590379821799459968115441974253
634439960290251001588827216474500682070419376158454712318346007262
933955054823955713725684023226821301247679452264482091023564775272
308208106351889915269288910845557112660396503439789627825001611015
323516051965590421184494990778999200732947690586857787872098290135
295661397888486050978608595701773129815531495168146717695976099421
003618355913877781769845875810446628399880600616229848616935337386
578773598336161338413385368421197893890018529569196780455448285848
370117096721253533875862158231013310387766827211572694951817958975
469399264219791552338576623167627547570354699414892904130186386119
439196283887054367774322427680913236544948536676800000106526248547
305586159899914017076983854831887501429389089950685453076511680333
732226517566220752695179144225280816517166776672793035485154204023
817460892328391703275425750867655117859395002793389592057668278967
764453184040418554010435134838953120132637836928358082719378312654
961745997056745071833206503455664403449045362756001125018433560736
122276594927839370647842645676338818807565612168960504161139039063
960162022153684941092605387688714837989559999112099164646441191856
827700457424343402167227644558933012778158686952506949936461017568

506016714535431581480105458860564550133203758645485840324029871709
348091055621167154684847780394475697980426318099175642280987399876
697323769573701580806822904599212366168902596273043067931653114940
176473769387351409336183321614280214976339918983548487562529875242
387307755955595546519639440182184099841248982623673771467226061633
643296406335728107078875816404381485018841143188598827694490119321
296827158884133869434682859006664080631407775772570563072940049294
030242049841656547973670548558044586572022763784046682337985282710
578431975354179501134727362577408021347682604502285157979579764746
702284099956160156910890384582450267926594205550395879229818526480
070683765041836562094555434613513415257006597488191634135955671964
965403218727160264859304903978748958906612725079482827693895352175
362185079629778514618843271922322381015874445052866523802253284389
137527384589238442253547265309817157844783421582232702069028723233
005386216347988509469547200479523112015043293226628272763217790884
008786148022147537657810581970222630971749507212724847947816957296
142365859578209083073323356034846531873029302665964501371837542889
755797144992465403868179921389346924474198509733462679332107268687
076806263991936196504409954216762784091466985692571507431574079380
532392523947755744159184582156251819215523370960748332923492103451
462643744980559610330799414534778457469999212859999939961228161521
931488876938802228108300198601654941654261696858678837260958774567
618250727599295089318052187292461086763995891614585505839727420980
909781729323930106766386824040111304024700735085782872462713494636
853181546969046696869392547251941399291465242385776255004748529547
681479546700705034799958886769501612497228204030399546327883069597
624936151010243655535223069061294938859901573466102371223547891129
254769617600504797492806072126803922691102777226102544149221576504
508120677173571202718024296810620377657883716690910941807448781404
907551782038565390991047759414132154328440625030180275716965082096
427348414695726397884256008453121406593580904127113592004197598513
625479616063228873618136737324450607924411763997597461938358457491
598809766744709300654634242346063423747466608043170126005205592849
369594143408146852981505394717890045183575515412522359059068726487
863575254191128887737176637486027660634960353679470269232297186832
771739323619200777452212624751869833495151019864269887847171939664
976907082521742336566272592844062043021411371992278526998469884770
232382384005565551788908766136013047709843861168705231055314916251
728373272867600724817298763756981633541507460883866364069347043720
668865127568826614973078865701568501691864748854167915459650723428
773069985371390430026653078398776385032381821553559732353068604301
067576083890862704984188859513809103042359578249514398859011318583
584066747237029714978508414585308578133915627076035639076394731145
549583226694570249413983163433237897595568085683629725386791327505
554252449194358912840504522695381217913191451350099384631177401797
151228378546011603595540286440590249646693070776905548102885020808
580087811577381719174177601733073855475800605601433774329901272867
725304318251975791679296996504146070664571258883469797964293162296
552016879730003564630457930884032748077181155533090988702550520768
046303460865816539487695196004408482065967379473168086415645650530
049881616490578831154345485052660069823093157776500378070466126470
602145750579327096204782561524714591896522360839664562410519551052
235723973951288181640597859142791481654263289200428160913693777372
229998332708208296995573772737566761552711392258805520189887620114
168005468736558063347160373429170390798639652296131280178267971728
982293607028806908776866059325274637840539769184808204102194471971
386925608416245112398062011318454124478205011079876071715568315407
886543904121087303240201068534194723047666672174986986854707678120

512473679247919315085644477537985379973223445612278584329684664751
333657369238720146472367942787004250325558992688434959287612400755
875694641370562514001179713316620715371543600687647731867558714878
398908107429530941060596944315847753970094398839491443235366853920
994687964506653398573888786614762944341401049888993160051207678103
588611660202961193639682134960750111649832785635316145168457695687
109002999769841263266502347716728657378579085746646077228341540311
441529418804782543876177079043000156698677679576090996693607559496
515273634981189641304331166277471233881740603731743970540670310967
676574869535878967003192586625941051053358438465602339179674926784
476370847497833365557900738419147319886271352595462518160434225372
996286326749682405806029642114638643686422472488728343417044157348
248183330164056695966886676956349141632842641497453334999948000266
998758881593507357815195889900539512085351035726137364034367534714
104836017546488300407846416745216737190483109676711344349481926268
111073994825060739495073503169019731852119552635632584339099822498
624067031076831844660729124874754031617969941139738776589986855417
031884778867592902607004321266617919223520938227878880988633599116
081923535557046463491132085918979613279131975649097600013996234445
535014346426860464495862476909434704829329414041114654092398834443
515913320107739441118407410768498106634724104823935827401944935665
161088463125678529776973468430306146241803585293315973458303845541
033701091676776374276210213701354854450926307190114731848574923318
167207213727935567952844392548156091372812840633303937356242001604
566455741458816605216660873874804724339121295587776390696903707882
852775389405246075849623157436917113176134783882719416860662572103
685132156647800147675231039357860689611125996028183930954870905907
386135191459181951029732787557104972901148717189718004696169777001
791391961379141716270701895846921434369676292745910994006008498356
842520191559370370101104974733949387788598941743303178534870760322
198297057975119144051099423588303454635349234982688362404332726741
554030161950568065418093940998202060999414021689090070821330723089
662119775530665918814119157783627292746156185710372172471009521423
696483086410259288745799932237495519122195190342445230753513380685
680735446499512720317448719540397610730806026990625807602029273145
525207807991418429063884437349968145827337207266391767020118300464
819000241308350884658415214899127610651374153943565721139032857491
876909441370209051703148777346165287984823533829726013611098451484
182380812054099612527458088109948697221612852489742555551607637167
505489617301680961380381191436114399210638005083214098760459930932
485102516829446726066613815174571255975495358023998314698220361338
082849935670557552471290274539776214049318201465800802156653606776
550878380430413431059180460680083459113664083488740800574127258670
479225831912741573908091438313845642415094084913391809684025116399
193685322555733896695374902662092326131885589158083245557194845387
562878612885900410600607374650140262782402734696252821717494158233
174923968353013617865367376064216677813773995100658952887742766263
684183068019080460984980946976366733566228291513235278880615776827
815958866918023894033307644191240341202231636857786035727694154177
882643523813190502808701857504704631293335375728538660588890458311
145077394293520199432197117164223500564404297989208159430716701985
746927384865383343614579463417592257389858800169801475742054299580
124295810545651083104629728293758416116253256251657249807849209989
799062003593650993472158296517413579849104711166079158743698654122
234834188772292944633517865385673196255985202607294767407261676714
557364981210567771689348491766077170527718760119990814411305864557
791052568430481144026193840232247093924980293355073184589035539713
308844617410795916251171486487446861124760542867343670904667846867

027409188101424971114965781772427934707021668829561087779440504843
752844337510882826477197854000650970403302186255614733211777117441
335028160884035178145254196432030957601869464908868154528562134698
835544456024955666843660292219512483091060537720198021831010327041
783866544718126039719068846237085751808003532704718565949947612424
811099928867915896904956394762460842406593094862150769031498702067
353384834955083636601784877106080980426924713241000946401437360326
564518456679245666955100150229833079849607994988249706172367449361
226222961790814311414660941234159359309585407913908720832273354957
208075716517187659944985693795623875551617575438091780528029464200
447215396280746360211329425591600257073562812638733106005891065245
708024474937543184149401482119996276453106800663118382376163966318
093144467129861552759820145141027560068929750246304017351489194576
360789352855505317331416457050499644389093630843874484783961684051
845273288403234520247056851646571647713932377551729479512613239822
960239454857975458651745878771331813875295980941217422730035229650
808917770506825924882232215493804837145478164721397682096332050830
564792048208592047549985732038887639160199524091893894557676874973
085695595801065952650303626615975066222508406742889826590751063756
356996821151094966974458054728869363102036782325018232370845979011
154847208761821247781326633041207621658731297081123075815982124863
980721240786887811450165582513617890307086087019897588980745664395
515741536319319198107057533663373803827215279884935039748001589051
942087971130805123393322190346624991716915094854140187106035460379
464337900589095772118080446574396280618671786101715674096766208029
576657705129120990794430463289294730615951043090222143937184956063
405618934251305726829146578329334052463502892917547087256484260034
962961165413823007731332729830500160256724014185152041890701154288
579920812198449315699905918201181973350012618772803681248199587707
020753240636125931343859554254778196114293516356122349666152261473
539967405158499860355295332924575238881013620234762466905581643896
786309762736550472434864307121849437348530060638764456627218666170
123812771562137974614986132874411771455244470899714452288566294244
023018479120547849857452163469644897389206240194351831008828348024
924908540307786387516591130287395878709810077271827187452901397283
661484214287170553179654307650453432460053636147261818096997693348
626407743519992868632383508875668359509726557481543194019557685043
724800102041374983187225967738715495839971844490727914196584593008
394263702087563539821696205532480321226749891140267852859967340524
203109179789990571882194939132075343170798002373659098537552023891
164346718558290685371189795262623449248339249634244971465684659124
891855662958932990903523923333364743520370770101084388003290759834
217018554228386161721041760301164591878053936744747205998502358289
183369292233732399948043710841965947316265482574809948250999183300
697656936715968936449334886474421350084070066088359723503953234017
958255703601693699098867113210979889707051728075585519126993067309
925070407024556850778679069476612629808225163313639952117098452809
263037592242674257559989289278370474445218936320348941552104459726
188380030067761793138139916205806270165102445886924764924689192461
212531027573139084047000714356136231699237169484813255420091453041
037135453296620639210547982439212517254013231490274058589206321758
949434548906846399313757091034633271415316223280552297297953801880
162859073572955416278867649827418616421878988574107164906919185116
281528548679417363890665388576422915834250067361245384916067413734
017357277995634104332688356950781493137800736235418007061918026732
855119194267609122103598746924117283749312616339500123959924050845
437569850795704622266461900010350049018303415354584283376437811198
855631877779253720116671853954183598443830520376281944076159410682

071697030228515225057312609304689842343315273213136121658280807521
263154773060442377475350595228717440266638914881717308643611138906
942027908814311944879941715404210341219084709408025402393294294549
387864023051292711909751353600092197110541209668311151632870542302
847007312065803262641711616595761327235156666253667271899853419989
523688483099930275741991646384142707798870887422927705389122717248
632202889842512528721782603050099451082478357290569198855546788607
946280537122704246654319214528176074148240382783582971930101788834
567416781139895475044833931468963076339665722672704339321674542182
455706252479721997866854279897799233957905758189062252547358220523
642485078340711014498047872669199018643882293230538231855973286978
092225352959101734140733488476100556401824239219269506208318381454
698392366461363989101210217709597670490830508185470419466437131229
969235889538493013635657618610606222870559942337163102127845744646
398973818856674626087948201864748767272722206267646533809980196688
368099415907577685263986514625333631245053640261056960551318381317
426118442018908885319635698696279503673842431301133175330532980201
668881748134298868158557781034323175306478498321062971842518438553
442762012823457071698853051832617964117857960888815032960229070561
447622091509473903594664691623539680920139457817589108893199211226
007392814916948161527384273626429809823406320024402449589445612916
704950823581248739179964864113348032475777521970893277226234948601
504665268143987705161531702669692970492831628550421289814670619533
197026950721437823047687528028735412616639170824592517001071418085
480063692325946201900227808740985977192180515853214739265325155903
541020928466592529991435379182531454529059841581763705892790690989
691116438118780943537152133226144362531449012745477269573939348154
691631162492887357471882407150399500944673195431619385548520766573
882513963916357672315100555603726339486720820780865373494244011579
966750736071115935133195919712094896471755302453136477094209463569
698222667377520994516845064362382421185353488798939567318780660610
788544000550827657030558744854180577889171920788142335113866292966
717964346876007704799953788338787034871802184243734211227394025571
769081960309201824018842705704609262256417837526526335832424066125
331152942345796556950250681001831090041124537901533296615697052237
921032570693705109083078947999900499939532215362274847660361367769
797856738658467093667958858378879562594646489137665219958828693380
183601193236857855855819555604215625088365020332202451376215820461
810670519533065306060650105488716724537794283133887163139559690583
208341689847606560711834713621812324622725884199028614208728495687
963932546428534307530110528571382964370999035694888528519040295604
734613113826387889755178856042499874831638280404684861893818959054
203988987265069762020199554841265000539442820393012748163815853039
643992547020167275932857436666164411096256633730540921951967514832
873480895747777527834422109107311135182804603634719818565557295714
474768255285786334934285842311874944000322969069775831590385803935
352135886007960034209754739229673331064939560181223781285458431760
556173386112673478074585067606304822940965304111830667108189303110
887172816751957967534718853722930961614320400638132246584111115775
835858113501856904781536893813771847281475199835050478129771859908
470762197460588742325699582889253504193795826061621184236876851141
831606831586799460165205774052942305360178031335726326705479033840
125730591233960188013782542192709476733719198728738524805742124892
118347087662966720727232565056512933312605950577772754247124164831
283298207236175057467387012820957554430596839555568686118839713552
208445285264008125202766555767749596962661260456524568408613923826
576858338469849977872670655519185446869846947849573462260629421962
455708537127277652309895545019303773216664918257815467729200521266

71434632096378918523232150189761260343736840671941930377468809992968775824410478781232662531818459604538535438391144967753128642609252115376732588667226040425234910870269580996475958057946639734190640100363619040420331135793365424263035614570090112448008900208014780566037101541223288914657223931450760716706435568274377439657890679726874384730763464516775621030986040927170909512808630902973850445271828927496892121066700816485833955377359191369501531620189088874842107987068991148046692706509407620465027725286507289053285485614331608126930056937854178610969692025388650345771831766868859236814884752764984688219497397297077371871884004143231276365048145311228509900207424092558592529261030210673681543470152523487863516439762358604191941296976904052648323470099111542426012734380220893310966863678986949779940012601642276092608234930411806438291383473546797253992623387915829984864592717340592256207491053085315371829116816372193951887009577881815868504645076993439409874335144316263303172477474868979182092394808331439708406730840795893581089665647758599055637695252326536144247802308268118310377358870892406130313364773710116282146146616794040905186152603600925219472188909181073358719641421444786548995285823439470500798303885388608310357193060027711945580219119428999227223534587075662469261776631788551443502182870266856106650035310502163182060176092179846849368631612937279518730789726373537171502563787335797718081848784588665043358243770041477104149349274384575871071597315594394264125702709651251081155482479394035976811881172824721582501094960966253933953809221955919181885526780621499231727631632183398969380756168559117529984501320671293924041445938623988093812404521914848316462101473891825101090967738690664041589736104764365000680771056567184862814963711188321924456639458144914861655004956769826903089111856879869294705135248160917432430153836847072928989828460222373014526556798986277679680914697983782687643115988321090437156112997665215396354644208691975673700057387649784376862876817924974694384274652563163230055513041742273416464551278127845777724575203865437542828256714128858345444351325620544642410110379554641905811686230596447695870540721419852121067343324107567675758184569906930460475227701670056845439692340417110898889934163505851578873534308155208117720718803791040469830695786854739376564336319797868036718730796939242363214484503547763156702553900654231179201534649779290662415083288583952905426376876689688050333172278001858850697362324038947004718976193473443084374437599250341788079722358591342458131440498477017323616947197657153531977549971627856631190469126091825912498903676541769799036237552865263757337635269693443544004730671988689019681474287677908669796885225016369498567302175231325292653758964151714795595387842784998664563028788319620998304945198743963690706827626574858104391122326187940599415540632701319898957037611053236062986748037791537675115830432084987209202809297526498125691634250005229088726469252846661046653921714820801305022980526378364269597337070539227891535105688839381132497570713310295044303467159894487868471164383280506925077662745001220035262037094660234146489983902525888301486781621967751945831677187627572005054397944124599007711520515461993050983869825428464072555409274031325716326407929341833421470904125425335232480219322770753555467958716383587501815933871742360615511710131235256334858203651461418700492057043720182617331947157008675785393360786227395581857975872587441025420771054753612940474601000940954449596628814869159038990718659805636171376922272907641977551777201042764969496110562205925024202177042696221549587264539892276976603105249808557594716310758701332088614632664125911486338812202844406941694882615295776253250198703598706743804698219420563812558334364219492322759372212890564209430823525440841108645453694049692714940033197828

131818618881111840825786592875742638445005994422956858646048103301
538891149948693543603022181094346676400002236255057363129462629609
619876056425996394613869233083719626595473923462413459779574852464
783798079569319865081597767535055391899115133525229873611277918274
854200868953965835942196333150286956119201229888988700607999279541
118826902307891310760361763477948943203210277335941690865007193280
401716384064498787175375678118532132840821657110754952829497493621
460821558320568723218557406516109627487437509809223021160998263303
391546949464449100451528092508974507489676032409076898365294065792
019831526541065813682379198409064571246894847020935776119313998024
681340520039478194986620262400890215016616381353838151503773502296
607462795291038406868556907015751662419298724448271942933100485482
445458071889763300323252582158128032746796200281476243182862217105
435289834820827345168018613171959332471107466222850871066611770346
535283957762599774467218571581612641114327179434788599089280848669
491413909771673690027775850268664654056595039486784111079011610400
857274456293842549416759460548711723594642910585090995021495879311
219613590831588262068233215615308683373083817327932819698387508708
348388046388478441884003184712697454370937329836240287519792080232
187874488287284372737801782700805878241074935751488997891173974612
932035108143270325140903048746226294234432757126008664250833318768
865075642927160552528954492153765175149219636718104943531785838345
386525565664065725136357506435323650893679043170259787817719031486
796384082881020946149007971513771709906195496964007086766710233004
867263147551053723175711432231741141168062286420638890621019235522
354671166213749969326932173704310598722503945657492461697826097025
335947502091383667377289443869640002811034402608471289900074680776
484408871134135250336787731679770937277868216611786534423173226463
784769787514433209534000165069213054647689098505020301504488083426
184520873053097318949291642532293361243151430657826407028389840984
160295030924189712097160164926561341343342229882790992178604267981
245728534580133826099587717811310216734025656274400729683406619848
067661580502169183372368039902793160642043681207990031626444914619
021945822969099212278855394878353830564686488165556229431567312827
439082645061162894280350166133669782405177015521962652272545585073
864058529983037918035043287670380925216790757120406123759632768567
484507915114731344000183257034492090971243580944790046249431345502
890068064870429353403743603262582053579011839564908935434510134296
961754524957396062149028872893279252069653538639644322538832752249
960598697475988232991626354597332444516375533437749292899058117578
635555562693742691094711700216541171821975051983178713710605106379
555858890556885288798908475091576463907469361988150781468526213325
247383765119299015610918977792200870579339646382749068069876916819
749236562422608715417610043060890437797667851966189140414492527048
088197149880154205778700652159400928977760133075684796699295543365
613984773806039436889588764605498387147896848280538470173087111776
115966350503997934386933911978988710915654170913308260764740630571
141109883938809548143782847452883836807941888434266622207043872288
741394780101772139228191199236540551639589347426395382482960903690
028835932774585506080131798840716244656399794827578365019551422155
133928197822698427863839167971509126241054872570092407004548848569
295044811073808799654748156891393538094347455697212891982717702076
661360248958146811913361412125878389557735719498631721084439890142
394849665925173138817160266326193106536653504147307080441493916936
326237376777709585031325599009576273195730864804246770121232702053
374266705314244820816813030639737873664248367253983748769098060218
278578621651273856351329014890350988327061725893257536399397905572
917516009761545904477169226580631511102803843601737474215247608515

209901615858231257159073342173657626714239047827958728150509563309
280266845893764964977023297364131906098274063353108979246424213458
374090116939196425045912881340349881063540088759682005440836438651
661788055760895689672753153808194207733259791727843762566118431989
102500749182908647514979400316070384554946538594602745244746681231
468794344161099333890899263841184742525704457251745932573898956518
571657596148126602031079762825416559050604247911401695790033835657
486925280074302562341949828646791447632277400552946090394017753633
565547193100017543004750471914489984104001586794617924161001645471
655133707407395026044276953855383439755054887109978520540117516974
758134492607943368954378322117245068734423198987884412854206474280
973562580706698310697993526069339213568588139121480735472846322778
490808700246777630360555123238665629517885371967303463470122293958
160679250915321748903084088651606111901149844341235012464692802880
599613428351188471544977127847336176628506216977871774382436256571
177945006447771837022199910669502165675764404499794076503799995484
500271066598781360380231412683690578319046079276529727769404361302
305178708054651154246939526512710105292707030667302444712597393995
051462840476743136373997825918454117641332790646063658415292701903
027601733947486696034869497654175242930604072700505903950314852292
139257559484507886797792525393176515641619716844352436979444735596
426063339105512682606159572621703669850647328126672452198906054988
028078288142979633669674412480598219214633956574572210229867759974
673812606936706913408155941201611596019023775352555630060624798326
124988128819293734347686268921923977783391073310658825681377717232
831532908252509273304785072497713944833389255208117560845296659055
394096556854170600117985729381399825831929367910039184409928657560
599359891000296986446097471471847010153128376263114677420914557404
181590880006494323785583930853082830547607679952435739163122188605
754967383224319565065546085288120190236364471270374863442172725787
950342848631294491631847534753143504139209610879605773098720135248
407505763719925365047090858251393686346386336804289176710760211115
982887553994012007601394703366179371539630613986365549221374159790
511908358829009765664730073387931467891318146510931676157582135142
486044229244530411316065270097433008849903467540551864067734260358
340960860553374736276093565885310976099423834738222208729246449768
456057956251676557408841032173134562773585605235823638953203853402
484227337163912397321599544082842166663602329654569470357718487344
203422770665383738750616921276801576618109542009770836360436111059
240911788954033802142652394892968643980892611463541457153519434285
072135345301831587562827573389826889852355779929572764522939156747
756667605108788764845349363606827805056462281359888587925994094644
604170520447004631513797543173718775603981596264750141090665886616
218003826698996196558058720863972117699521946678985701179833244060
181157565807428418291061519391763005919431443460515404771057005433
900018245311773371895585760360718286050635647997900413976180895536
366960316219311325022385179167205518065926351803625121457592623836
934822266589557699466049193811248660909979812857182349400661555219
611220720309227764620099931524427358948871057662389469388944649509
396033045434084210246240104872332875008174917987554387938738143989
423801176270083719605309438394006375611645856094312951759771393539
607432279248922126704580818331376416581826956210587289244774003594
700926866265965142205063007859200248829186083974373235384908396432
614700053242354064704208949921025040472678105908364400746638002087
012666420945718170294675227854007450855237772089058168391844659282
941701828823301497155423523591177481862859296760504820386434310877
956289292540563894662194826871104282816389397571175778691543016505
860296521745958198887868040811032843273986719862130620555985526603

640504628215230615459447448990883908199973874745296981077620148713
400012253552224669540931521311533791579802697955571050850747387475
075806876537644578252443263804614304288923593485296105826938210349
800040524840708440356116781717051281337880570564345061611933042444
079826037795119854869455915205196009304127100727784930155503889536
033826192934379708187432094991415959339636811062755729527800425486
306005452383915106899891357882001941178653568214911852820785213012
551851849371150342215954224451190020739353962740020811046553020793
286725474054365271759589350071633607632161472581540764205302004534
018357233829266191530835409512022632916505442612361919705161383935
732669376015691442994494374485680977569630312958871916112929468188
493633864739274760122696415884890096571708616059814720446742866420
876533479985822209061980217321161423041947775499073873856794118982
466091309169177227420723336763503267834058630193019324299639720444
517928812285447821195353089891012534297552472763573022628138209180
743974867145359077863353016082155991131414420509144729353502223081
719366350934686585865631485557586244781862010871188976065296989926
932817870557643514338206014107732926106343152533718224338526352021
773544071528189813769875515757454693972715048846979361950047772097
056179391382898984532742622728864710888327017372325881824465843624
958059256033810521560620615571329915608489206434030339526226345145
428367869828807425142256745180618414956468611163540497189768215422
772247947403357152743681940989205011365340012384671429655186734415
374161504256325671343024765512521921803578016924032669954174608759
240920700466934039651017813485783569444076047023254075555776472845
075182689041829396611331016013111907739863246277821902365066037404
160672496249013743321724645409741299557052914243820807609836482346
597388669134991978401310801558134397919485283043673901248208244481
412809544377389832005986490915950532285791457688496257866588599917
986752055455809900455646117875524937012455321717019428288461740273
664997847550829422802023290122163010230977215156944642790980219082
668986883426307160920791408519769523555348865774342527753119724743
087304361951139611908003025587838764420608504473063129927788894272
918972716989057592524467966018970748296094919064876469370275077386
643239191904225429023531892337729316673608699622803255718530891928
440380507103006477684786324319100022392978525537237556621364474009
676053943983823576460699246526008909062410590421545392790441152958
034533450025624410100635953003959886446616959562635187806068851372
346270799732723313469397145628554261546765063246567662027924520858
134771760852169134094652030767339184114750414016892412131982688156
866456148538028753933116023229255561894104299533564009578649534093
511526645402441877594931693056044868642086275720117231952640502309
977456764783848897346431721598062678767183800524769688408498918508
614900343240347674268624595239589035858213500645099817824463608731
775437885967767291952611121385919472545140030118050343787527766440
276261894101757687268042817662386068047788524288743025914524707395
054652513533945959878961977891104189029294381856720507096460626354
173294464957661265195349570186001541262396228641389779673332907056
737696215649818450684226369036784955597002607986799626101903933126
376855696876702929537116252800554310078640872893922571451248113577
862766490242516199027747109033593330930494838059785662884478744146
984149906712376478958226329490467981208998485716357108783119184863
025450162092980582920833481363840542172005612198935366937133673339
246441612522319694347120641737549121635700857369439730597970971972
666664226743111776217640306868131035189911227133972403688700099686
292254646500638528862039380050477827691283560337254825579391298525
150682996910775425764748832534141213280062671709400909822352965795
799780301828242849022147074811112401860761341515038756983091865278

065889668236252393784527263453042041880250844236319038331838455052
236799235775292910692504326144695010986108889991465855188187358252
816430252093928525807796973762084563748211443398816271003170315133
440230952635192958868069082135585368016100021374085115448491268584
126869589917414913382057849280069825519574020181810564129725083607
035685105533178784082900004155251186577945396331753853209214972052
660783126028196116485809868458752512999740409279768317663991465538
610893758795221497173172813151793290443112181587102351874075722210
012376872194474720934931232410706508061856237252673254073332487575
448296757345001932190219911996079798937338367324257610393898534927
877747398050808001554476406105352220232540944356771879456543040673
589649101761077594836454082348613025471847648518957583667439979150
851285802060782055446299172320202822291488695939972997429747115537
185892423849385585859540743810488262464878805330427146301194158989
632879267832732245610385219701113046658710050008328517731177648973
523092666123458887310288351562644602367199664455472760831011878838
915114934093934475007302585581475619088139875235781233134227986650
352272536717123075686104500454897036007956982762639234410714658489
578024140815840522953693749971066559489445924628661996355635065262
340533943914211127181069105229002465742360413009369188925586578466
846121567955425660541600507127664176605687427420032957716064344860
620123982169827172319782681662824993871499544913730205184366907672
357740005393266262276032365975171892590180110429038427418550789488
743883270306328327996300720069801224436511639408692222074532024462
412115580435454206421512158505689615735641431306888344318528085397
592773443365538418834030351782294625370201578215737326552318576355
409895403323638231921989217117744946940367829618592080340386757583
411151882417743914507736638407188048935825686854201164503135763335
550944031923672034865101056104987272647213198654343545040913185951
314518127643731043897250700498198705217627249406521461995923214231
443977654670835171474936798618655279171582408065106379950018429593
879915835017158075988378496225739851212981032637937621832245659423
668537679911314010804313973233544909082491049914332584329882103398
469814171575601082970658306521134707680368069532297199059990445120
908727577622535104090239288877942463048328031913271049547859918019
696783532146444118926063152661816744319355081708187547705080265402
529410921826485821385752668815558411319856002213515888721036569608
751506318753300294211868222189377554602722729129050429225978771066
787384000061677215463844129237119352182849982435092089180168557279
815642185819119749098573057033266764646072875743056537260276898237
325974508447964954564803077159815395582777913937360171742299602735
310276871944944491793978514463159731443535185049141394155732938204
854212350817391254974981930871439661513294204591938010623142177419
918406018034794988769105155790555480695387854006645337598186284641
990522045280330626369562649091082762711590385699505124652999606285
544383833032763859980079292284665950355121124528408751622906026201
185777531374794936205549640107300134885315073548735390560290893352
640071327473262196031177343394367338575912450814933573691166454128
178817145402305475066713651825828489809951213919399563324133655677
709800308191027204099714868741813466700609405102146269028044915964
654533010775469541308871416531254481306119240782118869005602778182
423502269618934435254763357353648561936325441775661398170393063287
216690572225974520919291726219984440964615826945638023950283712168
644656178523556516412771282691868861557271620147493405227694659571
219831494338162211400693630743044417328478610177774383797703723179
525543410722344551255558999864618387676490397246116795901810003509
892864120419516355110876320426761297982652942588295114127584126273
279079880755975185157684126474220947972184330935297266521001566251

455299474512763155091763673025946213293019040283795424632325855030
109670692272022707486341900543830265068121414213505715417505750863
990767394633514620908288893493837643939925690060406731142209331219
593620298297235116325938677224147791162957278075239505625158160313
335938231150051862689053065836812998810866326327198061127154885879
809348791291370749823057592909186293919501472119758606727009254771
802575033773079939713453953264619526999659638565491759045833358579
910201271320458390320085387888163363768518208372788513117522776960
978796214237216254521459128183179821604411131167140691482717098101
545778193920231156387195080502467972579249760577262591332855972637
121120190572077140914864507409492671803581515757151405039761096384
675556929897038354731410022380258346876735012977541327953206097115
450648421218593649099791776687477448188287063231551586503289816422
828823274686610659273219790716238464215348985247621678905026099804
526648392954235728734397768049577409144953839157556548545905897649
519851380100795801078375994577529919670054760225255203445398871253
878017196071816407812484784725791240782454436168234523957068951427
226975043187363326301110305342333582160933319121880660826834142891
041517324721605335584999322454873077882290525232423486153152097693
846104258284971496347534183756200301491570327968530186863157248840
152663983568956363465743532178349319982554211730846774529708583950
761645822963032442432823773745051702856069806788952176819815671078
163340526675953942492628075696832610749532339053622309080708145591
983735537774874202903901814293731152933464446815121294509759653430
628421531944572711861490001765055817709530246887526325011970520947
615941676872778447200019278913725184162285778379228443908430118112
149636642465903363419454065718354477191244662125939265662030688852
005559912123536371822692253178145879259375044144893398160865790087
616502463519704582889548179375668104647461410514249887025213993687
050937230544773411264135489280684105910771667782123833281026218558
775131272117934444820144042574508306394473836379390628300897330624
138061458941422769474793166571762318247216835067807648757342049155
762821758397297513447899069658953254894033561561316740327647246921
250575911625152965456854463349811431767025729566184477548746937846
423373723898192066204851189437886822480727935202250179654534375727
416391079197295295081294292220534771730418447791567399173841831171
036252439571615271466900581470000263301045264354786590329073320546
833887207873544476264792529769017091200787418373673508771337697768
349634425241994995138831507487753743384945825976556099655595431804
092017849718468549737069621208852437701385375768141663272241263442
398215294164537800049250726276515078908507126599703670872669276430
837722968598516912230503746274431085293430527307886528397733524601
746352770320593817912539691562106363762588293757137384075440646896
478310070458061344673127159119460843593582598778283526653115106504
162329532904777217408355934972375855213804830509000964667608830154
061282430874064559443185341375522016630581211103345312074508682433
943215904359443031243122747138584203039010607094031523555617276799
416002039397509989762933532585557562480899669182986422267750236019
325797472674257821111973470940235745722227121252685238429587427350
156366009318804549333898974157149054418255973808087156528143010267
046028431681923039253529779576586241439270154974087927313105163611
913757700892956482332364829826302460797587576774537716010249080462
430185652416175665560016085912153455626760219268998285537787258314
514408265458348440947846317877737479465358016996077940556870119232
860804113090462935087182712593466871276669487389982459852778649956
916546402945893506496433580982476596516514209098675520380830920323
048734270346828875160407154665383461961122301375945157925269674364
253192739003603860823645076269882749761872357547676288995075211480

485252795084503395857083813047693788132112367428131948795022806632
017002246033198967197064916374117585485187848401205484467258885140
156272501982171906696081262778548596481836962141072171421498636191
877475450965030895709947093433785698167446582826791194061195603784
539785583924076127634410576675102430755981455278616781594965706255
975507430652108530159790807334373607943286675789053348366955548680
391343372015649883422089339997164147974693869690548008919306713805
717150585730714881564992071408675825960287605645978242377024246980
532805663278704192676846711626687946348695046450742021937394525926
266861355294062478136120620263649819999949840514386828525895634226
432870766329930489172340072547176418868535137233266787792173834754
148002280339299735793615241275582956927683723123479898944627433045
456679006203242051639628258844308543830720149567210646053323853720
314324211260742448584509458049408182092763914000854042202355626021
856434899414543995041098059181794888262805206644108631900168856815
516922948620301073889718100770929059048074909242714101893354281842
999598816966099383696164438152887721408526808875748829325873580990
567075581701794916190611400190855374488272620093668560447559655747
648567400817738170330738030547697360978654385938218722058390234444
350886749986650604064587434600533182743629617786251808189314436325
120510709469081358644051922951293245007883339878842933934243512634
336520438581291283434529730865290978330067126179813031679438553572
629699874035957045845223085639009891317947594875212639707837594486
113945196028675121056163897600888009274611586080020780334159145179
707303683519697776607637378533301202412011204698860920933908536577
322239241244905153278095095586645947763448226998607481329730263097
502881210351772312446509534965369309001863776409409434983731325132
186208021480992268550294845466181471555744470966953017769043427203
189277060471778452793916047228153437980353967986142437095668322149
146543801459382927739339603275404800955223181666738035718393275707
714204672383862461780397629237713120958078936384144792980258806552
212926209362393063731349664018661951081158347117331202580586672763
999276357907806381881306915636627412543125958993611964762610140556
350339952314032311381965623632719896183725484533370206256346422395
276694356837676136871196292181875457608161705303159072882870071231
366630872275491866139577373054606599743781098764980241401124214277
366808275139095931340415582626678951084677611866595766016599817808
941498575497628438785610026379654317831363402513581416115190209649
913354873313111502270068193013592959597164019719605362503355847998
096348871803911161281359596856547886832585643789617315976200241962
155289629790481982219946226948713746244472909345647002853769495885
959160678928249105441251599630078136836749020937491573289627002865
682934443134234735123929825916673950342599586897069726733258273590
312128874666045146148785034614282776599160809039865257571726308183
349444182019353338507129234577437557934406217871133006310600332405
399169368260374617663856575887758020122936635327026710068126182517
291460820254189288593524449107013820621155382779356529691457650204
864328286555793470720963480737269214118689546732276775133569019015
372366903686538916129168888787640752549349424973342718117889275993
159671935475898809792452526236365903632007085444078454479734829180
208204492667063442043755532505052752283377888704080403353192340768
563010934777212563908864041310107381785333831603813528082811904083
256440184205374679299262203769871801806112262449090924264198582086
175117711378905160914038157500336642415609521632819712233502316742
260056794128140621721964184270578432895980288233505982820819666624
903585778994033315227481777695284368163008853176969478369058067106
482808359804669884109813515865490693331952239436328792399053481098
783027450017206543369906611778455436468772363184446476806914282800

4551074686645392805399409108754939166095731619715033166968309929466
6349142798780842257220697148875580637480308862995118473187124777729
1910070227588893486939456289515802965372150409603107761289831263588
9964893410247036036645058687287589051406841238124247386385427908288
2733827973326885504935874303160274749063129572349742611221517417155
3133618622410913869500688835898962349276317316478340077460886655599
8733382113829928776911495492184192087771606068472874673681886167500
7221017261103830671787856694812948785048943063086169948798703160511
5884108282351274153538513365895332948629494495061868514779105804699
6039069372662670386512905201137810858616188886947957607413585534588
5151768051973334433495230120395770739623771316030242887200537320999
8253008977618973129817881944671731160647231476248457551928732782822
5127182446807824215216469567819294098238926284943760248852279003622
0219386696482215628093605373178040863727268426696421929946819214900
8701707533361094791381804063287387593848269535583077395761447997277
0003472880182785281389503217986345216111066608839314053226944905455
5527867894417579202440021450780192099804461382547805858048442416400
4775031536054906591430078158372430123137511562284015838644270890711
8284816757527123846782459534334449622010096071051370608461801187544
3120725491334994247617115633321408934609156561550600317384218701577
0226103101916603887064661438897736318780940711527528174689576401588
1047016965247557740891644568677717158500583269943401677202156767722
4068128366565264122982439465133197359199709403275938502669557470233
1813203243716420586141033606524536939160050644953060161267822648944
2437397166717661231048975031885732165554988342121802846912529086100
1485527815277625623750456375769497734336846015607727035509629049399
2487088406281067943622418704747008368842671022558302403599841645955
1122485272633632645114017395248086194635840783753556885622317115522
0947223065437092606797351000565549381224575483728545711797393615755
6167641692895805257297522338558611388322171107362265816218842443177
8857488798109026653793426664216990914056536432249301334867988154888
6628665052346997235574738424830590423677143278792316422403877764333
0192600192284778313837632536121025336935812624086866699738275977366
5682227907215832478888642369346396164363308730139814211430306008733
0666164803678984091335926293402304324974926887831643602681011309577
0716141912830686577323532639653677390317661361315965553584999398600
0565155921936759977717933019744688148371103206503693192894521402655
0915465184309936553493337183425298433679915939417466223900389527677
3813330617747629574943868716978453767219493506590875711917720875477
7107189937960894774512654757501871194870738736785890200617373321077
5693302216320628432065671192096950585761173961632326217708945426211
4609858410237813215817727602222738133495410481003073275107799948999
1977963883530734443457532975914263768405442264784216063122769646966
7156473999043715903323906560726644116438605404838847161912109008700
1019130726071044114143241976796828547885524779476481802959736049433
9700479596040292746299203572099761950140348315380947714601056333444
6998820822120587281510729182971211917876424880354672316916541852255
6729234429187128163232596965413548589577133208339911288775917226111
5273379010341362085614577992398778325083550730199818459025958355988
9260553299673770491722454935329683300002230181517226575787524058833
2249085821280089747909326100762578770428656006996176212176845478999
6440705066241710213327486796237430229155358200780141165348065647488
8230615003392068983794766255036549822805329662862117930628430170499
2402301985719978948836897183043805182174419147660429752437251683433
5411217038631379411422095295885798060152938752753799030938871683577
2095760715221900279379292786303637268765822681241993384808166021600
3722154710143007377537792699069587121289288019052031601285861825499
4413353820784883465311632650407642428390870121015194231961652268422

20037112304643006734420647477180213530701240988603533991526679238711017062218658835737812109351797756044256346949997872511254408545222748109148743072598696020402759411789425812818821599523596589791811440776533543217575952555361581280011638467203193465072968079907939637149617743121194020212975731251652537680173591015573381537720019524445436200718484756634154074423286210609976132434875488474345396659813387174660930205350702719529839432714253711557666000257844230310734295515339450604862227649666876240793243531929926392537310768921353525723210808898193391686682789482811704726245019484097009757609209837240900747179733407881418251958425980962417476101382526439551352593118850456362641883003385396524359974169313228947198783084276004013680747039040972384739458348961865397905941185993103561684368692194853820557803957738813606795499000851232594425297244866667668346414021899159445653094234406506678519484177667794704720419588220432953803263105374948831221803912796784461001397267538921951191178365876625280836900532490045974109470687729123282143046353372835199536482743258331191444590178096077828835837301118575436599589827245319253105881150263075425714939430244539318701799236081666113054262539958338979429716020703387678150330102801200959972522222808014235710947603519255444349299867678178910455590630159538097618759203589373419789623589311259839025983102671933041892151096891562250696591198283234555030590817307351955037216658702880539921385760370353771051780212801295668419841403628727256232144287543022109094727210734741349755141907370433182766261772759968888260272252471336833534528166927795913288613817663498577289369009657495622871030243625907724122190943008717556926257580657099120166596224360802428700245473620363948412559548817272724736534677836472019183039987176270375157246499222894679323226936191776416146187956139566995677830682903165896994307673335082349907906241002025061340573443006957454746821756904416515406365846804636926212742110753990421887161276177870142588648257752238891845995233762923779155857445494773612955259522265786364621183775984737003479714082069941455807190802135907322692331008317595106590191212947954086036407573587502058902087045796700070552625058114206639074592152733094068236494415908910092202966805233252661989113118420162916310768940847235643668081821686572196882683584027855007828040434537101836510969517823357430305048526537380735310741859177056103973950626403554422751561011072617793706347238049906669221619711942591204450846417463835899382399465173955090008594799901360266742614942900664671150671754221770387745076735637421547829059110126191575558702389570014051178226469899449179083017954758767601680941001358376135785913569244556477644641786671153919513576961048649224900834467154863830544779143300976804868783481846727337584368927243104474068076852786255851650920882638132336231487333367147645204508766276149503899495048095604609896043291233583488599902945264002849942808786240398118148847673012167541611066299955536681931232874257020637383520200868636913117334697317412191536332467453256308713473027921749562270146873258678917345583799643513588009593508775563562488104938529990076751355135277924124292774885658885665132473025147102105753525165118148509027504768455182520963318990685276144351382136621523688905787866994322888160283774820355060160298940091197138501798716836337441392759736440170070147637066557035043381211135764150184518214136198234951596010647527125759351853043328755377830575095674254426847122196187091785607839361445113833356491032564057338986671781239722375193164306170138595394743678433926709867124522111896908402363274114966012434830989299417380305884171666130730400675883804321115553794406054977217059428215148861656727712409033877277456290971101348851843741186956554497457368452180669829110450580042998879538990278043835962824094218605562877884288021275

3884803728640019441614257499904272009595204654170598104989967504511
1936471172772220436102614079750809686975176600237187748348016120311
023468056711264476612374762785219024120256994353471622666089367521
983311181351114650385489502512065577263614547360442685949807439693
233129712737715734709971395229118265348515558713733662912024271430
250376326950135091161295299378586468130722648600827088133353819370
368259886789332123832705329762585738279009782646054559855513183668
884462826513379849166783940976135376625179825824966345877195012438
404035914084920973375464247448817618407002356958017741017769692507
781489338667255789856458985105689196092439884156928069698335224022
563457049731224526935419383700484318335719651662672157552419340193
309901831930919658292096965624766768365964701959575473934551433741
370876151732367720422738567427917069820454995309591887243493952409
444167899884631984550485239366297207977745281439941825678945779571
255242682608994086331737153889626288962940211210888442737656862452
761213037101730078513571540453304150795944777614359743780374243664
697324713841049212431413890357909241603640631403814983148190525172
093710396402680899483257229795456404270175772290417323479607361878
788991331830584306939482596131871381642346721873084513387721908697
510494284376932502498165667381626061594176825250999374167288395174
406693254965340310145222531618900923537648637848288134420987004809
622717122640748957193900291857330746010436072919094576799461492929
042798168772942648772995285843464777538690695014898413392454039414
468026362540211861431703125111757764282991464453340892097696169909
837265236176874560589470496817013697490952307208268288789073019001
825342580534342170592871393173799314241085264739094828459641809361
413847583113613057610846236683723769591349261582451622155213487924
414504175684806412063652017038633012953277769902311864802006755690
568229501635493199230591424639621702532974757311409422018019936803
502649563695586642590676268568737211033915679383989576556519317788
300024161353956243777784080174881937309502069990089089932808839743
036773659552489130015663329407790713961546453408879151030065132193
448667324827590794680787981942501958262232039513125201410996053126
069655540424867054998678692302174698900954785072567297879476988883
109348746442640071818316033165551153427615562240547447337804924621
495213325852769884733626918264917433898782478927846891882805466998
230368993978341374758702580571634941356843392939606819206177333179
173820856243643363535986349449689078106401967407443658366707158692
452118299789380407713750129085864657890577142683358276897855471768
718442772612050926648610205153564284063236848180728794071712796682
006072755955590404023317874944734645476062818954151213916291844429
765106694796935401686601005519607768733539651161493093757096855455
938151378956903925101495326562814701199832699220006639287537471313
523642158926512620407288771657835840521964605410543544364216656224
456504299901025658692727914275293117208279393775132610605288123537
345106837293989358087124386938593438917571337630072031976081660446
468393772580690923729752348670291691042636926209019960520412102407
764819031601408586355842760953708655816427399534934654631450404019
952853725200495780525465625115410925243799132626271360909940290226
206283675213230506518393405745011209934146491843332364656937172591
448932415900624202061288573292613359680872650004562828455757459659
212053034131011182750130696150983551563200431078460190656549380654
252522916199181995960275232770224985573882489988270746593635576858
256051806896428537685077201222034792099393617926820659014216561592
530673794456894907085326356819683186177226824991147261573203580764
629811624401331673789278886892290325933498617970219949819257396176
307583441709855922217017182571277753449150820527843090461946083521
740200583867284970941102326695392144546106621500641067474020700918

991195137646690448126725369153716229079138540393756007783515337416
774794210038400230895185099454877903934612222086506016050035177626
483161115332558770507354127924990985937347378708119425305512143697
974991495186053592040383023571635272763087469321962219006426088618
367610334600225547747781364101269190656968649501268837629690723396
127628722304114181361006026404403003599698891994582739762411461374
480405969706257676472376606554161857469052722923822827518679915698
339074767114610302277660602006124687647772881909679161335401988140
275799217416767879923160396356949285151363364721954061117176738737
255572852294005436178517650230754469386930787349911035218253292972
604455321079788771144989887091151123725060423875373484125708606406
905205845212275453384800820530245045651766951857691320004281675805
492481178051983264603244579282973012910531838563682120621553128866
856495651261389226136706409395333457052698695969235035309422454386
527867767302754040270224638448355323991475136344104405009233036127
149608135549053153902100229959575658370538126196568314428605795669
662215472169562087001372776853696084070483332513279311223250714863
020695124539500373572334680709465648308920980153487870563349109236
605755405086411152144148143463043727327104502776866195310785832333
485784029716092521532609255893265560067212435946425506599677177038
844539618163287961446081778927217183690888012677820743010642252463
480745430047649288555340906218515365435547412547615276977266776977
277705831580141218568801170502836527554321480348800444297999806215
790456416195721278450892848980642649742709057912906921780729876947
797511244730599140605062994689428093103421641662993561482813099887
074529271604843363081840412646963792584309418544221635908457614607
855856247381493142707826621518554160387020687698046174740080832434
366538235455510944949843109349475994467267366535251766270677219418
319197719637801570216993367508376005716345464367177672338758864340
564487156696432104128259564534984138841289042068204700761559691684
303899934836679354254921032811336318472259230555438305820694167562
999201337317548912203723034907268106853445403599356182357631283776
764063101312533521214199461186935083317658785204711236433122676512
996417132521751355326186768194233879036546890800182713528358488844
411176123410117991870923650718485785622102110400977699445312179502
247957806950653296594038398736990724079767904082679400761872954783
596349279390457697366164340535979221928587057495748169669406233427
261973351813662606373598257555249650980726012366828360592834185584
802695841377255897088378994291054980033111388460340193916612218669
605849157148573356828614950001909759112521880039641976216355937574
371801148055944229873041819680808564726571354761283162920044988031
540210553059707666636274932830891688093235929008178741198573831719
261672883491840242972129043496552694272640255964146352591434840067
586769035038232057293413298159353304444649682944136732344215838076
169483121933311981906109614295220153617029857510559432646146850545
268497576480780800922133581137819774927176854507553832876887447459
159373116247060109124460982942484128752022446259447763874949199784
044682925736096853454984326653686284448936570411181779380644161653
122360021491876876946739840751717630751684985635920148689294310594
020245796962292456664488196757629434953532638217161339575779076637
076456957025973880043841580589433613710655185998760075492418721171
488929522173772114608115434498266547987258005667472405112200738345
927157572771521858994694811794064446639943237004429114074721818022
482583773601734668530074498556471542003612359339731291445859152288
740871950870863221883728826282288463184371726190330577714765156414
382230679184738603914768310814135827575585364359772165002827780371
342286968878734979509603110889919614338666406845069742078770028050
936720338723262963785603865321643234881555755701846908907464787912

243637555666867806761054495501726079114293083128576125448194444947
324481909379536900820638463167822506480953181040657025432760438570
350592281891987806586541218429921727372095510324225107971807783304
260908679427342895573555925272380551144043800123904168771644518022
649168164192740110645162243110170005669112173318942340054795968466
980429801736257040673328212996215368488140410219446342464622074557
564396045298531307140908460849965376780379320189914086581466217531
933766597011433060862500982956691763884605676297293146491149370462
446935198403953444913514119366793330193661766365255514917498230798
707228086085962611266050428929696653565251668888557211227680277274
370891738963977225756489053340103885593112567999151658902501648696
142720700591605616615970245198905183296927893555030393468121976158
218398048396056252309146263844738629603984892438618729850777592879
272206855480721049781765328621018747676689724884113956034948037672
703631692100735083407386526168450748249644859742813493648037242611
670426687083192504099761531907685577032742178501000644198412420739
640013960360158381056592841368457411910273642027416372348821452410
134771652960312840865841978795111651152982781462037913985500639996
032659124852530849369031313010079997719136223086601109992914287124
938854161203802041134018888721969347790449752745428807280350930582
875442075513481666092787935356652125562013998824962847872621443236
285367650259145046837763528258765213915648097214192967554938437558
260025316853635673137926247587804944594418342917275698837622626184
636545274349766241113845130548144983631178978448973207671950878415
861887969295581973325069995140260151167552975057543781024223895792
578656212843273120220071673057406928686936393018676595825132649914
595026091706934751940897535746401683081179884645247361895605647942
635807056256328118926966302647953595109712765913623318086692153578
860781275991053717140220450618607537486630635059148391646765672320
571451688617079098469593223672494673758309960704258922048155079913
275208858378111768521426933478692189524062265792104362034885292626
798401395321645879115157905046057971083898337186403802441751134722
647254701079479399695355466961972676325522991465493349966323418595
145036098034409221220671256769872342794070885707047429317332918852
389672197135392449242617864118863779096281448691786946817759171715
066911148002075943201206196963779510322708902956608556222545260261
046073613136886900928172106819861855378098201847115416363032626569
928342415502360097804641710852553761272890533504550613568414377585
442967797701466029438768722511536380119175815402812081825560648541
078793359892106442724489861896162941341800129513068363860929410008
313667337215300835269623573717533073865333820484219030818644918409
372394403340524490955455801640646076158101030176748847501766190869
294609876920169120218168829104087070956095147041692114702741339005
225334083481287035303102391969997859741390859360543359969707560446
013424245368249609877258131102473279856207212657249900346829388687
230489556225320446360263985422525841646432427161141981780248259556
354490721922658386366266375083594431487763515614571074552801615967
704844271419443518327569840755267792641126176525061596523545718795
667317091331935876162825592078308018520689015150471334038610031005
591481785211038475454293338918844412051794396997019411269511952656
491959418997541839323464742429070271887522353439367363366320030723
274703740712398256202466265197409019976245205619855762576000870817
308328834438183107005451449354588542267857855191537229237955549433
341017442016960009069641561273229777022121795186837635908225512881
647002199234886404395915301846400471432118636062252701154112228380
277853891109849020134274101412155976996543887719748537643115822983
853312307175113296190455900793806427669581901484262799122179294798
734890186847167650382732855205908298452980625925035212845192592798

659350613296194679625237397256558415785374456755899803240549218696
288849033256085145534439166022625777551291620077279685262938793753
045418108072928589198971538179734349618723292761474785019261145041
327487324297058340847111233374627461727462658241532427105932250625
530231473875925172478732288149145591560503633457542423377916037495
250249302235148196138116256391141561032684495807250827343176594405
409826976526934457986347970974312449827193311386387315963636121862
349726140955607992062831699942007205481152535339394607685001990988
655386143349578165008996164907967814290114838764568217491407562376
761845377514403147541120676016072646055685925779932207033733339891
636950434669069482843662998003741452762771654762382554617088318981
086880684785370553648046935095881802536052974079353867651119507937
328208314626896007107517552061443378411454995013643244632819334638
905093654571450690086448344018042836339051357815727397333453728426
337217406577577107983051755572103679597690188995849413019599957301
790124019390868135658553966194137179448763207986880037160730322054
742357226689680188212342439188598416897227765219403249322731479366
923400484897605903795809469604175427961378255378122394764614783292
697654516229028170110043784603875654415173943396004891531881757665
050095169740241564477129365661425394936888423051740012992055685428
985389794266995677702708914651373689220610441548166215680421983847
673087178759027920917590069527345668202651337311151800018143412096
260165862982107666352336177400783778342370915264406305407180784335
806107296110555002041513169637304684921335683726540030750982908936
461204789111475303704989395283345782408281738644132271000296831194
020332345642082647327623383029463937899837583655455991934086623509
096796113400486702712317652666371077872511186035403755448741869351
973365662177235922939677646325156202348757011379571209623772343137
021203100496515211197601317641940820343734851285260291333491512508
311980285017785571072537314913921570910513096505988599993156086365
547740355189816673353588004821466509974143376118277772335191074121
757284159258087259131507460602563490377726337391446137703802131834
744730111303267029691733504770163210661622783002726928336558401179
141944780874825336071440329625228577500980859960904093631263562132
816207145340610422411208301000858726425211226248014264751942618432
585338675387405474349107271004975428115946601713612259044015899160
022982780179603519408004651353475269877760952783998436808690898919
783969353217998013913544255271791022539701081063214304851137829149
851138196914304349750018998068164441212327332830719282436240673319
655469267785119315277511344646890550424811336143498460484905125834
568326644152848971397237604032821266025351669391408204994732048602
162775979177123475109750240307893575993771509502175169355582707253
391189233407022383207758580213717477837877839101523413209848942345
961369234049799827930414446316270721479611745697571968123929191374
098292580556195520743424329598289898052923336641541925636738068949
420147124134052507220406179435525255522500874879008656831454283516
775054229480327478304405643858159195266675828292970522612762871104
013480178722480178968405240792436058274246744307672164527031345135
416764966890127478680101029513386269864974821211862904033769156857
624069929637249309720162870720018983542369036414927023696193854737
248032985504511208919287982987446786412915941753167560253343531062
674525450711418148323988060729714023472552071349079839898235526872
395090936566787899238371257897624875599044322889538837731734894112
275707141095979004791930104674075041143538178246463079598955563899
188477378134134707024674736211204898622699188851745625173251934135
203811586335012391305444191007362844756751416105041097350585276204
448919097890198431548528053398577784431393388399431044446566924455
088594631408175122033139068159659251054685801313383815217641821043

3429788826119630443111388796258746090226130900849975430395771243 23
0616906262919403921439740270894777663702488155499322458825979020 63
1257436910946393252806241642476868495455324938017639371615636847 85
9823715902385421265840615367228607131702674740131145261063765383 39
0315921943469817605358380310612887852051546933639241088467632009 56
7089718367490578163085158138161966882222047570437590614338040725 85
3862083565176998426774523195824182683698270160237414938363496629 35
1576854061397342746470899685618170160551104880971554859118617189 66
8025973541705423985135560018720335079060946421271143993196046527 42
4050882225359773481519135438571253258540493946010865793798058620 14
3366078825219717809025817370870916460452727977153509910340736425 02
0386386718220522879694458387652947951048660717390229327455426785 66
9776865939923416834122274663015062155320502655341460995249356050 85
4921756549134830958906536175693817637473644183378974229700703545 20
6663170929607591989627732423090252397443861014263098687733913882 51
8684316501027964911497737582888913450341148865948670215492101084 32
8080783428089417298008983297536940644969903125399863919581601468 99
5220880662285408414864274786281975546629278814621607171381880180 84
0572084715868906836919393381864278454537956719272397972364651667 59
2011057995663962598535512763558768140213409829016296873429850792 47
1846056874828331381259161962476156902875901072733103299140623864 60
8333378638257926302391590003557609032477281338887339178096966601 46
9615031754226751125993315529674213336300222964906480934582008181 06
1802100227664580400278213336758573019011371754672763059044353131 31
9036092489097246427928455549913490005180295707082919052556781889 91
3899625138662319380053611346224294610248954072404857123256628888 93
1722116432947816190554868054943441034090680716088028227959686950 13
3643814268252170472870863010137301155236861416908375675747637239 76
3185757038109443390564564468524183028148107998376918512127201935 04
4041804604721626939445788377090105974693219720558114078775989772 07
2009689382249303236830515862657281114637996983137517937623215111 25
2349734305240622105244234353732905655163406669506165892878218707 75
6794176080712973781335187117931650033155523822487730653444179453 41
5395202424449703410120874072188109388268167512042299404948179449 47
2732894770111574139441228455521828424922240658752689172272780607 11
6754046973008037039618787796694882555614674384392570115829546661 35
8678671897661297311267200072971553613027503556167817765442287442 11
4729881614802705243806817653573275578602505847084013208837932816 00
8769081300492491473682517035382219619039014999523495387105997351 14
3478292339499187936608692301375596368532373806703591144243268561 51
2109404259582639301678017128669239283231057658851714020211196957 06
4799814031505633045141564414623163763809904402816256917576489142 56
9714163598439317433270237812336938043012892626375382667795034169 33
4323607500248175741808750388475094939454896209740485442635637164 99
5949920980884294790363666297526003243856352945844728944547166209 29
7495496616877414120882130477022816116456044007236351581149729739 21
8966737382647204722642221242016560150284971306332795814302516013 69
4825567014780935790889657134926158161346901806965089556310121218 49
1805847922720691871696316330044858020102860657858591269974637661 74
1463934159569539554203314628026518951167938074573315759846086173 70
2687867602943677780500244673391332431669880354073232388281847501 05
1641331189537036488422690270478052742490603492082954755054003457 16
0184072574536938145531175354210726557835615499874447480427323457 88
0061873149341566046352979779455075359304795687209316724536547208 38
1685855606043801977030764246083489876101345709394877002946175792 06
1952549255757109038525171488525265671045349813419803390641529876 34
3695420256080277614421914318921393908834543131769685101840103844 47
2348948869520981943531906506555354617335814045544837884752526253 94

966586999205841765278012534103389646981864243003414679138061902805
960785488801078970551694621522877309010446746249797999262712095168
477956848258334140226647721084336243759374161053673404195473896419
789542533503630186140095153476696147625565187382329246854735693580
289601153679178730355315937836308224861517777054157757656175935851
201669294311113886358215966761883032610416465171484697938542262168
716140012237821377977413126897726671299202592201740877007695628347
393220108815935628628192856357189338495885060385315817976067947984
087836097596014973342057270460352179060564760328556927627349518220
323614411258418242624771201203577638889597431823282787131460805353
357449429762179678903456816988955351850447832561638070947695169908
624710001974880920500952194363237871976487033922381154036347548862
684595615975519376541011501406700122692747439388858994385973024541
480106123590803627458528849356325158538438324249325266608758890831
870070910023737710657698505643392885433765834259675065371500533351
448990829388773735205145933304962653141514138612443793588507094468
804548697535817021290849078734780681436632332281941582734567135644
317153796781805819585246484008403290998194378171817730231700398973
305049538735611626102399943325978012689343260558471027876490107092
344388463401173555686590358524491937018104162620850429925869743581
709813389404593447193749387762423240985283276226660494238512970945
324558625210360082928664972417491914198896612955807677097959479530
601311915901177394310420904907942444886851308684449370590902600612
064942574471035354765785924270813041061854621988183009063458818703
875585627491158737542106466795134648758677154383801852134828191581
246259933516019893559516796893285220582479942103451271587716334522
299541883968044883552975336128683722593539007920166694133909116875
880398882886921600237325736158820716351627133281051818760210485218
067552664867390890090719513805862673512431221569163790227732870541
084203784152568328871804698795251307326634027851905941733892035854
039567703561132935448258562828761061069822972142096199350933131217
118789107876687204454887608941017479864713788246215395593333327556
200943958043453791978228059039595992743691379377866494096404877784
174833643268402628293240626008190808180439091455635193685606304508
914228964521998779884934747772913279726602765840166789013649050874
114212686196986204412696528298108704547986155954533802120115564697
997678573892018624359932677768945406050821883822790983362716712449
002676117849826437703300208184459000971723520433199470824209877151
444975101705564302954282181967000920251561584417420593365814813490
269311151709387226002645863056132560579256092733226557934628080568
344392137368840565043430739657406101777937014142461549307074136080
544210029560009566358897789926763051771878194370676149821756418659
011616086540863539151303920131680576903417259645369235080641744656
235152392905040947995318407486215121056183385456617665260639371365
880252166622357613220194170137266496607325201077194793126528276330
241380516490717456596485374835466919452358031530196916048099460681
490403781982973236093008713576079862142542209641900436790547904993
007837242158195453541837112936865843055384271762803527912882112930
835157565659994474178843838156514843422985870424559243469329523282
180350833372628379183021659183618155421715744846577842013432998259
456688455826617197901218084948033244878725818377480552226815101137
174536841787028027445244290547451823467491956418855124442133778352
142386597992598820328708510933838682990657199461490629025742768603
885051103263854454041918495886653854504057132362968106914681484786
965916686184275679846004186876229805556296304595322792305161672159
196867584952363529893578850774608153732145464298479231051167635774
949462295256949766035947396243099534331040499420967788382700271447
849406903707324910644415169605325656058677875741747211082743577431

519406075798356362914332639781221894628744779811980722564671466405
485013100965678631488009030374933887536418316513498254669467331611
812336485439764932502617954935720430540218297487125110740401161140
589991109306249231281311634054926257135672181862893278613883371802
853505650359195274140086951092616754147679266803210923746708721360
627833292238641361959412133927803611827632410600474097111104814000
362334271451448333464167546635469973149475664342365949349684588455
152415075637660508663282742479413606287604129064491382851945640264
315322585862404314183866959063324506300039221319264762596269151090
445769530144405461803785750303668621246227863975274666787012100339
298487337501447560032210062235802934377495503203701273846816306102
657030087227546296679688089058712767636106622572235222973920644309
352432722810085997309513252863060110549791564479184500461804676240
892892568091293059296064235702106152464620502324896659398732493396
737695202399176089847457184353193664652912584806448019652016283879
518949933675924148562613699594530728725453246329152911012876377060
557060953137752775186792329213495524513308986796916512907384130216
757323863757582008036357572800275449032795307990079944254110872569
318801466793559583467643286887696661009739574996783659339784634695
994895061049038364740950469522606385804675807306991229047408987916
687211714752764471160440195271816950828973353714853092893704638442
089329977112585684084660833993404568902678751600877546126798801546
585652206121095349079670736553970257619943137663996060606110640695
933082817187642604357342536175694378484849525010826648839515970049
059838081210522111109194332395113605144645983421079905808209371646
452312770402316007213854372346126726099787038565709199850759563461
324846018840985019428768790226873455650051912154654406382925385127
631766392205093834520430077301702994036261543400132276391091298832
786392041230044555168405488980908077917463609243933491264116424009
388074635660726233669584276458369826873481588196105857183576746200
965052606592926354829149904576830721089324585707370166071739819448
502884260396366074603118478622583105658087087030556759586134170074
540296568763477417643105175103673286924555858208237203860178173940
517513043799486882232004437804310317092103426167499800007301609481
458637448877852227307633049538394434538277060876076354209844500830
624763025357278103278346176697054428715531534001649707665719598504
174819908720149087568603778359199471934335277294728553792578768483
230110185936580071729118696761765505377503029303383070644891281141
202550615089641100762382457448865518258105814034532012475472326908
754750707857765973254284445935304499207001453874894822655644222369
636554419422544133821222547749753549462482768053333698328415613869
236344335855386847111114304982483989918031654586382893537991305352 2
283343013795337295401625762322808113849949187614414132293376710656
349252881452823950620902235787668465011666009738275366040544694165
342223905210831458584703552935221992827276057482126606529138553034
554974455147034493948686342945965843102419078592368022456076393678
416627051855517870290407355730462063969245330779578224594971042018
804300018388142900817303945050734278701312446686009277858181104091
151117293748736278878749074652855654347488868310641100510230208751 0
776891878152562273525155037953244485778727761700196485370355516765
520911933934376286628461984402629525218367852236747510880978150709
897841308624588152266096355140187449583692691779904712072649490573
726428600521140358123107600669951853612486274675637589622529911649
606687650826173417848478933729505673900787861792535144062104536625
064046372881569823231750059626108092195521115085930295565496753886
261297233991462835847604862762702730973920200143224870758233735491
524608560821032888297418390647886992327369136004883743661522351705
843770554521081551336126214291181561530175888257359489250710887926

2128641392443309383797333867806131795237315266773820858024701433527
700924380326695174211950767088432634644274912755890774686358216216
604274131517021245858605623363149316464691394656249747174195835421
860774871105733845843368993964591374060338215935224359475162623918
868530782282176398323730618020424656047752794310479618972429953302
979249748168405289379104494700459086499187272734541350810198388186
467360939257193051196864560185578245021823106588943798652243205067
737996619695547244058592241795300682045179537004347245176289356677
050849021310773662575169733552746230294303120359626095342357439724
965921101065781782610874531887480318743082357369919515634095716270
099244492974910548985151965866474014822510633536794973714251022934
188258511737199449911509758374613010550506419772153192935487537119
163026203032858865852848019350922587577559742527658401172134232364
808402714335636754204637518255252494432965704386138786590196573880
286840189408767281671413703366173265012057865391578070308871426151
907500149257611292767519309672845397116021360630309054224396632067
432358279788933232440577919927848463333977773765590187057480682867
834796562414610289950848739969297075043275302997287229732793444298
864641272534816060377970729829917302929630869580199631241330493935
049332541235507105446118259114111645453471032988104784406778013807
713146540009938630648126661433085820681139583831916954555825942689
576984142889374346708410794631893253910696395578070602124597489829
356461356078898347241997947856436204209461341238761319886535235831
299686226894860840845665560687695450127448663140505473535174687300
980632278046891224682146080672762770840240226615548502400895289165
711761743902033758487784291128962324705919187469104200584832614067
733375102719565399469716251724831223063391932870798380074848572651
612343493327335666447335855643023528088392434827876088616494328939
916639921048830784777704804572849145630335326507002958890626591549
850940797276756712979501009822947622896189159144152003228387877348
513097908101912926722710377889805396415636236416915498576840839846
886168437540706512103906250612810766379904790887967477806973847317
047525344215639038720123880632368803701794930895490077633152306354
837425681665336160664198003018828712376748189833024683637148830925
928337590227894258806008728603885916884973069394802051122176635913
825152427867009440694235512020156837777885182467002565170850924962
374772681369428435006293881442998790530105621737545918267997321773
502936892806521002539626880749809264345801165571588670044350397650
532347828732736884086354000274067678382196352222653929093980736739
136408289872201777674716811819585613372158311905468293608323697611
345028175783020293484598292500089568263027126329586629214765314223
335179309338795135709534637718368409244442209631933129562030557551
734006797374061416210792363342380564685009203716715264255637185388
957141641977238742261059666739699717316816941543509528319355641770
566862221521799115135563970714331289365755384464832620120642433801
695586269856102246064606933079384785881436740700059976970364901927
332882613532936311240365069865216063898725026723808740339674439783
025829689425689674186433613497947524552629142652284241924308338810
358005378702399954217211368655027534136221169314069466951318692810
257479598560514500502171591331775160995786555198188619321128211070
944228724044248115340605589595835581523201218460582056359269930347
885113206862662758877144603599665610843072569650056306448918759946
659677284717153957361210818084154727314266174893313417463266235422
207260014601270120693463952056444554329166298666078308906811879009
081529506362678207561438881578135113469536630387841209234694286873
083932043233387277549680521030282154432472338884521534372725012858
974769146080831440412586818154004918777228786980185345453700652665
564917091542952275670922221747411206272065662298980603289167206874

365494824610869736722554740481288924247185432360575341167285075755
205713115669795458488739874222813588798584078313506054829055148278
529489112190538319562422871948475940785939804790109419407067176443
903273071213588738504999363883820550168340277749607027684488028191
222063688863681104356952930065219552826152699127163727738841899328
713056346468822739828876319864570983630891778648708667618548568004
767255267541474285102814580740315299219781455775684368111018531749
816701642664788409026268282444825802753209454991510451851771654631
180490456798571325752811791365627815811128881656228587603087597496
384943527567661216895926148503078536204527450775295063101248034180
458405943292607985443562009370809182152392037179067812199228049606
973823874331262673030679594396095495718957721791559730058869364684
557667609245090608820221223571925453671519183487258742391941089044
411595993276004450655620646116465566548759424736925233695599303035
509581762617623184956190649483967300203776387436934399982943020914
707361894793269276244518656023955905370512897816345542332011497599
489627842432748378803270141867695262118097500640514975588965029300
486760520801049153788541390942453169171998762894127722112946456829
486028149318156024967788794981377721622935943781100444806079767242
927624951078415344642915084276452000204276947069804177583220909702
029165734725158290463091035903784297757265172087724474095226716630
600546971638794317119687348468873818665675127929857501636341131462
753049901913564682380432997069577015078933772865803571279091376742
080565549362464641260024379684543777339026472512819416320076848736
251764065967540693621758879307855916478777274739272002910342949562
447661308200729250734529170764226621047673037863169954237455117456
522022783324096803524667663190861011206745856287317413511162292078
865132941244815471628182079877168346341322362234117788231027659825
109358892359162055108763298087993165172528938001237817434896832151
590562493347370206832232100118637395770567473867102173212375224325
241626358034376253606808669163571594551527817803921774322823436633
772811186390511893075901666650742952758384008544635419317190531363
659724905158409106582201814734799022359067138146905116051922301269
482316113417439944714833040862484269139502336713412425123864026657
258130943967621939655407386524229897879782198637918299709557924747
320303239116410445906907977862315518349593035305923789817515891457
650408025109479123421758482841881950138546165680301755035580054944
894884871351605375593402345748979516602442338321406030095937105588
457052515704266284600354402823678768550982678161765520375795655481
677896038927498355608791541177749423573400764161093294003899982199
267257086957326068774974224802023307525187650255968420760693229988
587579898896460744381788170081548895226516722834045277219106991415
764639485231126794730865803195076455197675628957428881796812090026
387145257858315277615109088631740243695680567873015235427804793414
266495223833707117511265375503942372098784668049139473446530714079
622597287130503077258714875570502582573466866613802351426056116197
405543436548698005444879295970287590352258409782683598666446586045
694241390729095266249932902973440568160683805726626057277088407073
471496060064561454070734432782514087474275506722304845357006092214
390002992981608211717047917614505191008132670375214930740567853311
106058352912781007391749949197845112915913681107394055175208019630
539350740248509553772500367054665162330430425087442324262404632115
078997336929985407041656261041976700202415094892411856092409637604
429612002364590706449770627207919019235964807048923636979860198283
087284228564752353162882791324295524814447505521909672046080689545
181712204930321853740627247421519740305769043602686360780792004776
232429551829473522027244376339027721392087767065716241639751785859
254426923428535274328856336850789651962072519416556061870370550218

462845434257850383000095374518292958440464918838685793483961151297
160581665745096703677495836666693121881763679644943617130416037243
050658485131749264055855194018005180908475211868224616976149243238
319486434415908558011073070311201502243416073157929528752936835820
397003389112114170685219366589789459503154389589015303827143001929
589074149943592894083097077078362875914484037045038618966975811201
852319231868659968038583812370329156207578835948780941688205531605
128190152647592807574958154564221341459378167056992868299895611982
353837157880480478704584175394665497690173220310890070303362911767
308448450372145669644401469545173857434157810158618783839278552609
399130570255575559060947051498093487773320072797573038245989466809
680822221348485873822999281794090825665209581655472475244566743697
594474686376332428904269776106791933910983300422310293728298798903
209391092682836306173610173878123679898645149311702437128285882630
486298884492207415640607147059137405524665756971870217355287245439
427714809179364437650637861861324348635797411258520863459927803688
792498354363298457687650165065115345008695721239507544785683173631
557153527046524235259737513408825461609661440746675514226836031959
801072152463551069171871335731685485631280857834435623670959650949
946968820661185118086034202821331801249410991502601435450017432730
793625113070298250499417994284451146479329154599555909587807621636
668591791065435966065253525320273650725989121255686842802077246487
722010996631829559552903393312284364864475973560859840760947298389
542433932623153239918981852264180831296333546356874828863465618504
810632288805596737844562000941465603499280879405115310057587129552
571964111506850340773710604380371259575596985949362058477512026354
947347534748189262254190352671614429284899857536740692165271630086
060654373736823556588626486343689153218095572204456777137368310458
075584529612832832606319629728527966674362974800821318627921869044
284342630735760703999669430789508147269730253817375694922751795354
326156912040594832860949992366412287881226419148504856328072066418
557059520375030322916894489427578306090910852410601400683274205583
969773823150734996108758763704255564964086855071942256344966732430
656259250474581762733281816017019698166542426378763601453035946538
450325476674999737340835665138186025156520283637389171016545414882
674448009105704186162626837971120886141357279611099088292970229692
128180978798951391504270936786444983196420134566833908775943006442
485623012124614511697921939634409508083229281294270436599146482749
984375942113020418297308417178813090379558545603247170819195302771
465794555475544754284434408139388908609776017857389307518661906505
018077165001840744325854024184360501118242990702323417243674525365
349594799063334540754371812699399833719218485418735979845348934592
268515068182662490078029335012658824974226241885352526636702827662
499349829488748331061764208429016923052899608978604130065109028179
805040587107671179041130217482796682353001960220253185576789843317
586806378359968791601538922220236575765581586611409199394861599209
159917553341783033347643131635012705390697079326567812415906434284
721360235218236741214733124499944334155915274315931687477882533155
092770336202901222597794809855392200064527162280855398278906584233
447552821276517650572663267691141075034845871896996434875775138479
148183635100621466818585096348887081456976722020167991199462417776
688907917136865945960726468538810778783002161368276697026223459418
737476733537998884403427046803042551694127158739320398444374604547
816113056625176412759821181939661101850562880555942566060032312116
180994622129301002470913347150682268430458680300904242861682025562
140946087900065191099495570815816505828983340739466084457565780636
690272843462018587328252924796505286681408503538519837523637451925
622795490290557907030283950104854835929834542814487304358047053315

081510503001521428117175393649133166172621235405527863308002083177
055630294963594201654333094094177196326234119387105161570101798053
551679370860291366756986097124120368583812957695307798141365700174
761356966986146068491439699573837631695824602513342108072621713601
943018087209888551415024163818325975259593165531865833117126857941
527206612218422661411825154657484878312610347834546749258308729985
447421206445095233245050877431496166555251797168020991720026409374
921907569936896330281391647208963581771735555848592706524504862516
419540550801343510323389813378302497701822754906381499964723334079
613041469739476372650869273347108415685608430921316240434629863920
841660055904598506491243505264766067600344441618186403670083774114
101094320588955598658670077863671896944089622321374034113597199133
135946553685446692367652589012108413777432482191812747847892287264
892970032371873456157981599834839100412601050746964599430331978810
634913923812490503061433407918328004063907098672596197098311265960
147473725330526853717742146554005873924623727617364905198713368067
723952570781360686683261395014329509474851594724667527201684316586
608807512768584755541184381169011622005552113484488960668259227431
319007963011587084670117654935393046563356225311244727796669005831
190616101972663073970542531439818457379449486780134618217875939076
999602029083965677287846905736401564015047696448993947541474608339
918696889271156942345492651246645507792554028105037622035967530558
601856492056062879090769453339208808849477828894851122154743230191
383245562993881020614490266876010207753210915684977830740859649857
967152617010039475494539917698791323546550106407355816999409756248
149967443278429202762644189793918158394562708173301582160225519659
898769376164019861207466755048861110855726764507052622446130222335
852072273620485057289238815884938754535229186399714380884061757286
220950122506515863104258884134355431973729856217753072022629475552
483044445340434888878581170341345342522354319407877972846760181583
227097745180929342193189815812482832658950040704855206099893783900
341914163044639163880549658786501375046341695655156618298878630705
842306967660254053024811471007899784211830489010464056896539702885
595530925558636052158957375114089564905844156774937105859648014315
874614491250549253191164653821585197370093280194530320572628452658
046046337816631429933076646646530760590548962888724189716060225882
617577539922055131509377200624863085562820493575752724995567089221
634233983602565328731029194007041176919220850015116735670101958971
001797019578120892910969417754369904368202563024054822625401905696
507710581574240721496339560365270283334407305750073674562260584649
886115101689612181119058471714461068719761017456587373796740697137
423238753839030317200200207205928488785123911746471673743737923283
881966201687622191346233893762599527025672138622112458980212130501
407288904300322535504095866818724139369938193069148744717186646183
111942603161664070377316487001864799600243044003242241809402278533
309011509880870678268835317200767522553138008818780431690190072804
831799287414125476123089606833095828377667688287578688683092976001
011974533898331952588619630132917094385816615374171794496319177154
312506959853481285684619377669894277459170918802520012749905559407
289696594793331672243621567896776966708035229039018485730806275670
867658627104769409203565593025352743418965927002227049233186829991
560936413757004988537304596396152734629396974951748062696451793018
719986788537581415975799314806608557232568374305282764175670050288
040489429899580948103534833934144927885925262192415547231997143385
086637320926632728243514933640704589683852345624744361175256766987
767597223439206357507471552918102762614012992480422883990297879925
418517499129630283990729635588579890593317795908769073905646025623
533567221552259468838298452882922966275137162422172954678670715840

9241840841475575825393852409633020513497047406953995678979817278609204622868397357798151118681526598846069497589654813146511503926263777495137615572481951161198772503445647107385134359273555387124623755981938132142384415819290700463897716838872079163617414324970791096581627464297170728717251427458983568970955346268201690853561089448984071005819203021769451207717745887955195104733841847399807963067678858451675757299043069715426423834980098708699336709121083944535062459224323123482785496603746571880148929379451478705406079245759006012196221239287200172155886663457349714095337211516559857579417244198890261670161016115578343150254603287811984240274846085107224066767787608552476177738330895026100643883505502054563243461678594519417956698749685152448838475136181806671083161655642093692705206118985172926171417144346555087063060635510129494003097591677991584260491971209543227026784326542965724032720887143219996453132025871096771651285496699625526986073117637182074988273997706019913620930832307368382064557325637659829125781314922242204279712414416299512659456397927593803838047826231604243253991328511230322470375619423217330478540785762440132917179929792407833907157579814268168646553829468473992058886316559349198678969628404473449680240770928313764081033522552427174041076735654244410044833474401017264410529547872963458986405012036080244511903509949744939736171815752770937802092366681358416362683192634067141827974213425462207054156000509596740456168404517717479527903532549325891204833857465900967817304160005210889346107687540042419778030828851812001733695591271377141950113613044097532791905048915832463991434835316486815485791786329351239255525102111827885736960602769313014696614334496423021143824837056335327938588952676720766889712744358156320881066501495681435587965769098577659027687074536592763649755534496173080781609871032480137951361703677634575949756862080139963745517624251477806287222659714554829067692957136435721526744689878894188207512922257565091435528288746141950978624275278815715664007637210378031940430958442725492699871692343318900221415031139987652606887615667402101972017196023908610829749276395695411530322754601738707956259935797853024434767163995914623179312399899869284379757024923695515872976838540052276514956144471059719628898881571094151717015181147435136438540051162462021311748007919837497001004713634325232815789113554504533719052750682291561850033284695679262262081904424733403625038927920715859600393631533688427243753667996986479347411331983286194414606539227840999031438403545650470567895520248271760118743356436902435030856313095590552503904927316133117349225846446090245350791901844112993216997704518328535864804285568222087372136164905863032563689130841037602156799270200053223554398046531193397754590440450785680213984650096934295473102692499475864660580916699841606846460872939438082743082858174796941728729903110131926755738979840913642534796949434803777033646349584768629825901034707278612186230019866079877826842459338356389195702068535216032116352300649887446002001704130569853651546687520238593751832803728511432748116996836928492204473805706334966187112409478359158696268586435891413598542535776887749327436345147544886408688180303696524317556883002058607732569597160864854158344684324899630770113713446751569302448854820771241335577323069494580672678452359436315078727281579015730700331787968544362795257190236232746142628687327380094977411228562376632149046532940720261975390717404222595392428881645597965700309571413891069368450362682310539867437532400527015347458933256795149418545378088270634572959621690853835353703814181155738163782090325615198697453576464121254980760051561417072980469948135934831505681166427932193352798227147157673401860887215187996693502527007575560997198828630642854481282751392806947027501481632897273143473485285295046048832716739783

9815636788047804436021090073207273697493446304997314425715604331336903876181009488731207134827108158898574832658542075100779531183268617080370709359276149367825308583404823510036321663789574262025503501168615434073795045164828967556983589355220201736795480757819095026979812711487034311903631122461282953038205128704309294719745946908210256347889954317715243796962112812245034260663992688521330791963702777804488579205730469908009234401866381132520971230964760599899479257598510081730396068222199753273016065826285275825766950785472603493829813358252817867060851265600226887178112535978293373477914127362841886561759208328794474109697038798547369840254580632948350223593935435874802239897609162962501104739311694491006669072306346931301697118206325352692440438400937242844282097093648569094689200873717532525570305435398287278123011398080938670154748858034456318713196026785487938933162050076752641120443902375833427242986996547863685341028488573702547255023656634186809190383886707879072084036194021646701215348379781518328264725786288152071010814995589803381189615694417567613407170465385121709021237778843336496518721199054075818773943975283641439530442459139031788130041887918871145531482674699870555879310402403888840838506873416250716572741851349520849636709555424504394839480459791562282824837879341527203622633695618055563710768148888936192757426599358235594315308879330527675587475123650658439694756042971920023198680243517199378681003611023125683642560795974105741536282971800464977485737183786390370390153973749116546854997164539416112164176107171454017651905650525206622778831290457196932059902413753959838619826032054958395016755525096441371182225614960140030230354078992096986775078672000380742679705303071679322960156486228085184033523501706085895129122232461178302531636289439460736527713365116316464461990990212249224123151689927678558637363155260025034884878132330019101893996167027314169996265119457426367619650024347371727290284622097983948710659822700099549188776961885054326532118022194442822284251525561411874340180419461413945147128725275923912559644373568339728963312676782349103563329612947191015157143115795490933903261411918654752376247215311020793691158487422058227473432017355850771224379698579654915806279502740977168861148076163151618553068566924571717692204436684331273989337941116297224516999854685622157024175947117699529165502116855001089857619346394559088262707753114657752238846343519376539734984802454976076024403080844890106838786972612370978357824516680117148598367940552904619826216566917202742628548239339600182545994092543081696910329784112340228856001905493427502231852947128296096939768137341977042781213001473286776057194059699792755124617184349569856417128724811834654206423187145518241528676305675131162677177350617511245463387994265291270105789956718057214365579183506917779307040757329043974949958224106238105149176502385041827300966201717509405908054089572837554063551522199658207573513157075923615398639459211155864000988097552610538382568992721584785041746065161511337883360976012114848700556016581249247068256844272045472896309420306650445298646223594226008554991589149953606498428034579492757009497959450602378775019470624632394954957823082283066840818802521076639074230973720916285337176806216446935432317917855305833171420847988630340846572642693955700268576057539347888587094600582723230519108117514234912687336585960799891732928915896001815091816337400806035475200051511751029012299248709615459280262060761698272181029167315548929423740851967433079166078499055782101935713662435990883613859808516156417476946054785540081953530670803089697630452946868233210532878237438944115685176271711636309401479909649456354592950130739003626821007326370082356150691269643183351716254390304698989314261544263595113634660573786549512445747526216789547036289048304849968040377225134311

937373441236618586944588064018584073147633792940386340435919419872
355263015654608051868676068043160845128459160424413269879125385602
991599672787661951950531764883134693257366894644382558139108486209
663742674579831301222343872583124422033094571457541470479293875858
238997738515213523723895596643122356432626286011474890868171592810
668727084008203377186921535235269263472268090825989889840026208152
178282611229313118208660070996860365409818326807558247767069504109
975861436243552161945353029200254667367996485043373133495208210751
199258926638995647569858707901856123791578864374469037871509500112
550210038845311923652965599461900474846620642347942329670060529003
709175578188708193522146871427235277632559898086948721113845980014
123842163827824412736542446748833381679716201128861914154019367129
094789902646664431560983729615019686242282506723061667209435465714
251493086424887785986827595887490650772602509518295367651811823686
169447243607837642947624692263194989219646440683169287661615060508
138463194151162025779078630718012311594586038965625265542233462344
545073947886902681594975131168851436945210216883190446168629763325
229863851818850049286935727647668238555646365544964006317648285575
785866610228551564859908820958689444362546986795238226861159699100
563660829267915337538160661122478695313261585318717638859893779291
889029987938798100036973078489592706254104848593158543233956831042
390299070263443797875691855434089764407601308444819786265079476440
830134942435834281885915259293471436317533749589701072873501270788
980481635045676667693207553051840432446100740321676471836083708475
065126930707660849825299000317850305853682139512735038638246056425
103377755809864643398017186208142663074172592226000511091342681074
670129014301654101064933212283790827515001003530015654597508323772
965439697382047741626571065740821649960626227496187953347907065988
974871779564334064841745645747906925170149499810095353413548908754
836327579522407206986291024671703579251441766703886609906985726260
581240825336225218992000418975745765315123000064445715931701771688
635483333051921582055946117357716321132233931965320386199005116178
171334001070576652689919708169202219464704323795356411866063920558
609034457064151797782145054722278852987210197858846070047420028468
873795844228949974333656271877991721137916164492541329715652879529
532639759538535920950138633380507561369530899547584883024261962758
985941513780515805025767540401785795852448831172105089277089227273
431973823884687307168230248788688585510108073522781405371406520758
107270848167263977098731455162646911423286103036932984330300323676
162714264067587806731883971515002798163374779078775038307986759404
591073921034587404219617034925808189907205961291586420202885734009
114955238865107911371495334639763988183948804530075074740372280936
820535430494951948332833470075161979008687285439962981575605891637
624723069162871111137676086480323752459664930411753946136464337804
671165055504670671836221285795048067165630427626711429999113487698
447050370637900181096888629721757951732433802780617470496302042492
916619171886243355599282093243919445711886321556320161654247055375
938696624656334121541014032286990930159132885808831241242882876373
872742838038590710292748633351503090445328052597795658920554562434
297982794134891756382400771612173324736428540160610044337641457220
785921715591401037832020132133833096380778904095723810558829392796
374381660686835195059277019515361601722158904287856784820682919441
698718192862730827044416303962547130532843883379133747687358261221
162583602728961624559041896770247453827583966522993712351630489833
012421417455788591594256059792427721819908556279848605617453684478
923796907975594555154646853163024462325674034895845462256744858202
042457391994253094264224504202689038150152683602412559807597523648
162809304891274615119623154611400822056396780658535407668688227542

650381225999162076017089556747446524234452017661650325945665912966
786324621379919222961458671422482492880647680321086477994100410060
033906792752373625460277429600734788038356687522003482457694908456
862696057715701919174892260635208129738797443835483286136939562450
392976805783223402171676555917766840375723484409461762931288492689
936871389838822271060279037990019045583360079739277410926655739233
147025909233890654388422351324115388018559234956139930223919645050
450369352927011566305153351918641864823442499919272027295345959906
304872360804159576002966812111683172366038110542803591445720248256
456105714055462420821343520948108417158289572445072063546816002305
120140848054358742526171017681853883557558717415424775449772221419
261315525269109175563331932322243218525422182729149159810583689702
503522813002141192486014248068079753699647771939490680468355280834
732761030604940973309169031678309793463661183278453186871646268073
883365670456601042376850580139507443647963922284112697945134773004
924987864965636794909929132712528977651918175427962806084932375520
815361113240339713165504391887960198382138585000773242461778849187
581459642642337889793330819488160040113126525635693244659398400636
890315254722923991414474377069633893576192603918924793631780083102
611419548543605157787160049557886565797066588551042882466363057207
778902266777042512681571979533225107638903681976284402861025880539
233932947467202408854127649238644760216116262082421299166036229918
492378223630098347811952291382184732634228575912097980547828525059
183798336801787411242644746002256241498069140074097972102327853957
561512834580616541111792671042799057939449713494632895045651286884
784187175802050458328387485313736911351025506201027753458094391050
010218339732456504728894768792989259450198750767122363791875864720
121496606115128048709648863056228440839369443872169212084920085155
838125107074195518720809374694245973117281172105192890389637039423
577686212766821093182763664984042124938144097959863114225436483965
499983479084307021764385554351257436828228153032222380834767951113
557014806318200453220723794891863572149106242526993994671015366846
234105153338142684770627585203524099207972086991453730109551641503
317628200196916411546026820723669255275141842996992053985343307306
805737238050416719722112737405078927266340638850686734458560773266
648384578027718911475801323105519878413365218519071460681389868867
103147598264611293795439526672867275994833590259744587868768496462
683484434414135917714587766088077845357183932937193739323640835633
757668846821111799350554102085561884901020160050563954168745108220
603555410817666460524124966224422804545243216032036019464135609792
001959024049792923673298924553990101980112140290868699920575891777
188074146122205024728585715367530747814389730571787268366360157613
610077228631963885264623512553807731945956356796538236249992655180
433079635962110674552852142902629498265675533527310046878865731047
246649332656792733134512295505918623293739332608607745135077530901
574443829487339779605322849358301361837958626480321297368474817516
476913662110360369509106666505171711508278200932788358722598394046
306837631811808904423626219988123682680785795262197216687201745517
472627818032683058548803970977047934831035439855907843552776676033
139884605271503138856332467688927104595851932895139167823857735772
658100479825639355193520055204080028705967824973937478860528356493
591497838037796496000521244583477900175604246586665199807702883943
851638095504304921960324436090340085174660429627430976838715194598
264473594023424821104475729111777958773134155360952759570898612586
771456252399450075938020609355024892008476733229308574222255020645
569023912654366357852427242905605320575403082101451238209021746697
579765347517250146583747884808053773515042222404295760361375432486
199655891939220504699982106293160967565179075132296077785755331026

5858425760866867645355209277482755675451771699508789411805936305249944967012375980065534998739666395394417017059698101512719333118407679232718539539809764048527846743872316432910029065495308612833302664007580129618499207022002555972156957588376168784364346792755863573972253564884133060119289574642809357858081132331433115287482179766039712579528900364071989233281316116404169377366280132597382222374268189176489596422703380390592959649696482133114473166765041976781108490966469425717069457007871264014486522428469488976172567465352205061621073001019262483146821203551699501522007316384004132030333242312167082685468931758436630430784350785928104478492663952652398718644173380085681692321347429754583269402161253332837900960648627785494126679513674045877416945596140762656625029900692267267876036587137932796041848883939339346926354341548095183623323317522937035210291464133127520371171667548720634738923293785107290295144629274154676194794274716691603049782928896147458702649979707920638724082502300642554499590401197410853516784440901880646293748354439614400353523310304041178457228902958180581032123743825898702747370401068377771592512645357065083009214792583498924751274536220061058545759973693135297078143742841340551954446721489415057452839171603715453082525558343202512542416624457524562964457910769717152147095185055003550543906316882581057850746356562047914667680556984384552027709969719889807233714869563567031776877637897432734928293439051455670607446079704769316462781214171381827437856146219708808702106421105737785147135883737738824076528045191427137488110559744718310093937519765980210024101251123081368260338474491087716132285766026393884928495989823656572720426357202637482564949491262914191713064628059566982549360326132019252804346170439028926027993140436137026582012131285148815857311178210413103357288871817295262711200081475064026830464189887697478791731737038139991888242416994212152776045185956711909418073734793310997092831554681656395271010461137625406644958618385463898220899677832955011143149959368039822230371363295742321735744647342109741491743641994731958840052638726959231836423254918455955045343778467094704509594201202114220864191279049359945213739248711074323149511380429379365543637217263481907571135312709307952729522112479531498969908089466574769556512436056114200866399056099000380302506124236077503293413472890501316772809713162683495963409292243031195084878867103533520023712730202916592975252657039210421496349523857085605723434621576956985134068304548331545907536471146996824209102321431171769227738534770417794076441001301048596092707211320523185382227444870243327103987811479127546080836115687792151311310450083663631007517511025900280864277150209627136623974010752884454683316182115027892643072976355761055112462033248005310599511150543148482955343295983057427245173788652719300073232173623758732731489091094553740270481185557199051683938745352067970859211896407854895041094056996598871598863362077955045219321563361246853031747054439402941829263552401554523160986825531389701880153970459625016917966481250155593231148267300563383579726032860177847414960045697257834956205873287301245145557634523029864814954410090788352980120701265410952518460666201767420452573679946907719084537874820608029048251670176619820730618331239219353569004070521549893903446593880904750772416954365185807506649045944318886297872357160302248135220460109063521450828063974927551284769435499620339916448879197437902095718886320024750207910237907307296374632633667459427556378453569136734552401489712590948036856628232100500394007310663207525728314711519263328928520696723934717509829526021254947643301953574383509258283111339115390633766173730772363027988986998579945016592376906754883798892940060516282614004815046948281403308391643424865093963545890913280595111633455036563482451915058317949808318272813479

5050772717335949663371882149192837871164639035669257799424573943555
4730449355593968480327902086141968150826064810924688543383329866639
0745478052636291615627988031878282707451630327863907566533621975066
3224248645769459753596673200603898262930000761251494798008956712455
2569559827585485769012463686594942242277271771518496417510715984166
3572072412243719680672039270647894278942171284264133427118318479444
1334606472431411501550985511712414668243312352062840657226926069044
7479196447297528322749569819632778728162595401202053807329582500499
7445930809782409529912965423318498798800771681631986086512088315866
7256506594414061844683749631892913745993421603484822883158289730944
2161473689255851699271553115588887600721703410244587440208443428277
3004673097955556668115013003388895838023146431382900260076322850344
7583078087889518031398102076278898517435347822512084675949743002444
3789584289568075266320362769629946018083494199491270655913084000588
6265639963911040685104128200715324625642637145635575769452849271122
6355771963250658965455364821254592633552572925952814993415878776511
5692231191510233734407169916564763982000896984629843997759385398111
2133218103281989699457926176493582974837338775235285946403513823822
3062694536345810031936725020698280738433341175283157314342639896411
6347127053034775699155800311815918091137880268838547576972923398888
2860323029977043066628869553012102727057633959897689410249968479499
8168420119925613480756440406559462383708723688812548949148794873488
0861416810552114001845517008444484294847550732736642827222063365822
4017454988082913018839140156809050000849546573730003274779720991755
0746178595157995320223728523592040074251522563861667562031883981177
6186119602216284743190797025036745928280467817853664739356003540388
2782818457669478233745711382212193261672950104270694095202650280522
2898590935002394490874562620534522173119409577830195360518503854966
1406218253061820365182733706211198939024488975386358180994491815788
4878336528865436542248302027892417049689651104172759475017812267855
8143917486494243573009091712648771605959209744581146295542231002200
0851205225897647781148270394267766642782746259395117438071986187222
2655865040300284691469278646800318360346381726405702707422620342977
1875558099386871240465622333891464658305543013155095285109726300500
8051882652726853353729373385691826937171677303161186474948104242155
1279159101460656979533313377409593674932644146370242752453933503011
3099283364854070698403439912124524927558029979882409206644640425855
9662008887419164987730275403729204215810937814713136226288666694544
7421244955284909149219337193623402943371255755699886529662364503533
5192026777637942482082860568936231521523178850145213132149146986855
4835944706865850109813142058926764161151621094053567807368100897344
2458729327052108535726763805642288409296658844777952795467107351933
2954747130150792208403282322044289446782183965471109021173407251399
7247573570085553127432199967512595825680632358808838843662032622666
1914149347404364980002473983320924118386674296092694607014183881788
1107142824396577963884398647823137154249894725830411451495268724233
6189967630588168208463274374412103905527652187107355645257133601144
5580455856845586504328599176765196193271143498665407774514500473077
2711714795712227572018128864464407775174603282423173385337652989811
0442322404677246320479517980971576025800885768975134059480548268777
2884776293846454960402703705085394190927699370668045517194160403766
3511801855136575451095247034602260020741742823849481782254906365999
2084749037583205744677959106755660640775009347129817005818769408022
7992690460594987211763415191488225186704395573100179371000466572922
1803728487979715692278888397041982545657064289089858279586256599011
3759687500785698534209443995971523667673559911557090061413018853955
6006933050826115788315979018829128777653969640675392080848582290477
5561905186375490594176472080908485239299663653777468709856801423611

3707637046742361802921867959247697776529262929041798392750534329433
3844765333985012282836279851502637454279667177148419757339065728711
5430543215752354493205346537542382048448508846345908533866772925388
5204449844131368637518941176848626136036819373635133932540806852266
9214743073291344676252932264084533084493864715156181394136343503644
8177947550976339255988278690369632386330342579445292292377520328744
4890200405326681393547528550174645317172145995081455613646925266500
2271153373818175978557950419880754858113362891549009039080607754155
7573613737559880187573075362487370012912238261134381039234372313533
6898891533749493786324984941764281417045284082969399172432328677255
6415048376577311449335215538523001781108276163630370902052595037799
0925341104705700465652519779256793314108866326405926231788931260311
5285758716424211903337987257758742901290375936269727234314893572577
2418837941862768645667758686920276014398050163871435204776738809000
5789283633817797388457344100149966433235822257925351711059485607899
1824015219982852269465095876314924712795201644676474027046895454355
1030698261799914022340728548915468068420957432075066211544876266444
6757986364438802325863608869187594422715214296506641613849638150277
9721730712659205782660027847181400342092656930703090445702459646755
7649018527813931481315092036410498459690602253144748229457070252700
4363040611144551422276693665012542523720743940182775250894143291522
1517059974545931259468212143510622763303318504339488951276720637299
1512493681935703191046935729052762887687825004850548005973230753266
5227792552419913159617911522069419685479187341566997810967025629933
9932081645071741734905643398652199866390557093521198524390679861500
2144862392843873982018760228547123039494596615725875096503200712477
6657593813721248011341535506167547203695791055974610671125417117455
3695430147191419937319722797169021161357262524311647228936664414266
2124385498136236949635712821160368544160710823177510780129830425388
1419089224920859536461082139564811320531607370777207605599349815033
4240640775123315121589992462974978454743857855952270892671024791999
1996450430401660056217629623401492821816115205046438140512010176322
7979026932712227012592708163045794086959388503088585777767698805777
1202774618583728185859970177211160371098273932414719793766386484311
6000841579272530611640850151500165203002001427433763904187886226355
2747022589848494690776947476132763910525994056603823823716369435555
4706581748273071824741827263627240462399440284444736424586444751044
6902997652674973443569857085390578191599585996096750612830910194744
8865650751261397136329276415834913042083009508511004140745574437844
9278985760726105769741819633696790755188383220173443764398053682966
2687328518939530815972138409987536577466354932531139362559789543000
0911914267407538592549690157973419183710401699917900945678359628577
3224471479073204569647197863154908628412333251748127848288098487611
0221009742783475164627905539385196688956965108760628729574590889200
1702386720740106024538941519547393281424662231268923626502720564022
6430217769031895555206112711463146717038915773390065452869232720800
8111578757374991035324446693616535175221246886608059397380546894866
7556025887068710308118989220242174952934582195353009915613553607311
5909567346990699248742680019538217524621053498627010613215907572600
2408043008278683562931983842710521983547275117642330279958926872733
0531183558056875276124091974244476335680956874844410454670283523655
1415276562700804363097477453767809820873498038498259924881067029777
5494953522829951654655985068742831762852085719613937978285057790144
9962321392204623415241682380388944662426737300189654337647650363411
2518285095120888648562947143987795665592807491648962562185926715411
4692176768396054500821642162605610642314443579823069196578047057477
1484600729681823722879775604960891581786867293632379024157920472833
6469702103139751800978415985500070553649387532125749616748758725833

259925957615074339186228437988301346044540880817809685491194541193
470268965059919860410997653211196581062966550051161836517062029288
087760914984616731644268641970892306484630567545738872024760165257
760852937721093358445387107402729259191524626762353817978693064215
340131633701135735635111098141821129662210736726269615672674830775
248874448416766573702400485083937025583859101226694835806839154547
916601645691486305239359779324467255886717416048550387114903176075
537321944728305822191558078807524536969327446017473605242058646968
697577061218677619720587491045165142715495423853920232526975123495
465463090613294600566507283098728033873735155375223563183570253700
649409263808031737463485403611466000484687624231089472379165007451
797052486284672766337551730368736838564403704980661790920083171078
821049818331552614850537354075035108223939247445630109692042278844
737169688950911185736926890336659718522537770329622016708106551812
675800940852515068477579219138932138092869611953122090503801810765
874883683178827814252786261879667606821977039093260067296151275571
252786437069898354444096139173790354548518040397333137480523587910
955583040481534804539187854038243236907304310274062641777762657301
034703384021129669084818046162496487394734584412155302581522214994
582224994194195472564103175021144228086523028022134240931939327276
78195990608112598623967339458989619071679777802595116314775762640
285882625148158216439944135061960811758904619511585390826133549603
880323713522245169681180597512189590028591797390866524495280407827
130270045377437267855532504850397463757394646098408565893018482234
161498658315034660821862236058019481145549035154742662660612950268
784097547798140726823956931472487609828034508118938340409615343148
630112486764653154787584549465222275318773560890835043837081120882
441759938586466309397048117253004020305813409044745051156377054103
501416686191248525269493348297851018111472329874045396127540222219
095844050872306623268888497042234567000119497518597964940991489713
853622794588740760990432854228127730581830402494510870633698694686
740089481097539710090849476830410711529550638887652490545659994260
773886347394552511448972036104793757254472396602354774812749416069
835101314764023641949146105980556375704465155667123652568282701574
452847602207817539723371640969862649205576687615644577446446649254
773467297255570538828590789231759706768639824966294555601938731527
103627201242931201764252246448031819544683337639946131383614457041
608883422253715587835807016115602717754142472333152781356694009898
004445823899842006407489589238923892752289147329455312404247755208
380523795101239384358587754549990012720682866599985790984293038460
073296238426290797218233372747669464015269204881430422739438838386
988072365034008809524512726001361525704157749789546427459286696216
415427519072078965765676204708762910259298887712834058061317182068
879509627355230802280366588530930270461940061446449186278566424494
208162102038327611169622442138639731157130118991853169915158165025
834281284874149275360507355014927516496556894986881445782807241540
090116176936589862811374592790322578489093397688160867085700299534
572157942098099722053214575142715411220939886987456280116533207925
455196985191038428157268351201092367995242906867995456830838859301
366721852113536417244228370492060364815444971779988618739061970126
506684370640425124459951909006226082179845415139874086156189246593
08440274701471016725471601668601739769199766201119989301553540628
177813282386798739883185480936514175269040502739923269532293931036
045698425205947108776022321016774679279356253076833772206929809952
133275493410764068293696256538097982992215020076190656713233330719
175311095376967431445827047452191856565617305618532166042594645538
561688375993453276738278878122231537281113417355451707355320827604
407745254423078545374811259665463557459604327036854215738696224447

9609259367500830989140006853836358817787486427106882578787407992834182519771408422304894979155179876782746847540849289938647634983917539244593293129138080738765005052200666662727343844540498968011834325534999762501192176787558098067233241678261782570891163017980881955837910754011805096216010930804225701805492976467841153876914307088247531217231379403723659287710434554469626659999262339332986411371001268040811602769694022871365072981064452520165517338604686504062129245789271472274267638614268236764085164119476626514371013938556806427007782965968048607751794922121562917386716354649889853835751532497431583541399132213650515513841090309027554332364412022530077042821114714191814757096183313782294342072543410315558281866932838668366072691638369677932010214202904681337049153438059246547114970835401227241006503949742164188669227447368995062528945027771898946913296346758587926423521163354647468642605485613157784036114314902695442750564803847888794329565560484433918406020270451468278242315140650702210485195920723120049337176738352370930885652643448419467734538241329688543063024778255435028195957175433268735831728279337741010263471725258000551089980879204274477838536427497206543092247960572140033066159793981569706136609839640552028766999172254724020639606096429945427059154600073536731549880773908300158133516035730111114109280154122806666705878555092703338500983115676285161649242550929283039087709889349460723490286585602054220670371568046350038260527637108239865979318483093676416563607907066052334341113779312161202058809514614377394768353883950472129452834986548086483788501946767694562326701998713318455453483736084512767180056787542358871951058956527978045378344846504681469516775381369518451030832390374965716214330796386015448161449552393511121218944302382695405786011646737366479565206587250815927530571313438356992004899961804325495020521955502061792779930564245836658721675351928175033449923918332562361626502081490355786124405183440403815991358271738433734045297449996405991865666415356124243080016261793375092142965808828322195705784317169794628455133096838246000369899618059298795066037607124327255975365088203863609588090400380017604750786697443325877232154383259983998643950114495415077009728226536958394380850912841104162909663701274249881761634410166742340050683616764823271038894223948202530869672229252434075060265129885763587813750085100568868743282747187323242898477335425815041625895502385448906849676764892829707281158435116760776172604891355851098147895084298498360559365937105320205997904436973534016628764532063718869382189780157321907629981036125683876483872698536012944816073176186580668059683733894119826500873262426696002409088320762261178399915744021058427898450630360141993392836245540276835099897204218596209020162101565192235842119488202091237839275571856055416562054553471969786612350583489628212860820840349731199881077259045458633766108505095823850307512842596428597494715967542592403495586097964340196646672175723723707078518464663837067170299541698329886912472818768027381254962938987607223408465709509894320165487604793394679468513437326303922309331790687303169941800740480006872513659785795859947801994965234272868898871781351617155057783915871386404057895659182321370814005871380883652304716712718220060186088112572603398624035420675212769089210815522603293004441018906372365919571195303028824858684782564883005251812608103542135181224715840046275105924448705837095408353189752152361034204084507641376742347300588220343231604746330435062814232108294872409025947644118910322337404979474085782776220482618219514282179811243726766258468951951069986737402273230026026150597064215274602326999497006158235928282229783286840199729036537816816002884117306733244966283840324353650413975362055091052197490957998605957269413840242675559674863774293085831406648031844531532908153215494343

5828804429373556800527667018000947887335886091364949458385268927913655943428817418645559410296179299581260809706454746509023426184034501081240335390006107346941209783867162772161370836145151105007720117042140575102955114913702554533502068141165244769178458694354034118791350719472868333896624761011830170049726189561183989816053909200891172772452827329958680838010737813140018760672501269264546450976733747002367678201352356732426247888048234362900999633010976573057107450862132187796828074343989648355242714487573058303218024945210923199120417862983211064561898234504950543971618030395685126538014922516948784795547241863827862758232782129939782074286755471092498218244686147958081408355004668755962615790617175902192718697237845472411298557573179374795351829558429913369281405884804215715380746853113023354946272141844005632397445875377275180714660165706503537500007800054761003678636991113239858621322182246246434350103632239859670172899284252341131543432629303907359534291441393387428218721484186131279071626858266847205954664035651133279272928367042153333781564897878724347231657710811890588115922053413447767521297746355065511098018114547089217012441063492394924242267383494394078654658363868597002601991541683855861557896701272200232200316861954197028924757421666766801524808240221111561909829095288293422784064903953396720086499569654470752118461343409778577773642631658691698762749541886831332475145315900233544095171491408135927319114619200677579215856331076125470709339611644150880072729394563684925327185891551688147209601141540566400389210281186485459504119005580079283947161996760030187700072991661348781038991897992779330826033333833405791933860125992663543506471009126063462523857434635268474929790657800172876659682562194685410779874218445504710482511389936542799445932024438989851344256726693278613295048517020426704168104239887877662828350193125454951010870376696381206031276179962188931877783052045019481204742705204573212548733903930286680853928985514539518307016773725339156792769039073362485903433514761178705177976647101075024507681616557253954820094809110586317329891753118416036402195034635732195947558600832082926751237884955166725064922072060974120312931357435374521855454983025804156517986227801646893748172397133811236953637358110573939105369179739293431977518803252435258608082755374099972101540080046979927943422345447689707580313149065499764572719969962803326920908915583817603213989264488023769100827420906680800437399250454122368497194097746704673167378878520494165644737071325437283139540962318133764738488941218277568760582754721153484064111928660919806142282295524907588525871140721341401635238119989127477891313975746828093424728231102189843007024439996429064445084478802766865394635783597863301435743073855224801180578551630030594803517023052917619376680448974551900622981417402254687938598091422858374494142946684056784478629968730373668633975101391007984558831971893984042058517831262556099075164256666091448576606836793744806529724037099333962928343483326610413687134472596294417153661683256929874607519349004367548712450125173882289594264322061718377059516656649038896234159034283659246762389215431621094739650098692570895075041141578197189457994851682923997676852605909408476925555603209473017988926182294738346886884787742147478211246290050487616242097572295178607339598869641860539956912742611053799648648272882147298654479372705114310364153995043024924890389871904738048121737057256637134651471541312220563195699529710744845423257854093196070374806243288730574037414313238215835562671427568755755136182019176330108628379725855115674172305047190608736162770832629644295804827975636308237643616154555406169800458196444670667810243347845988069248477274895298262045169437003711201912953531129197138017595577974532179706899810786979967116140647258355731385280378144794618645821634745203

985589751231713640797468385145592041450052177212291446699278647652
010036539788997094195677954229000414384548714348852855651763080299
251676444247682186490621512191723425686851600605859780896623668832
012839653122703074654818211999482253881430040168114450362116720244
462048282967776160165637897576349795548725510809105781339420347277
448474876989841921828085630416492602991762303626322504418296296521
543856287607037421868140047386309450159109132542103032561351107575
582873478656260809325645074346337233422408558581633853715306945878
269202052395067272475369001398011496431659458297164868632204841795
219642449832794880631346462010891393287053134556150378876921145927
268505146771355995890632238650764778282690168036013061708569828863
363533982166411661335548040370382100344583808150558303401797120822
493909503856609585571395374634762832404217519342656686392559177433
783255482070386105633012623762876981734728224250946153189070215082
050421810397748940765721499083247852854595100246795973930841106272
522541569649389236827358143460772759803346264312598278889441818491
738026870449603886707186477083156478758911780354308201318656582034
354073422928347455769651498683915039761412613360789480997559164824
906255168553679482474050984649608568188917203699873757964398001165
295270277237226019357555720232631014768692847626362851893048492690
926409854724936481814128316893832831257956621359883554452066740895
840923148625755911051962200050308020425737002899660124136355648802
803399956946560958857632199260300046853975598028765558317107063997
506660476148677763563226116127152242671096736184025291082552446153
885776660277960808983028370687781398492381254517178987577906769165
132460310875518147960012167620168554361388753511114464644596594898
628685003842938167759796191272999045913439604283622782145743849108
066267372039815968331145831327755737193964762139470369487134483796
533672088650760949443106748938628101668608093548762040629531426836
790162232434421625009619198865282501847807500930929896168789351440
485278448521019497293149122933664283836109583591179266973210503286
586371961913064985733208661524319891775175613307253369060628944014
036246735791686124190767973072153896099260914778003921829096605678
051574245394812705158278656086176628088767548528264353457929751091
037432431480490509972013400938712099679922667327456972199757397498
352955663444532434557032626027829313689388962967691490051117916415
739641516223459624143879984997239721062591045242665562829601459679
012861764153524786433047855814962571113956032515036318374506194258
790732974799065403378129323435496477095994159702169181036814733833
330641513877132215173398409381746568333237521245212042635149480179
573706485748255881296241114146469266177478173860156155696776808063
542808133926222268057358604395739162738771435084847701866265316974
888647386824309419601892875891202138727709615384880950656532073442
058984978568214481099344327143794129234072975479326476182962040361
443641127465240436917542835856614059594332610091323144864164204976
494795520171710865170698122416084821707217101649482479807749180166
663180760457163952518386095827183272086570529825589266492312740506
731234877203497799829560941063603051658168190384801114703042390182
045758372731652085922539947510938900121122194266654459086779269137
115495078966657667654609628827777519957055450729792366620852350781
689434003204754374040076217990918813510949939669431342798599215806
292704213826756214353405924672023502064258541096859551282959888016
794748534882762322608988214260279669494883399735380911531026157275
260615166467574723112673113045630210164427562827821914879246698975
320978326529216825843304790854783365426975843307795571952000101207
872401988134949844384367638270411742100369511690111801683269994661
201008605320941579019288976139784035165111599346420444148276820545
506341848306161979946027048964895243897025843417177319031533093214

7980542020896195125075929364901627814740773224772573220191350456805
5599978569277543054657879842859468408586784134114538241240720656755
5982648262576190303383417425184853854038470371006908765080853508640
0217621010156728291435673677110351164397836344042830234780735456699
1438177047450894587211787839154166530924726979519526863928233003711
6850678762078775481783910819732182904787993291396078874176833081866
5318199940659792678221322713459632471409529463076197396749984634933
6360975806725366155180785981453495358216014802602331762520150636633
9939135142877511535321241122515057065723115208537650284322101584066
1898257004704391718649072412089171456120249173004379934999420658666
3798578734606048061922811946433156292568671087969712349623640619377
3881121802073791598180109759080113272257843002501113788034957920433
9189928830051624292176003376410793371968133192067582991826078485244
7571177524201683493481941400539164639352182737104891500365804792599
7615834365135534943843191509214629308199501835916709425302654032988
0324967615843963471143532471437039221486178438282611386688552159844
6134450580330263691439417435599175378716668814004529689343519876522
7230084584655015656598952113011048528816939415686706351783192218555
9553050002986483254447747771995501650826588967139640889880567958066
6916065806094048513928010222769761561382608319076033245484652866144
6494294839667733008070732006751042625141429624471453687509706878500
6600593940265187786103276547028063257299061968975918873866723051100
1237949329259764957482625519592739447176400925561852118577244308888
9458931304570975272586707145565142360341819890315195457218862114911
7103453059657845082618680743649773583175770086475879964322744548955
0078096671196162151367695085308923361238666283481102939804607435533
4272724428104903280756767003377271120949128434487450813568822156033
3050438835175410814830375344342084122081683605813262345767754279311
6198604543050444851055580041167943376713205581470587272088253604733
1064967931847963735278844788520587318286600656334932560235908889833
5377725079702005054144021055946107207649244091363372278973994663977
5123411788366312509006141623227657028541048506797449812718146764300
8414103002375256537304952767275484545999787163325331050619024021511
8146810014651262851039759839412882369862113183152477649679577744199
1332394798552871653023199869802398398473198178817133310344339890833
7958000051319653452338339010909704447143479426502628574031518152033
5465072823118385198658029362135224379754319380198343291431250275777
6675431686988860286567701350037258969644586868341764738783906654444
2181923585773107870023191744542871416003026828372404946436034787699
0357332618811431010813218855279858973034505344033037227691514045311
8236187832171998898905508290896624197658559805783414287373064809855
2907821459411264949921965113612567773076994605802064072391808669000
2017569564175955272113593375897911604759823155872535644568257143744
6585668898203737054970452907158469737635558706092801201769780532933
5796783807950227922001052016768988732541083893192509071742888108100
7086232075510184800417696968262903923998393811623663847871308193200
1855592678658980709850229537394942175424696253547043954732413392477
6485210376117773112313850016187130471064778393248758500636199119677
7877532680713924689844038826593605108546523692226192724034991210388
3162262972411440835568680499807460483713592520753901701446937391644
0928648639190537573932945565367754356329489530854791973561811689433
4694434436430308714442549106098294828815811595635629933779473922099
7851104067216644803205310670913037594884345787343984737076537474044
7930809034382443397058305326958562998479383048081779750890193239788
8196444747281348548648563997367907690393025212859195095945330313977
5185298186626201176126095321392633918271825632758305911893721069155
7764383887227842285290091226125140805231508120272624773706671615377
2979623651717118309181715228052653759337375581282348642969322667844

7133869598876915809508115049936337356905900842892007054825254617689
954164710778011758607143286624044830552364259377579855244869608072
673059076502488514081476189179998962929079540606916509862750703309
100886611993183653478110689500553232123231040994315669757128432110
589272907562665298306834612688174350276344573487313081278785396682
594804502445089945385062622281565720665625908071060090719474158064
342896173131515706055811399896076568427723954812062465492792246644
108673930170526784065224750410536043235086881525438218840578152295
198789560649956069827453289227327038537584520927092429466734689593
377789658067695128590449057399130794876253979899894685344867084276
328476440980465348855120943606428893738371053515595879507510368199
958600924794052205154880777749983061313790264128273715757106128173
624978364745020722775619521267432735816854961196988825831126166950
522240218811466930625749538470869958657459988789278684738719864383
790480463746222816126871276345113094783166175997075950853325746028
493740010436450345565804494429503453183381290785088833385837869771
084982066510206279570766983344517793452718037691141020755747743154
293290326295321149788262035159874125464228843952777954992895647543
471058985851590055084900569690369399463805412744078272079588120610
950182666750528291004286440115969091560260245872117456045510940768
469797368274814597904045521904841801154566347833534380881534140372
398178819077576306472338368480766171878875254440731865830501186475
632030171398339007898754244110262777492594557872631516087487025048
062038162606284156754299711008457236079436838831775697116071774760
197736299860847092256124190334430386806075160778365027891666283609
317675969553014936812797935466652393898654922082126132776378982029
467995816243987059362391705117507050493924429371228752072100479003
695203530541747026881003131427531174456246407354520013033515441611
612845306363822062031827141203471057333057060956104199977441294378
972333619529368071162946497417467460616194428419577150642124491154
067073122138420641412696715456643891594717779694935195834684336783
221413037431073429174373437635441590737738077683355454522041607558
324500141271289974101494705488647243415899129609228298624074551605
891496310210005958713471920979723983683128011017526431868611183517
017358675406492657915137405821629724201883751029772027692280780173
235365852486610373552463605197417587182349087381977451996041351604
688086082725590610482822257576746358191662903439070547597034809044
004304333317413461423454127456779872589232409091510873059202427900
149673701513477215142571480238781897278909931923211885180400304976
289387311988687633977056903190741451762975055829507905515712897726
034354672225187519472277750347806298881580276408830585887321140899
356256544526325626293042854399332825503295028936990777054902947079
622008390293221444112657382089568543447852253558437312693375479376
599430699100569908215603145081988649438948867959773652023776380526
949555871454270658517474445964682352694105685193373700491448623760
659795743742949313762849542374769629842362040406990322328625482822
335420165228291288443421575127060202153831784521856484115066939436
436446339032946286921500120033173722315945993702440466546440107095
463778627366769045642599775860341423376275925853631264370897307579
552699685031320690918306791326542030640031482455986239265759757317
759128625308946541251662284071633701497902673843253016190101372978
864695403425694557263052203876294232648064996238163085500312651680
544788556819973108967957554426839220485130919026882403377120177863
986046398002560372060692953460153673513009351664904759969041534844
228406494643578396273959796970119995996897055007139802671431539123
914611613581834068087605346672553050422397928096566221091111847789
650335190031281930814047064787403671555521140340703039890722323391
594235126529717112144915912874696964544557092280434733841013858874

2805072514932018367654986544261906876750303979569390242134374752592
2028444937070321982409508528743929412781595864754303669533654646506
4381229553856960187081463036000681022319353567758842217066271778757
2289539374973944984606881958992605790426632428181883297682570878308
8901643540546417536779752140149169816134993044910420427417299073189
3796985131245595860639919965966899610800504940072963970989595175742
6349501131523954054363842477157673057968997809351123101270006068318
5601347056168842081862105906584385468535226530994080955068645181969
4310455005698528640369722726449640722091072805065651759005363319428
5718826190168520911094446230493872762260130096650980181502161161898
3149917554486648451019396408924245351858629668535880723702520862906
3963751354424084167679610625407745354397187202203898292588150488176
4626321440193245912638467764538782149003218736052884016158146769340
0972434249669596797455129521524754130038382417596775542251548689039
4984675846106631594198812117971334525092753070140856142635030152716
4737087979022966356830017879902880841938939224918844891176700803807
3875888780169770111345334911534802106585087570025517363256882009754
9552748712253571825516978753150955690868985464837948430351870614922
3135734029631365279127615262306104314092395653522974932610180235748
1449400201075752924889589293245803518893483362322662110704722279515
7896431135332221513311128130269965705654236666071242736067583376782
3519101251099443703046290763466149644955996730321258522840068128862
3206013843915352393209115790604734139362973234927591808942336565260
0939483133648102906435863118308259658785978471502344907874767879955
6674248520510401039997571039402206306917347420208969917560052888983
7367593966293674017209821954183371282339328624773171964386256653148
5512699222367766777081986434967998415260451946404590163957896049271
9139291043490275683817268404705229814089067131491526250441754527101
1123578680129893682849339139638336607814229179455434491680970664914
3188453781202579621552321288398882948309602541545158301556456231326
8450317418576979979910789556567996082529165537586122338380700692199
5796394198374261176767691005073575014710412739177835963479441159243
1607409649189238641623144315098433799999574123860845568790650179668
0465904011900931064914597645508744169170936159780546717465899301700
4137539046825444198493069773963033614330040322637044138842456485314
9600091024035914860434195671889198585615655465775089014431777128612
6456862190012845946074216074295710458314004620124639011021019323682
8746231874639063905184609082474661222258683171698906360640254048934
5087506001935383235847779777771841566927112240044551677054193107301
3883694387978890462421759004666609104016362270064506716725632981352
6191695775856133339039827759965225099040387093278986221595494379976
0630648070964941770800581227055933091621184400463589378563243586414
9106154006820478790162140445787717398102952607173000991217971137541
2433348822666186718059345350035979406261016945589487985287382394613
9259273830058657862369012719296382659263937819596877763449192781383
3915273468510317128350116775412896963401763368803347613242500654791
4483551600242312566460801078670258603760993908004517562600906555542
1309840427357430050066877433135282061707299033893705322254670042056
8896404652393614283079185404169666783240709559587709423200941561099
5553433443491385438840861082486242898596197412565717042400678767126
3685935372271091567040606219434726020409941395472013175524491583451
9442749191929135023558344041872069743458860538337018589765720622547
6686389914740617138440911140542444892418125280587378444375990337027
7144323207852046419314755947583142919416971906297697904498821308013
9258759048587570102804989009276466743181741931273879879190866705649
6017414104518365473639211201831272642135299075075316741860411390850
0791740441726580092889664003508561829937247213684341314956929570409
1981300160608754127957466419031797332593924021074167670242353517427

2118285715161832976814222607309029636948713308807785556632397283 3
4225270656507307251890309039502297514554508141344442816541436441 04
9217506227064362861017571711204836658149705824635780075504562645 37
4462805259328415678857985069010580452797562628572208304783543668 13
1330317233238135264707525779523301528916639528654318999573174578 01
6782672814602226403818995669379948424210982489742008823311140013 41
0440951609308313090546550315955515473977480221462406761105271613 75
7998628354396966578355245670079360497518767958500434785944448348 74
7345525999632392588201044528789576723339110852081379948423471531 89
5261281875108905121354654969246066557673451857177405113980900507 49
3228007094056920655442879928976809138532882392312742649637907197 99
7852490903046095850203281301188191897987538612770509831126796867 21
1781006094288603341607408020448532441414457945472105469892916649 98
1941599750811708399758552531253479307072377194823738336760655418 50
2113337357535711604984088630696264989015115562982779223043334984 49
3678391519856268504325200844798554629691262997883130293630646337 04
5033155263752040473922415707285777998802963532890069840076781896 97
5635402176619429442475372564984572265506787359093405723897937819 14
6080319827113924804979411004922981431759499199310328089795747253 37
6814606174543313264489248037013462642669263173424435742705177475 65
0675563413336005917831376373759203890204265171691865422448413609 46
5986362677543332217290972806772121812294501776642327167331091927 52
3377242450590808589275655643441154844388895321327028560540600643 52
4034011774394263831492694203667677312492334460461527922287117473 73
2068207379504077538070248751397369748722080794183627245792671586 05
8756843738256966015288450159363605799764879554666897963172764144 58
2671409839302605844437311832195273578242992380530460979253217595 29
4376466967178599565547975260104811160900192559877031603692890535 46
4219693617900985472078551257259753257687803418428239419301167647 12
7802001324490541706871940608748642509924900412437779902567236238 75
2134487546868018057002677165904571741675093585374663278466147170 22
2912775381648935814037542041316317966204626016830058358402780842 50
8470610285632146763492164415596565399512441115272252098855763180 78
8086283744439537336386628913943990459186935629799316913207431084 91
2387826967081737983802760073283523713057038353881201921780745570 53
9213125048219766934063942802345335096980545219196416496966420519 22
3222933252249809906809438298608239338686773954452673632194440159 86
5904066528670206510049409786713024508960506363114749789754318632 73
7637932022795300108771768440261821800385909060770293459786409329 69
5112338532614945655859677117544246194620867417898814347780270237 89
1838654750736725070123164595421042303682530154992729230497065006 88
7445529088936646551481938480563745544703197171864227209658706300 49
8343665718303094585252856298925461326079017237462336013820894764 04
7217671021078363530632839185289426309708352418426880666397245435 19
4987459495272368823511064759383705313614945233262990060613544202 07
9008007844359185142381095406246359288007491747232601428291506647 35
6469491490753130404119487461275842517254229933765619132283441361 59
6884512305097922908093477806100012024307933675460695671886784758 90
2916716281510479109848199687950537574761034383928929818347559135 33
7283428884922853939659501645942296849021635769846036566677068849 79
0614937895866238978513950301955252071159479162430380571339104412 35
1279771742589499718132089939724094576305043817654202377493729292 36
6664085826356304701889428471366217962807794758141064720396869005 73
3588378323839385156436769291095321263095302372341887763775951325 58
5719886841563511434654449213461836258200177391119635659736209174 80
2895107119119312161615049356614001989154067719147406045020084890 07
8521044898407155872491318142412374531473909585928549192619551275 28
1540455554894860530439518305516386529635114358554267895788433224 70

303223984629694037003638667059755189622821668494721551679940102372
605276192061750456049663717076263845953005334443878994432854366301
464142075150267652998714841483859242046843515052858926483541295999
641906383622255006162029851790807999551716142892743322162806963512
162902965050345455980024292038066113122499875748777814543334957813
655800830045879054556552375964308994728294156584689806294311272597
554693021887912731035300216864227633661031890511086335963986070974
737419552934178507801365337868776791151473833252513002371910235875
886798039385529704998321830389985333735331510345804434025730422758
682609723983422315017640803327317622631967565989772971839422971652
277619673408573444137475914779317933398924359940581396032281346592
478558756550575159428611613176739552834815080851852071547951439266
728751057441386769718902087776119592459359290863839696200576865099
629303818145491280734180972040320363366766499443919936264145227145
073503705907609385752440000947482297136237721592630836022139158855
909461407406976301297086569690667624421861836355204727903533175309
360779778798470802390218595874487896074523749056283747418931026806
280481743381300198221277706092184700815252327145986723437854310497
839043649305860764535756025983828162540984963998318112971881438639
542654408280086193057291799568887988182572440923086077708626351313
609463097677409970284727668296668542908454052022919003034322471898
204993824806075163872794566894084973666516228133694882858339531350
502170536111817502101016960937372882174683892618068718872117122032
067336498312812545726927631056482587901068575208080633928751364817
583758109695596993384841991062692241136561171129547967778123862393
493414008547053784583782851512327987308840937357211004586036545449
453018681073293761108679782828034366133339780538614863594371635008
771193119559848021828992677797694091393084926892397161087516259813
646427951911630339179612919001760953221655734911037471209457900418
602899221551175915683036249471608650518632279742956136519830351897
142876022375071605999046852270284824202570096299382791478180154066
416917925969650230616749677247841947414200924297729713888325116655
222730338964447305270490241477275647154092376806641652822611221660
551903510495321695382999570174114651029984898251643905699093945986
820498552583128764456838984342109365894310511407555838492789369974
095013891988212479916209888392435587365554523753623890641072663136
573386887150143766765743928298320734613140394952617095308448965800
565584630952329631871245183661663042774985273620234906785915576936
220534724261112532638914302528937667373926286064609916642583999874
647434141639526777322469333134089892282135236716095628251349234859
268040735518153607196567395027069135357634343767443117249084553776
703266698841449114808884801324930258438177011878566635729353987811
406465883694172838736570843375751044799123597365972434455742718384
733620516409860393102195921211225720343651001396389064945296714205
608908617698288316388283825229707658189611895457298258107339454017
277497834540687764110778004407429296639807958026689312946890891500
618618418921853041643322169492782133921182771902167520208059672627
494630028053887779459621856830743842989256463024089063366760647389
704968736267734714319346437826952783760286146583892789336232361603
686965888599440717090143857650085623703570747288123004277647474703
779463200055437274736584724026183902508185020399413109503970812481
221077612832459563995640730378422829494179041799135365337060929535
804128449019567717436326558733430284401481499075465103281813878210
907214339837454150957218087233216393141184885404882476133156499354
030311313197143885666833802176668360829503236040595136775927155165
679680295859733803613440693078137573011613002657970242655917863431
943626466230186872587963057556366078289969536349814522388660073014
771879198641606614390805777255192448707082910976735549911200612317

536184781317654395572950385452923653669413348562178789261547404561
504523088531184389783350748069944802081583047802929139502742186788
691980176555468181445443074191102292721931864440749790317259939771
362180997111761468900013772407009234849963308320304297321967582789
400084665235071155111834810320144835294744885188632413303960676395
857662392727435386476553325926113916010589720694912160419438436276
922502040836018348118271585534392555374582362825523726253314359699
646366678255933821009175987444402718522512906425157453594796130527
189599488847824353172256275413109599850442774753526388871189264977
071705502201056823253071154389564758303122551162876396883514326272
862981524075588799596209804394659688932951904490105741914199814985
879000053348961206916311175468253485829007683953766264145205273936
078651380572241506897185582776538952151358565897640730148811113373
869245888909822760229733951244135075100387258948206647854453929055
116049254655683017922363526377554626840904947800372684716102649508
826206935748463164396897896270083376303743017561945738907884810424
309285523631029835517451745446606529767081994743205995165291596008
715652115461296735413957327751678443484864598339137584856250554601
750792209883587719338601394896751483376380213903415290428362645376
374580878281537904708544575967614210377236123296196402292022895146
968811447058289309803570015041036948417054728692275197044694697292
034951969766217664143620321098749717299081440003715739281891528408
319742294373367747782587092187172252984069305556935225836786253877
699037108329877979505169169629957607026624877256111532102452287179
703093738005633540592931089018740049575133056457554686458819253443
722717048411076045834505257242917684423561001394566824566042880004
740729261916575440953365000544583331333348665819727492760012423238
225171846893069408572324615542392813887422027661693797935663441045
037115598236757714312491279623141150164528880440942390051734645637
803629395327787801636044104275918642025167711823591081249478448954
880725855125745496079956918012971317205403842490960873636213513109
752862098622426241874243578376772912641791880137675200320563326189
241018616513034810244006856808606110481129427409009375274857858586
840922973157793668860861996442907361574905330345107467921392139357
341469378267840533131992354433034607195155205700661001169301061146
455648916805009145005561378494045436931348591061841933018954548638
521986008082004028633222685779286594749900693607510578023474201503
531773730083649498916728935090935421209752074727250436661880061334
959093061140711046459924475971754240199654073065408416973523925041
563550039026440029227515519108629413672360002694120712781130876365
316152244335162266919012310171362950133169807700976703122592334099
542352764898442089109243902756481617004072173927256602429583715571
067168541418713003571310230544725937706454206437302838147841910218
688218466567138326128263780697530988608290663303966984683962467476
885114506491315186155246294792481115987311090797711529805880920928
581622770652756777195393112035731943345934347337295189941572147227
619036780043087959047999266424281962016300988837148845043980122446
245560266040861613319972328497876159317126814004045056189668690984
708239413708551813261996376889702124152313781873313001560112199565
703541410653535638452439655642672721743450531708970862034765475867
412814640719792280574469540684927959945904582090187317658661625178
733129693572624876301827480530656002946241810151431318690240874938
411242158215083730741283373100322266839576907350698817682754848177
304995391310318465327838386562671747600180627887580492540088784039
279856464964515527892734502001541003131905096271080962889523768982
872651391408194511671959166631818853373127982279478096384186330812
324934368273270884716848408230651068049840198996614184868219292371
243226293284424830236101798391004269904077447991908376102111123960

672507192997931365177067316450479862322551979702992566531510159660
459669015088706888298252728640598951425047655646438613971390930202
571945855825271392719811327758886955446292060520268767522136796682
746877587452876077863449138269956348008254414413182534720494801421
265432982967846686054377906133891020607653897467837990904198226428
291713569800434724666969930157511495371520437403191810795486894324
062290945862326245220966757449528566016465787368842465402656045769
732900129583787420171151005742659749253328682586625702458378115218
220127757607872278753625441767851681849197994949764910003693490955
081945020556381122496479676624965078580232712368944622866979631971
539024990109911767205329659201024327415836646285103519400541457190
714863882469446903822456883885007824410392501635971537484905694524
560531254037917603310165347750019986495805835536614207169974117331
055254042055211077318581045894610462735580709471466835287835222442
439519510959648019339972822544123729119753352333978820050032094830
778066283364606324667100180087066628897715761318039445308517785997
967916175623642457991318747995295187367560206724336078627831644655
047133342557745622032970583706520846148146180327955657231128913791
506107878236724170631574279086027582680483282048253059594486535530
533557360894366837877887790883577331658156656404633363117896557755
386745135965474379288244327761776652997753788443212262675878961266
383306843849005800577613730946043245733141597876165553722630161642
335345100237463536829894247824255806480766433618052377415631403789
337126999008115460840814240586928446408742389124577519366466994637
359158441193177950085848065280520451386178972329910964611770976297
169880547414864040358883927950040568096688268252678332587535835160
050579458531484837770296761832636064913660564711850804916359111816
805735686256767574836279625954231444084268694441780846545900109830
083247012732767325186296528101198756674251237185471917419644610996
381436922527648768652429643328488026710488044888015591064476982918
336443256383798347892249924247347347492558557293151861103453413733
567227462578276718755128522961571935018632517217599994227794412512
769491665964117645331130767839435875570151126833978807782308932767
292196739065650167909884959899971836201837724669791646815888400401
508326413390170244028639070088310664906834976762880088097131577264
334164705251536471773066139272240563257100139729989909559374773055
963634856006159849612535183107450428280599101135615276461371873230
740548644387095103762391293174413926799644747323618213633118585804
069936583777606558414953328326602877854696894300229268531019343019
873705871735821809800669389125076625708474659506289918468346949911
962050562881006235243400502407512125659762183568345522576684049165
251570758414614413289520970093068729022716370563859061059216969457
351312296992925835675318834452109537570173563261816644245918307191
732592805373518481830987229456262172540444189864039750384513606111
006210718088689290538855653803212319776645007978808922913907197183
215533766071468815888614665937080218111848640949124415780158696473 7
239095958580311735493963934232398121883858322269062273043691547964
773290362031023158462282118660828589608169094090006189644213461734
468252143386306086410764913030963038606156122694775672705661641983
226612829559405418526700993894418145266998151219653967190513843135
365503213138068242451734889475925031241924841752557403818235113902
616355370936864688471015259866820062966604332671588470284672528273
675136369158934985721514957696957393793129333387868587015586438472
121908813119471337087338232750005623992374477172103479216899958700
150469805895623651885426829398566671272305833174739467989387917984
475726396699725651509335049449623932989411838095115220273859361991
620893155937352131938012702984818829682456924664015891024522408334
073529472376766018719083566257343946835470483622445461993712921994

552160770522653798347510666769463255511566494911680705230528173086
908826823801294125418146730584593435812734334074634710981016973378
451137000361466614777973756676762211878253942360370654923722566475
192700250824888860406223409811545113334223901773684113599153372373
184676634050715689668193810358548079907399613453888826857567243504
591899740069104470411162878652679201061613247199984482371523349978
363752301431351328269553952901086494205818604396159053058259754001
573475299749827230953870577210005396418869704874528973591568792707
994416581049442687939022278200261738842463895922113926387495411419
594330027084671423706812813778229848743892258019606732295576646222
560726500832043734636892069742573101488777832814597005506211252970
943955134820697067809204578900900556359931930300746710425701791847
467995201645098538151015396961735455277804306267757948771097991362
593662234937064837059816841914409008692838417581369607702626526373
984217927518655855340018024947388424795076359369625165815980054901
107970726953244886137434993884408366146859209020138746292972909384
539568930915524703254564948483425584353927500268398089195124385714
727889228818004727979105941649371664175676509443374654097289014406
328130189143863392806334434942400260228810471699972553393957076410
706789505905241632902212917615707202813379630649859882322426710298
264264548227932715480458764992124132126818276723090347559579303115
848248948301417317193431046635619982934226608452419772743000894757
519064436425070401157381317794809599533926915579884005478289536523
956367416572964880463463057367371172158099902098944537332554066492
445556570479772079145812304506188806693477316115492135285980811109
640356420103206503138783298144430856387206579389407056232795868744
608528406980628390128319940403175369817291011930274216487446018619
632159446845380755709872212964758426105804371014414489107488133766
721383545414247871366665387182071284847617070028023077139862001523
285284674980517160094177008483060781630740674129158570458579809143
541609290613494597096889257105679100556752900747504379946338211192
119990091221539655631726332987359358386665001897021037681056553912
581127425650365892142910191935677400796661271382307140818828418649
325456700504789023579983462966520539034526722973679711222964757638
427953370703079415632893117466348996286910518604722726888778758797
953654811330971852577488362549950780896238311682394650511685470862
613640217820445276226218509468771458466676588999479371028457027858
288649455781921024708840980548840494289202758632513512032768369165
509333757568774231103616106683832158080256433346454271722024956218
060593586057783683982546182236449833541991908181754923962168710528
049514224637589120113615979984380345538987436863794164300305130378
895831279248454986839906586006407899335281278519409840167197297270
699322133907184209551782475206802684636165397716512345743403044324
661478177119961085537282430917112635195011915381033226170096078197
922946035526018787669236212486362488512903544283973792325138955506
401423913076654675381145244024706837652806414248720891345137963859
994493516086771074601432747723851028474946663633461941723016077362
976288977251283002580846877265301516820292508730013462199231565387
199041060550741930363390184442397874423384998260696760570205353684
564642727270348943923664845900245979494739486041667113357170281209
226805278156883353132643317590299465385748521847109720477182480567
215619231319966276378282067062797786438225580872740355388755763722
582999050673591541471494743726498397870576633115053342116121745340
896541521554977462478886291183035260403687328220250708935308435234
580815071956958892412605287571839649630550766286009111672617530072
817388845881237359853726929926264266600217297690409322916645780080
286157310501383405996052151802023374674932941095769139999676638521
753746488507214642276836486091831973323639215924903900069678881210

112974635837340525868785445702221462087368587279664145301762633541
558879405907321222539467073782654675608107464960418043395795387211
330646467992861229485713933856329761617850891155827661197902337999
866357704749637968223993509579545082055051118930347793570244303528
304428347024105904612246808113753997074287434351207241798271000829
331913714192887714098986370546271136142170603160388771587341075626
603462603469320575746363265306120596147410967864366328128489246217
276990604400356483137270171843261107628690706296287678248337252181
678495208701873888835266818806885615538210291793846881259759223871
757568737766365217279182935988891248129048499965476445965554595153
192301986734214396269052453374634498603717927205427968168892955587
945755341314658812833102455747928050200866695716939577801534143906
770746884443719972294731420962430984645053185396521906026711006056
621714505652396167629158214100393073338929186256703337144724170924
079448220819579349698115524925573254088088316481951994849251885979
718179165071886497535369431957636602624261722924254800560595721748
153559340925382832433344777942345089465946829548015616400884023550
373234965498786621710766801062510274472340547773872282337063244223
465713099833535636179045129664535920772793879392700954660146105091
802732697555135713654909405170986914333834373538622395662531670508
132212346736878144276185478830585005810785155567880769397324212208
730661826200908305041506079878672078008638748314710467962218043947
575563099086244244382809070751636039213609739671934940819820051893
084634184185137758694213859700251923572103523597814756562837006498
935806194277478376736716568604401242535394254608374734622249608298
272472402187537341510443884271408930329039663170598527274357572249
198043964068936908330704606903363403761135669272008017206016525870
166209246565318321783590348318466849633623177354463039337934892379
583823380148352466207076888417756468257271713619148355289440361157
962468253470999577854148164846673573561133803192065822135496782962
945837948992590906571508585899240368777247095602252060304105945472
235734307619920200387034244022234909496718095119479811812317662161
328126574188879267178040238578005598529232561568824676516359048340
588004483845823024199841762420397502821442033237813646956129181609
088070522692744785023579437156142854961033099970139477214606174500
788247541700679178188133730735538786796010124219243417398732897632
280986762293745343728998117259300822232462437598540018372660873832
647120725544913064336449951001947825445255425611985444689633861923
341088611902366362520061671773407268444876708707863399288518785748
868906955952057560806553597236255486657680659973002696144997913863
949137643343951178186561697245750119555271398766633102419936496159
367342333676593518995108210508545586590240452439501495865709751688
017729800819922259725289161528326432871330191207202622505599302105
520059364272067206743658081959198389468624150753802751656622826042
558448723696342315273704964736012472937470582351894637772876085862
713952359990692232587035991070927536077178731275481509403512701381
707948704002794636433688427716924012826404447538300216806055597399
111532756743042507916896649365346106649030339264547982624507527529
703551170293895493926050261167328050806361911350415038722553548052
495030725922083212991676993938578960521919040226593296893201528053
855848832676736575685837994286855431488484598780431993710784840893
374197790800338636966596327000480075341073313028695828601359287661
350885694130726895270622119444657090135000285078170081732969360699
447808011650899774698383275335446223117890041424456125659236190671
377822188309901262050387138637461107547138243333426066111911249960
431197487300355784675385580931940536564143872408715930702800223362
024034209266924841036541392460325281513910602580690086792469478464
151377425304908113337485925659032521084378705836901803059338532970

010969600087425044814184589259856965345569808272371276255400483792
707641017020870006758524443257752645790361826803605262387899668756
268684575871134948261702787264207740532779178396605930246805376125
287836224216318147642047643334565869242915619461741479293032672745
331987962759051058255643906412796059960516294105583577003536563242
856713972433093599861784854409718181725544777914093209591840501649
984386128073788718816754788756505663196319767304705864894624045949
269766453285109198744337351215664488814513250978299799856828301830
292718126587579974915259421426066384493476182366943613010007783474
504544383940594638253164174696216796375403949152116008355340458730
070341674476885386353724175911919125730296287576998669830602845055
125425577813049196573537081097538898051449828195851720963288792497
596687858557626872836385771428233523466567958926948548919544874241
952228540275810132725725884846046549518516222527214858969727263289
515266100741919597178328836594559768570577262847855954488372407579
162883631484906477914553487265755850111942264869962439109009594842
195050118285457021898741034571838981790486364649082967773150836776
997335515074170081220258053885245195363983453187617781231829233845
161492018946787202174802452815919012422558651698747259021550762497
491226737659456330761660211943484032397991440702488173434330292725
710928657389894240649561810909797654985118424771133900728809299016
301869411421261170343722496674766825988818337774891530013580023146
024260472055275799319899409643161144298528316114869897317486423082
626493416316845278016222869452176524878906995560991510960158794169
103884595635966852936125991245729283769357494996000637454051029331
252353719421156501331547537362809071428157731851185927633100144847
811577305157277411636321762995559710643742787164040829830710463054
191559937148153916125547811264374389039745212073575767777507421150
508298100857375238351838357539933202975989157824880504120705904644
073227688483087435345122645470626940975444513697572570891505730232
425735672721085168470673901113721018280580460322479166007383269145
415931982924325433746486049633965342248172938325475111403759384377
805581002679023593384798958654860787419414168840730434173496904240
691742895829113388157394322771015619624776355190240217127468627824
721979967626629009191769556438105385692640185914761669543194077693
489655590603130341579094455197560296624875454879091117532699370937
126380672256751463060740233445983148205780778552538169343648053568
079452045386888722145805202281371652698201162506165729737974807500
297233921909750122049474941700659392967296028738767195222550630864
385036022864184376624009174719032833908399536747468613101093275450
853701032488164563575489558603679918936112978761908356737312274823
781827018531064263096134724871436054931890377026133291202074118570
203965492336859086327290034787222376598941863189523397575264482623
228467814741678394175878779284134112230198805548371919662085993112
969783033886516758544622791340538447608394235553449033163300012475
799652761685263194532930959627313146726192661819832194566480048912
724042165730413638330422266489785155626456553219711447297326058213
214861528010097276801510402989652020638606860045981614285237499912
085209347292307977350301533905977647834289074748778151734815736268
828728847309608641886645303294760075330935853802770492600732884329
411952086482971131759375253443889588142555483851732952836011377915
329011335759811747590808295590475065765845686049895198699305066250
601709707832987606304681600952109729778751360186320545895578180059
587571719117232350915787137515399125515052606959476059331576350909
179773320833613680719455156407495330357188422563693171183439732516
057365037747321453500563595638262474936382247058368475214260727919
955251074551304244338336405493970033337134880029978594657544942765
183034121119702102998733619911776479300476493264915219099177723625

580527127257799284186221272202594729578364241569518389042618629319
498502398088279018227708670870719353980183576382804721762702614902
491846340252361295645126001797544961312208728483738833193857402018
908281767850298505262156335275083393941014555472125637636854640790
946476538165508050017967373743140990894748416914300650811821039900
941917142905544287434869177808284127163283349333738760189805192382
363730219765007029919984095395364081929393544844337867252570777295
959613871007157921472183707580415005441349860049290749969893790348
810208279250693005742360174677126398250444798794775513836887538882
077572120635319586500300839106544714907549279711557218460901553945
733251863898128582468776959800827417655549925565263787184749887063
291494390803074172603675896802548718373999961968293266122412170677
133971378809252017026230147178208006391621359820529738550555820959
403332647089156195552235663806264125747913463825374925991288014312
614436201171810061047225858418500286341562115688441856620282727660
065536243416531861727054704601829523329536489605733307453064730077
394581740556221801029658695455429623213626808519358450025873573058
666519561744637181113447756361029316422929997712848473992479149975
246975746167652401333988711899350291995072594035417767887842758631
733862021231433122324549995210226464191705902063721563648704410426
983363333138216958848319812093693683591904114931623478727636627592
154568410702417405297925694298191498174063952714478690511842343371
955926123191937579062117858090932058847948363057956121560105651820
752164895293647504997836425968788047609259997018653611113136048448
104343137267327149325176406779591282704180928409930214809574578663
493779227121375537125494983646132191054790119508054816377823175531
880540483447456823482955282130638303595464797553133860371316576407
833408859359457376719674086252518097818178803698601166138834712999
153771123834154528740489956469028260300688542763451896235735546181
582222140719678668431026826557381151949371316182349253043654779188
727730395772916676035989068298497927532644579302056250041982158178
336797583245820167033440163751943613079360668770605961550745818730
074058855418570777129376539546111235520177374526753650277123601026
263711408502493754519972388118497200485295407605375755048633498517
603934340365895296086073448055312295533568821456711805760475884194
205834963384542165377020226288732032814262719241911346980715350623
640604880106101761396430655066646671459774792751275013133465967646
396069944057031186056087812280632896781657653727506296728357626397
482838464729501891798560482491985007991609232399766719647678330126
384650808804283111098502554661296861855650350012361068529743566446
561984920921101266375831195462401126619489300838284386599999928333
379487659821355883933309759653943516874770254203805203373382317893
878282543047736859273772357478886566858736092568691056377446851131
559478673651648492178603892046920573921373965976293426617993875988
610557138473901586953800144003377394259635248692636896090870539526
251096127290887376798222410747678488299026259214172065135443271991
645998333033382050970236703791891297771139000217964645568170138089
418258462959876893635924393798037012604370001954537532094758565668
626186913776932365553855337366401811426040117126345320537251244688
083925045538064254766280934506039108615119488314246477393594536113
462632539790305310615515404757043183580698889116850882783575406260
470081334894277564198811046150339108196697436038560732670871560877
665885891060896072087471582697016905626687199268158483351741024109
506049761333221030468320093162948196664178664108925963354038629241
301524764041519532761824707235278977276957745431491457204054179953
185788374981085050571576711105815852167055220110024031214717157984
645854332489073410987610992956496761565447188062442849337194222747
404498379859647558413348941074260833361521207750192980151294656720

842115507638816459889661764643697603289243245105302982505122426807
037312128083935122022554084151320294799980575137686493365567616784
994713349269562577390784837182482831556792981972877868629036310560
685800990722240215387661471364480196561481124538862716542334288756
197979204855730192999750041758818622035508435269374224183477542357
560534672554956154189883817856029267690850613136595288039173555602
456879717723106032174576049759502322562941963790630937955810449095
876213591677865829539303065765309230704398675706257606714270638526
055475959525321304780063261071076808321621001457946409776926800691
390719372725319228526274289573895041376854774592960335922726252666
683521707031894962827245652845824142546063037280407774797988541294
635397999246474691335524337231830453538489080808931525181357684852
728589173285917464503656120068829470503204716904153768678001929305
063669577855088550542369890122298087791267061052356297358060222018
294315807355521909375865774736264736992888812797829333934998697735
232413759931554636311929820706537274786072589973120693062721040157
239438426087560393263870639290221903085890987772201985593853726881
479322882922369825904643093397881652299859711143887919168112556374
983131611093190611563255289261205865159851493976127055624087676714
060590627593678972863204655894075319271591295117018443755758535236
978206034603081114085616222042904289052870934871938753668199421196
787160344751165632170440416053513413901731366894638873855531386368
243369975985970616457062670413045912643728498914835689045560909348
101158092318073018459984087990904615749310986143133159197840606356
831884195057075962103268508407539511046071367743150631865568117504
568429109859360948634686959367227758077306072883798814246810034268
587441953320342225922259113156871855129884383997718184817757527652
868727478679975609559814433269798023224692517480084804373540267386
844464825094568371986966198330889858783525793232810047849800001659
240729031466028150564724110345203157652765771714505108046030512975
963903369048782270839013310400538514937353749729516134897226397902
119889634448662018819029576929504346472305784526520058067990645390
049554274873960333111513343423239392815392857552418925427533689936
707673603270769534071539778317693299858002902473809122227024700301
497321483099349332418808211182569586232946518575636897541635746895
986602665172871063731782115440732830840958229371768628036856451591
525703290275690368571298831278118747345960741731009788473156283864
948619310435016618122663037695937267645885383809430494530230302680
142109755025038907214842460093398754399153838421377545972464098687
379266027941662047086632843876627366087827215003598927765170744547
706538396196028343102852384091338723785639795368257883705830489472
663481348213171908883396336724123153639729520379956140542026523557
331822605360301516107672701613667753472021089952406019019073107167
115721315313139910873460499485588793055573290748667569249917791477
776275257215331530591915437576402085562431149445372545956809702564
757642444230904740701449387200931485566126738641899425494931363104
759618933034909499307284324090098660429647764160636212894769517265
674169221041267919762026291755853059616058835981509438139888155464
739539002210859787185924059647802767889239242804773232416801150880
994290751300672861497273785041600155380972786910116538163760299560
019987567710528743417964863494875902284345081024845232428506194564
649282883380246745314360076653939325316906934715341110259091559509
809996077710819240434008174019090499522416945936708415512633504468
374235408291264653803549416953846871915947864482169071971882790453
741758978656539635436417496421138332391272660853829567746264220437
486137508696560381441154467817463182415780125489762580240567221816
519025525646655104178403139931552734970128274640783796773431039575
001167643501232392187217369395615725612096294658612581792259971229

3601560483252932466059000746753828911358876966050230432754644157772
7204135535343106923020990409588280284249254566092255047367866335357
9776701147547793789512216395039174883700606920832143131056511403211
6591497160545033152608756244303975120162704447566549744508291084499
1427532865125788432014337191619507424345854267127681102600799697733
2731091087404071388839859302056854770568128370032410609948808912033
7233751569167712944767701057362851752692267386733249041105761883633
4334373993174057361935369077705806991870011038755068258651233963411
9298473309667875732032904837005690335362168372869158682248493164588
6413099556128076135431583947979650364579844225293998032521346097288
6226953626724707628997177963276334614112070415414830530440196754588
1623598606346657273305740334246756825399885570038420395650977199544
1002683762829751197071569287780588231902617101475800897373783464999
2100430570761585953225073361087295702715074312297920313721103151200
5786945818242017418320565151175338128457998173296130009722859113088
2820909053314760119678185038367530347047000578748360997590909129633
0344182765505119842942611742125017453108837615277210320916233088333
5710208577216259509925298643641820689439656908564775124318290301833
5339510911467513718534246305851770744343216131691304544562072955777
9149889048547850294942518699230156420482367299678208543277081713999
3729713647285516236910280943949049809571114798735326336108615449099
2136210719578627182658984645459587009069249248820523435112868738622
6912529335695565624401533344756716240947811826571155954756699368422
3562499979227723332856784786245269498130382957671588368253903484611
6714968014138599194055597917917858281975784812372478022962734271322
7380701712131593454022544168614641620641854955622017580271717419322
9604030724285575914037487524125583648684782653057902112930150460099
3009791132893911020928422212628874397239879299987221712680244269577
0436408269175123947288580976631735219034774020783010825008230686744
8165992916214204378559690700839634317491570400704911133097023046877
6615857483135080144475992852020727860406246909862458183710566318255
4920666633928689416422316813978537417455898355023981413476275686611
6221186367561134540185061230145050641464766200254793727370169115099
1057005880583855287751553568346135550888143137449856363777369433477
3077922369202328195126019883348531930841391296921034511566461558177
1845160918653048971195380110248525749893158647233999267453725219144
8787799788807562673750638723780564697643526861306774761161564030888
9810722990061362029138553864683684245835443420724906526943131926366
3064557919103281746224652305086811453922379034699935761819228384111
7831112734266093171716054723027485870001047866059835368762042349099
3563146793544370070867604441608093430388964169122938462935021661100
0210761640546614532826133025098992955391927596299462782632632116566
5874319551733594278724799548287227810793149777110353425543816635055
0218200475598457194707642967827158772684836236111806592445159528299
1523018180897167227176349652283750680731317414453350933010558621577
1973367591051672048856745415728163217259397927018267765927879072699
7595865244447986278487669539491461017760577603607110750866034557555
5712962345406637758448773140658050218144414570121613889442942543011
2726143996039751548809684175388778709977105315689605779553635967000
7806998565011955361699581910918533374036619990661867745865365937822
8951586192168358385372055171819669900290622524429719647760765792122
0834997981483108425338006646056465462844105959758701053837837669511
3414411711576580152919723932831823741907241827055621142924812595000
8621934825451856553970125840647774590941610778984486679878798360355
9430670508264698506509650714242879841665013303364759597132945835699
0587596970583659840237526455951428415274309347600284805973744511544
8230400857745381944142354918783809292229783184414022384436112321688
8505624335418588432511544720643284962084563281194108270588318935422

884543650548453563300884266856935636428902027669230848663361182991
429872638798806829980861239497632951046359133826912525187946694508
941539649332734549729944898362994739917547441647197173177987268394
360240105216610149815265541625403854517795215840024958798797410495
248004753558164544116079649674374767184221183581573767370489681657
618646684473995745738638952849566518957447866597778195075225888298
702478900964065318520474237695238933550121847859966007408965038385
951470180407234577687838560758095616453392168848975425983059917537
610132320635432534424048860000309082261900373063418486886143876373
649417887401204826095051275986339050977024247252980175882639229387
079367325221116705792644140908543740148530459025037169637477458607
191405425694381561170144378884418883091592292719203584129871622866
850532460389435650023073416708375186459536802527582405209237446765
733512706016011703490806822232723412140846959667332516156575806659
024310130320641153751168740775678740603592587886171973634936771114
265430484708113330323186633985509494314397480484078764776783277053
488015967141016984435669780845487805182319957564073978831770271135
643924204452033300760976436796999004095854955620131358480587537494
725693403309091728323941836921932491518687235477393921275611794664
018511800138075010277721713064204253265553611432390788203509453770
750843488923010206936485172849761293833257931632804024023662247707
358488505586196021481895075688961464986471085846445373294965523337
264188383262127117827240693226571570786417557289614533829164489186
520495527295263300281049823109857339430816022566981711150564218030
749436110781361389682204877365185667020919787109427227650347063385
085500842117094040508256992457562828262781375133270805294552322160
845405765437854007179908127688366953749752286406714615345649011269
387426711403621513820477587549428565722785336658487290869174951010
237587497660723016951857365090579491818691542049514818950633136723
233600179192443975940164167719835945106934272172934837133152708252
285878147644954066168266066328173859064681708480980195630954019100
230303837721074832278139011682082582389277936139561206216213391578
640790409627777430623945887116813593241244337109448308742299489657
270496966891909767872956785683749182662280759470730876390942917918
464672898935038166571603238341300482214907355731011475604391076423
070499714171792722498893625118537718445653611243536680334158347109
999781275045931072949201640040438736891084890000220658968949509883
554543303448063469068362642692622526048050382229656658564454638172
578720242239306031674501605397755165542460307432569145384140667700
093348172625337857836954968801819714207583047902504544932943440806
547069667092081966871809574518223790333116866601065885464616222513
680755807281783990499382032540352222147912787357337924050581704793
436111604657520350964992030094306338515155701039654361560042502091
754083680251075696272405400706130739148399782154975269620067771746
125375177474080770421469498072465669210313803655901391446319337852
495607651289588470395683600524056037732266484889767598647222236870
457260025131465330278949073668317542852793043641684491309014822977
944414539776700050476454539441997442534009022064970795065778667625
625790416787951719322821604842790422281457455552585011050511118532
051282481704493408500651110585967966113480543157990100271163704146
255884514695315016137653098634679351398306442172125391421048484018
069955555893386469844709722072920441600174464574485789885219133254
971330254820980219920946867055130885041123215989403060607764070886
215302252839630610614984492974704512812064392509526839331630165354
068929280565187157265787411940217478091727995418741181137373534823
204924028544437285424144786673531720397284099921075338521376852189
920275476375155088032382034514104490336878610551139745556445344133
528058933149507241545365042536863587651146455776385286184222500373

544338608419457202578083624670516135441219360521249265478557979011
265815919933225542147336102522035640035827908575507305278835431594
674179374264974074094794894477957316609623021732397288402601621550
899074510246296718368591603789059816357439266727829502991817957028
068636510124544515441318142965418452451978873052020028802043389552
095212624250682073625164648296888315050959701000226437213534878582
602533578984284992642598493826986555915745522772230447836700451292
620325907284470070718264639429939710579650492402721513090902016322
578929364662069079114189091709554858581709996939845824188862304346
386468537094692019086644250014237049070605479440163636224484204946
141454073340772056136753779947174346418696144163556429471591970959
124572988939233815001041229439585288124290316381893911829364047567
480132005483777642241308322733790168055134561187865263787390846029
832484496777676526714460909842724092219442087290507772474227128491
998627528840954536122442608122367302636241666463676956582340509347
865011435452230172110431829674611812712477267475584183473918296468
924243908358983041077861222164667413927458084410934467091407688908
115480426990464476617903706913186431644872934811624753142709479512
183711895430801606136867423308652068568392614804784456647494574832
329837112783484945756818482357381296729860250944563100213870768049
043011088410435606595632913551363659537905774508634658418379378550
213855073066062032361892026534379655424091388667805176486602355686
801024443819982174081868308063265793445013660695883116352765901963
710912216830217994317817811597562569334811817590163704539548800254
386919502939484296333878802324540268683115920771472660964081472974
256413523770713265586567292609352131356326973863345139232379491272
741604407165332837276663606992078289885158189007406817883560033839
550249105442191369494384025928975768041647987388754419071010073882
502600250529371571205988217997519052515481351289265070350312953887
973951968071463129797393988552240677107478132966112514244409425462
058656056386484117697376509322232005813738988859893022336308095219
342652281506753067731168349920030749784495333173923562877249889011
049829135380994323467387064792939183829847365091741599344224180136
090702185376839482371972551488138816352825082378087561773037185933
102376901551814895668026451066955667635627033163755042821846935526
079312867717163008152297052501399440411109952375878216898707228324
155404378594936488165971060194170111775308197796006102061075809541
843822637717441589308934402454807763589859838646004481913063291821
212522007280634089056273136156282514259729116909696211674082471631
451891747360069596699142308087833837868659015986702232142869157014
142480704589721910542004790420726183894565916757662433748165233431
013197777875062648144789623796854491833393254452263282389839955214
350864723998824618234678333412034969696346523102970980070312729811
300298748758845155628443101315609908946158784058400383614543062750
283843451683679399431155194067233688033261838130190651593168620191
839636438811828697041164945876942211365769814951731860439447681922
394006701455127928254056530324642352419083789115209165207534501147
751337617613160303463500158304324119830345045973111548023529147267
556528539615498251732218702811891475582192510975188147499627018320
123866466554470962703221196735206682568834873759645072512079691451
687396399872950892928615057450939183524898641711515633710772070437
194298978525854106512202087219851152011968200668515495090775699216
193168057612255084107995644735723621151384426059118785236111157667
462461676058949088473218825118818916537294130184756365083622904096
877270759063075951737344653812358167205699861544933744135511580828
599979725070005425695844829042157032963296954183720611253277818507
824353239187267379753901060421898213335680014917629276358973974915
103361029448548755412659458830826273087297415813599878505897081564

2932415956520572243886015842078104750426281129044255263505482966l3
4319834755788519322226718693036456672710264959940051166308663731 72
740445456949737487485211033177549364625380611334474310806832630846
622039370773105244279995137450193526614235225514186805510400502143
876778592990110859251867499131314500087258371166936982497699408416
160624284063083328979971618705057651962404924316599951518966497547
503900114739890318968783264557847453725180452235972687766876242850
753816616792488000823409032034807146522890222308061496574270447722
125026619237142356260929122601825058373181197103907517533857713780
776213177245287947915831714843227314735068371778815798520230352800
599998697766693700822670880420433042717610360443602119574053183239
775082537624353359925874480669523131409508267297420082719591871616
960153406545781475710124329470340498901172403145627070070858913555
130659474830501092675331050476766851006872795324432368964938724349
140188685802176697065515885025617415207031509272651458735885771669
074118956676294168134057842406773388665298433582820992092796000256
053731611957486517297171140435836830233310269244755634963018267857
351110563974947335708175806329870766803421309668272612847950604361
526544217036355406583290195474112632161794143686238782446810885100
608798206571969473153168872765582925484100600262887084707264146369
814546760230690648480001950891529208834752002948330118357071474860
460032318036646630113783461481020801040824162464398628580275352540
541481178772578449824401215358088326311157679388344399416742552671
812706870485790500170018827661154025989664563822695284086125700000
312015134146214627435881881137521596235509096186934825303819680850
849675713080265221001754452150438824469635391354522294838227521939
781610063081571394734757164331002885720115617471919226677195436928
312826604396069925463721960291425377797398316744381208097218818831
236226603387075326789425385591691829772833273126155084174849512359
891579860193104630204088365812328283393282877527485978705364732951
561411429853246103430255531301949643011670379286563766956985479637
443740469514404752486274767380255896740849630272538858173832095777
727044265967645023462419588725735933861552680812047751364027860596
714899368123712011862123490548171292454815430238041036501487535674
543111800604500426130787682215885144267302962084048226136949742620
817609999350033446197688418790304159595153926411196546477482084960
353618894576122048571862646143232749719188085841721650249255612284
867044407945280918253914446987618136633194396064637822450816138177
872928278397648591104634556227172221781769229741153867862146057242
015889821754945547494863631767227436470898021546200732501302370572
121626662522005303961351678831013008568016798771386008087441449608
596103041041197485369831113671070824797474197170808243016916661770
771312763331363815453158913375254168398408478643177506675039488466
367772146792112185361223631672188803806610698593702379096318692240
259119146345846149741712192550199254747960048460063345981864608011
593744703731663195351890879205648107281187772402039744024602129739
110134992696648989782233646553651294973293415434068946943373818266
377860503474934332702908375618011054934690179339428739905663796976
347810695528961987646189850722086345874757753558684468723357249179
047654807751039237363961854667533349597089174705010313969438090236
340457990307072485296328514308887866880742498163585636339314194762
523066152520565896307037142091574467866737683351558224442263717555
290549395328823666896153326331493583928128224584932540555941071950
713799703563742340097316130986462139379530870947165361256508033157
850445730000941413946001474525441403816920993360411596583800506303
682545663080628250094880200341800214558417554634801876535677644115
164771043843669008537061169050325303146835437133581809292400768050
958188888031319229966049866511923553334427159951307690820852662967

740310259473022591776820132591077731585784477312075886450933987756
187266253938362357576251588056203092312138665780721626116181270037
560534462263494983862525666524229234436513969720823782599576261080
998493754227356751224109232447930724282802917623537533863708763873
518155274821112448002459124640511151114996644626198433900579254635
394962288892436232521864025248104905959554083650286893574890542000
912533867434313407342265195998144887626448318552732774941228785613
062258218781200116285735213380860436525201235079083015059632454682
818922475989132871694359851422675732581509249821248990518465907278
237639649232119042056438491725564318734416229620060447190161161278
608069159705072338317990240010621164747758439023757467891316957011
822646217702894571191364126858718686358249327174656270672807513674
315975075657747583764063380449448206683521783321333278967763836574
467462017288395723672110981540162132700681687402313661948332501044
648564646036412531741333237960756729373305212297457933352566168 55
892004375962513420306383429430609715847409538019741154953001028216
505595925945919485334822732715544487352136534472942394955964530478
805317945586293418901077793490276022180849918514125716531651374508
750314014667742519764762046166931133260453878964516572908438615194
431140161514230702247163939901004379068641034162367907418506463768
256603895503347734896731133431362942854314887603124731335419670980
008452642740142097631369587622585910093111299737936001355335292074
829853672042761269847640066766986610534552072872187381806791058162
907487010767369652166873448787438277199732718649255424806684238330
274106960918550071153548924174440794337042318254560683867024205233
933058031730647788593322929965546621687057128180663158107596988037
954190286710515896821839986172264565237272159212726998561668843085
968396028717153852669414793173289354584495315021859300866891179713
664949241053953017401360785889154713408500397680364538111157208612
956394709645574270823873126874988730970590053373183461689693417093
000008616802780058956741522844366300229652650701385626568435888629
758589271228973122504501939753988019599295859466744488527923464103
724733413533839025948077395517640674147646580145330375512587839152
060027305459805828008341586750878202182980291241797731523538577064
067711668452133686650109064439918466472914384152284355957780524178
692213439026209703590303502527032839798676548711129716415065768915
393509094042163002921262342347128521083954216649117518876848901601
635079499087251459442840907695196996180377128279292330631394632150
965793664885286718536589854282324046387338281784815302092030883156
972673439255833643216320660898884580711362776399966495706481333243
008044307069228179629683286131639498341581788714262196654990514044
999490513227583290203973389028542575136640742837719838951375846035
685933196763654229787959796756828399831018152542366659857278588886
806485189459707162034673703516804567897410832102068776915310505668
766877329334920023893505744369544516023429794578060306718931576795
190895808112827048686785651794949425317989898545584635110166292415
067016117622197572925577322229957957026951427313412587036021325937
476429476772338553939496080349432963081459079933815943114610237436
482609052748926091149978175992425233969728695252416687315009238204
121285426136163532491366251378662874417287369277732668533899905091
442880593169617682577285592777855488912248808866962902222009071053
198672733203501256083276186546860690046121765511410345328312712044
352295100167947903133505342535567838691922343124905213327943612569
046803304540642593143348598935298788225495318574248810376413754148
449982952274890279695089814986469076164438957523435665064979825941
525032426325529441165969405598958665076121533992974864105280830988
791971237287616972907302953015863380954319401820266910469313930352
663628358321962934195022055821562811510082783702191422318615775289

443074012512069822362570413511621279344747937375070858534490402518
946776914742064913902473152404739223757035683312553974447363697759
131016724855642522704985587132991847584382118515249153210866087093
894774655589097681500909155245318437110167970439422720060659347278
649237655946958471716429025786327183436043870606152679931992517807
196060181997889618914413296815327355365655317827878987704548492565
683154048433686635893482791153784996014629433017853591892226871356
021156380668887360245242861517707711106712851439717394625668407770
725858919518657200283026878274880646248625804514333344541330861637
868233257296257953800673509106053396523255759682415048279519619749
459051008217962365670147705645902747898018100630951888962137903769
365337298726812820884788701063082554158504213341014958285427718069
494633813881682451903444805049224355100033141429208942257683134801
951041953956483428383168994699706893612395299336477360596737956301
617803184226182619920816348676196602758664471180876032530070874535
085357549089483316670801325348249711806765228158023607082333904142
811702294135253600330633026112455168649227533897653332750883730873
546591411189798341977081211090804713744235632419974361958142327674
056004446749156949455787149355479222541764298223075736651596039395
678729520830762129957290564633327979056087360196683806841521600534
098228717682054303049482964071437795896778917852651344209014796569
969586033217610283983223252420909187497569528250236244494235687350
103470187419905300293809698609087614945672871126806871959924240064
653277115700461234695506725963015667229090544556889669490363819793
746846586653406795597194462977563164582434386240379348980473005757
098395158216139214440418894226816655348954143282061553926819933381
323414313987908720655644117610051979103079211594464124822986954039
586697896296360224807663263111856093817090755322596581714925458095
004864281930723758653310934741026846088351017655232979279258864296
905772257139082911909071964170853845945443359918962961825813795766
195253377709395930937558695979150585469590600816003435570792205728
418485855996164771561906337685043293655454747429793082284034010421
477940049481806545729224483426104801520489332597893682357594775848
939079653986132009777388783890023066496506731865265056828395821962
580338070209708988714146215856544262375254313938425321275734074533
191162955171187913699270353917235081499866237794428418843345714929
271033322663099327159181177798427378975014789433268497205154307237
560639987729616687253234709907174640540240739876530764999282725555
733397102244685228197440635674154423398952240404254833976955371473
159903911519958160949598512103745365994424396455866218951207314020
177355678185319574500159138619106408997869328313648390096137571062
723478005228242118426427552831612858697601566046431833533610397233
746019991538893157302858826916092049488454130092262588377714048796
551601554359374511078984718088470096060778907622069368407378496336
096342509584708257256336812670064291029822279991576193941230501066
561932438529131227088307156747196820218627201948474469147750995873
774866029631262112393626268432315339171935691378989196606671277097
343228082519847506195406203449333070378426798379941771882384778573
049239862558566116335286152795713435314524810391638351705507787722
297623979208407088711586623991923319336495574109949375410066796880
142650207310666332190372968824698040807054186317885193804782714122
565417999942520847288328203476858489725525747181941141110041741566
799999641975328403240933119063192104713467023378515181682298661343
846179559222892272724792951269711902324963913804404399574050092712
081861325429437494680803495274028786638624393417108857657456509859
476694892184500640546563007857601863379039611427130965704638609176
346038756811696167424770017570120962241599529760603853488570014814
031370011280296945431637235112508802119138585426221056899489951830

180914171906159263693473649530715417590666788072282014882919882051
557077635832956721911220357704249516850618829530889889133774280092
605574823119088319103131939299334559231342822908244952580052392312
035468409591811803767004110412429520600416749760555822753840278557
228994429097079220373479880867350017022354028870748724156877915062
146524891733255247701844863336042379174274985534336281951376593862
764032817426362481472009657057617273393219713701624994376072232561
327874249377778589269330335964016213344136498402711391338427470757
769543778601175664910861942707182917441242654445981363785943440204
322865897546386434827291483675790906124620843234390391923443343496
772773556111421320014394443227320381369085729795736326744778943865
774890385918099259886296977925891374705285779546130320543303677522
033550855052641852468351949293468352432860294168994575328382103070
059714264453901409018029918233366474407788470720216230623856055975
822134483772962995988321194341336945834461478359693702832682714104
848145288290526166403281494081840243768279808314945204633401314793
187522373778064144956575621060530337373631466749971428199074239705
585981535036662090465058448358290370627882179517010954976396032910
465540606926458630212687402703333762870900863607757172312759161950
765391337763291958221560239574342934468871298084612180268971042434
170908330991098588888835254085942276917768288120756179439690119075
663452417006163200810141847533290811300309310975867707303631842545
293345309766615291752366323656474216904228061697515605333059925079
176825022364645999570337747610841475018859988302655204068322532391
058724489413214920420150763661972890004060592720424962760719992999
765156898504788208519098035733115741544655500524131490124398995076
737791147971421276661555365700029980643522358559463340291519655744
773725774525517368467724114822876372680019635844862426037986498657
582130805125486775367180449618700159104473879342430418785461787045
785436649442843850304116481926667184975252670736583993025400618865
946300442593498642188736746677914010289921935190341984732576022585
319484839338206114648070364899786708653140531734815143246518534005
640853019289907636016009140767076874864987866144724164384262549228
598167910812920521882291519447434704103619261982249688650183287881
228655261494487243355986406705534886676421607699601535508232824182
707156181963143431092962804052569380172100643874560935856366533754
096152099368441090064234555949678992586527173749829803717638644155
408339933247328130954900909116944267647099606051366703401744118303
662250489910202822410449800530639392246517643281963200444786431071
064518182924901554707466301366585027750507967666944709231116950742
845792691986465479689769857442471202502619936276904918689385379697
748241302056076304338922473675747538314713417546782974962447706654
093819818294053395278667728983884828299114239277363245716014373375
263048026324942165455657671976751934720546499442516009891508526537
508002510756054326553772723422307196967945272246615973866021741689
039122722547133825915532284525152266946972817303175525367108519113
588765425443579041298241035431744232764343271370654209963215706364
060968713845246245663352691301220789207803854120376020634119553945
346946694930916207958199116593075741982692987786665036590825853102
107170150184413675291384847390819223564708665621950319865198555690
374767109471408761353154871815930278188382078139400086999967045174
005890292947204951246680739095172243055169301048038278147544641937
702694249327243368125202460157153486104406075905633203741788371475
352143957277788274638618416087213432549823690048373821826384009251
021559976282492414832391100246927892536253848077699875241682775157
981445345592180912352016230923561872620356180637137437050124625681
248863511622694756689681361908739138611682781042246641848813774916
377582353071751093363651592078320285074878177329456795722880270292

533039303562960965509081112345990450064098314626001133976600372981
3388131614498624607384004103873895233467701560476564767743753091 35
303602773064948548181815798555845871362783153768046482215248418050
024360485920424819532883678403638789956319332163183177839752991937
552421965960896506553739404460898228965083088605090890249651264721
191096922940386605909137826663597944840783267636254438297382631612
385127135883189511072580994198572394263896590594982782417809235047
599580728228798338367066100204159537645690875936082090530464645605
498351090377847790876519746000574937826856962268923656473689664002
376142139140408853022853014229240229342391847460728918244015840316
196570370051165037483283612130517927679202949584496107578731219312
362792490708774704940277207668639512899595810379182555275737019903
563985512824029479351343047014985331634148827241470870511372210732
637816770795704244352542402658784910323099442185047657104762926221
526379911773502945540414719797361893916413646795825081052536221009
567308707059953511023282255406880818424246132905503411246368206259
564429291757401920097057467517837870949738346200035152023509658220
513234951881288097417013828007277492706073842957867654565122328069
601873592793842250298239452654566153768909500376120416256518301073
537003907029150204753714277894368805917320302711857789917666634257
164269537166959331831417686469920393292873147806545499610556358587
858035598893825325627842779752774869590582901784353170386419677914
076504812980943838768811633599534749783496325840425665564883523023
097152638961085263284139935517370055701579243314557133926350649126
910328745743368016847088321019831805725899635641749947999141176464
087830985873887601262243929152513512743163114240916595798544231194
074263914199573700819368632439542888918921590733557117772516588695
449446490515695732422360494291061139881878797814569230082568163508
933768853608784915097614076727220176526327006304042988298532360410
002404018299071505820953486678658549147523103917653043924444519665
134251485886593572530618789331732908416340355221647410415435261375
821818187912790652821080566445817008188462120953276418823619373937
158454165450046137634755726916725247610255780211198221219161676924
799468148510221108354686977604706507970232697917944664058254587841
235137839159878685765801747173575840055450021699156624893432775705
316234398574651211255669761595794165500430927839580643678562017610
936953432274420323727782921264107279273315388054265718719523146147
608512412071162145370705234609873528525352639855517375988621522832
527062331717710576468344207112184896971626329212490614166624188760
417868396533520813403993199974584851636867649088685910452680787306
216050214958591937822714926533322860968536505039861403799578393235
926809107789054855586108859242822422592774477365117818270019813885
316073056803365796762717845777429169997919369629629072997268103049
709697061750361784872804915714553234024897008651825057184139097089
981443210863274307629534648301060291760317398316298855807697144339
567729015294792494892573053103628809298857109774203433903894241774
960849678531158757524460721062635221799957944832824964981796880877
703560490697406097558151120951620501327709107803913461147510049698
677195780467282368221758850855512187378823843550239713535647675312
848875111455843944130756166908021940470540250925616388730579959357
100709542152424023897386614498430269643615697593835035800086525206
634482325093428912815946824688131107670648072715392133808549088932
174463059788581127442534488131962175507453904692292260778682863658
751566809447504786267227357076953714897264860136280801508442263265
972211471187217154458187742615869707938869559231035534774484427102
772791812654193912554760484431809343679664633404282833273374185062
986549946001209056686091094950352084418389916340306963343519971372
234045101839365628394905715741199173881420686449188564896816333551

9506600092884333252480673558417133749617150550934263718940232530354
4259938439418771874208814554354356164303489103148152057658869444478
2706449109953352128432519104912469054321738051067941859880544401289
425123258990996231232405387739821014464058496559741586595232058144
988525103769306549748931350603293607448181499898201118274927781520
1132404643038340009302231080547259597551216746706659229444385710 75
829356865159801179019948045358247172345030176398914902214494890216
019868415175873791916826610983857384537652804189009337550323487675
887576583508168084898048899461346384675835827589450046648026022470
795960731123470870190122939638421992508876853711199854331293724294
847578836115174083358437533109066594270132580329543981526920681054
804215521024796511454543319711530574099549378383693200170656410239
939685203415131733092513860829839610344837564348547094563741106045
616668328026369760559410786005301485403212528253223272517323249355
788226593959508373340095059845300844861549376083077293236978053902
069489843652286792858078158108085806495326331730564681609178514712
540008807225793713598591960203211769851661813820572666448797145605
056476417427368418914506734245675641604829030981897917595674479970
441848154395604702337843568126761771579873748731652445882100164106
192876715295197730961257950401327995125123044607173765330443488975
837750220067414677801697322800545673449942537241384582367759639957
228554593078385191403950474413617589100741462268192976969498861286
529855178802499331966356382483829419247431923558426763507319858030
301534307486182437832522793357993835685378113275565386473002476743
067237584455570664332239670583789750194011098458453031203974160814
952863365122483951151426513952136199492804776145672284844312856596
154497313827859533076736960149415863707036217565867010430358696114
579171483445820548229597116654702113627728249354079462907060140372
016903577892399326303272607254506004036460502838092961007600067621
093582161548809682798180459087699075582797111496748587103659798177
900559920461992108621883339386436767545357823633698908816193564218
210955951100939837537477554658607786559433062248491278978754508135
580009553618632247789455782167285821565583485741692055782234361503
253551913069451960052894986940468655864528839323919612404399594779
057554351905822581270246825732160269935312376217316256389732475711
628596069997082939495981464546812429119289449321675789363587752365
870831262612976895221403712133337137363657009749611146795473894021
625486684146352498148656843712993256610369032098432452443637457892
832745325401013787354608708578491533913301848796502158881092990371
435011496211919724372703633189011799293100919897206605891949918385
269867800580939230917378195429850851684668129923342594667076177767
558862080126141261464088615606370386756461288814378861816884069210
573731007147127556028255238461049428731994983801419274943751006947
906095976275704074256052792040373513225643720532009690271261787884
195824392343316525246682094254662729793482420950273277702953598156
498247338180616393871547749197535049321791743206684340920620175808
477830518875496124423952011896490704766018606357333213987937346739
149080881235134551377407155868222354558845754468634333775403138713
026260714622401171706024010652254911986468430964157219449244602828
173252536670353723004242498660648053112750195435652322568738263515
606179781774903631475049573203258272282087901580037039472207847114
408535302162674050665051256166950057399083273250689952126975261606
027252472836652446769954693569475947257566855811894258537772576809
839768588064964418575387172908716236658429546000645728360531387581
386362944104313462995374182776271853061591934261217713201031002114
525657669028709745553310907385811128955140392716687522479984987499
178502589268902148245957825905186445482530867096054152463916486748
199695691963759719630398096105803354613278943598184350869745259200

054859003037296168307195762268543536417311817450957993316477690774
640274025905255388091862936860458673952113310468554748440381710725
066369101455731473828250535705756613939175526069518689649250344686
647491465261560855050379139420298199422219947354382317832230368471
373302474855942982638040651298489197127731269793994244683681397971
930894515313010228207176024113229639122806181570953761845202287863
361784261035310734175920397829165437023953430922591085010805655918
772527775548002700192946141441537672275682553321417372014074713478
844343631675916143872255943304949771961233842226616048964396242053
267797041430802411640119680891010920634290380279255156967953244161
928348664108382860544153699643659319691377778700593603048202291302
651459229613455802971827243841968767243708449263675507056334050268
371994493547326526563206627438083699582633516760708235495298561583
433119524392297003987910675268314944224875870597119750135716880807
708801638578432778181513027786831168589194611430108958921839289713
359413928885648854509163725939835974207671570746071497524605986398
966957462315777686048797201480357679564845898219702887616123194701
320955924214488355505762723234434842642601175322306182230356585080
470110118991932525717205499629266412977350428504370262289723585281
626756378963020389847435594806121738739066853543845303929312988193
883304118342377836147805975750584066225413362933578093194781966392
974235039084805932006978991767883396869131974825886474708627997131
325613717273081653340613946256855059072754586450686465652776825553
429721408833837278820102890293240313242102002610635664244369661208
304176869322010489934515597321174663009086712008355724205292251062
850302940669270580504400681819227351425634658435481109593207340127
496949000254472079736037916466970319503383284835516767605831036545
270857655498002823947822313718870396521642078414038632005016875592
892442489164321079620031371107462606935918955818239988365915310970
042358174294600735961247432905721092909762924104106566209235037924
431392689030306220340787058475213684434981400664399682817772883283
068082967474851072684228563950311923967939970227828083290403918794
270125640317319867054809038172901093826770327618187333823329928735
425179121467416968444384160995792173492547541151695503632929460672
187983817798488683627829099798430217204175362522299672743257163080
332626794270088346679931237227789280490726906343593863344827373494
687180880694508882406899726165871343751874071244353589993574950576
391055026023488483193010977628751845555614279728428487603938721304
909025418488426977514011626937613955045856899047300398762225695695
285227027007070022363127827564720918907236614533831506450866015716
672503044253134573076142482529934735508200948111074026427032879613
545589972387692438810975970444457279722559558214831857922116838192
022376660147053550332990566389961139502003559003953143148531999733
956110064596295558214961621580455163249615249846254913386661556613
057471073066064947612592513473986724042947052713945870057114461774
359248919999779853985891554580117570754584198570746444171573528708
831815566490671161372052484212406756883333463263093946744059153928
124346865274150763671083329467993079601213226236297192288906112943
956865890674688582258888398916501883553307523319815790355358685515
578206546821833215907429103474695675663392485415223645371500388621
789026343137853026622744881799998738533234152500505075994452916010
384924296473792314485199676400312042619311018390010745597693245743
996519682211157017225000078018520076909279952748195722352249009245
510210083294350604709038217623401235278483873772731431981235331216
735074162478419546325344615208289122378046922908509386280752677373
364891675275108867186907485731511798719112758973717212220069790268
627015397703337623539168573023532778080515008525981753295550807877
886672815650966691615839112721698699388759112688648485453452898384

500172007531788096127347744030045241675032393038367061707101305504
380587173067566833533745378303685599937759086951306218465528579235
933917419171205417969987256132453266577397569709321705621938004614
828574998937523164351347470736588209810605778865416514732489817870
094630138507925592226072971522620388194374843914310594095992584334
465657681739689326110459870100372754352511637744161227299994101861
956605142159694120635513144859719545286080974868254874524459036260
473138064839379734468186624970072155471060193500238648389343756227
635012792584941732643662372023278553594194930450011152493701147634
634157542640955747394306944563542362081212241176373576970867776359
301935638364440288936305078333228036674743943248657079895085250872
741832683527199515779265271987637499790762084389463472126203607830
817381428047878554978289786227472441770030163255013397053724176828
153235161769069219970255699962054642437265357754725102403129943553
864594831470194940156026684943031837836936554661866566254708258607
489483972825155891603855349506451384744221188275629862063313569213
435053541753254622942738570185142216047979181239135818570233638135
445357112771171943216604661431015474198215549290475621090189572080
606234908802904067854566746372417724868119007420655784822192950106
596683535208679087585534490092713251073537813112328600410529188355
048282568212439318097857966641441641974384650359754316704183852145
907794335773149648457421486085488674529131457458931518483420505854
272116027520701053028812182044257185040797177351938264441514303400
003896508354760695211261435151449409699151517833258517247989474052
420610045984073638435113829829335370285516415328184689878043592175
819760111037188260115715212198992803575460838874094737522040639123
362898280661873195323552920401422000951548088070610074538656397258
970803032798551240570967529948775250348381191484476396069023998008
588751011612900600807691194381030260949480659847619690480593217852
139982865901636139729473334245297578429975902328892122887617453643
431583753143784957460887473734258795875821990193538981422942391794
151561315397930251464137986089598876754136943230404870285554197809
229580446989901929045589068465978383379944925127160494133779070648
657858949675759940506175576329347568082892202911154918648820159214
617765449921182725498867656896225170636148321951406030844868842947
490817141227669895297665284671870107291933779292824435324313828520
635706158076925928260322211940276877904292408365323232151023540753
423210947605321017167804788960416851071973969399186187946346189679
713546786722440290516444396478293266946358491866150401165503213795
823884640354533706750014682450893963508407963388339316440021557629
487655451496229849457357045563984582865390103120311995586329789859
964274241654564022155269311761821934040528049770013958185699504506
269083218442209858065603603966505204050926529449163112247412243985
455233459397360215848895956457603560112394722600290911023235818327
077603181928957893191200422829722719276801057856446673340203186065
797599897673630045155344121222746492117841921042993023304754593408
148695733885585311878925724359962470195810494083427130065971636437
516574637102498705362916929006099797979820814714713112899508518492
003986404614653025099491414340358369556884216151820006672539858532
035670787534474101821344997039591787397534962114723677715107506443
419320670978481013906119468142996565946949980301501505043949916581
936434061754712006023253305100568566199539885210969917968103065156
627611400123939441274050406560022170985477796442468587486319694618
955103513339164119590397189387610544264230246441278596632014847955
273227434092863462009840598245338576386151933644320983391819574962
950525271704159411032941605520070797004452742655032910680168291821
505488657297908306573320056671104039316642894607397427613272069913
773588877646840767260164503796909067376249123181569461325842146511

243018088628385018727320829304932053488349080225799962089315382043
458746206252362968122407755716676147083325307435180280156466152412
335772666545968950153512096407409879933525112423683932555080407795
389065135984869314857268748994901408530010625403698440243398573821
267762945491927728192707271007595054194545037909180516361580836288
870015386724394500702749898432185567647403247143923664843411160620
931019601825031780683539857258391335713344930361449170866597972333
881453092174031811747752032581674338945826496752752520361126273672
109764543134023806587201125134514611723801638359472687522817638356
655896188613216729989394014941251035646583363688776090658837696741
829191920311945646978094249838609041603309637645292794234193002301
740054343225854625094743545517096835436975603565019923851473718492
670597233277579791173815247435316334241172984589412910750455504288
587776273734066330416039180826874172606596159893360778633070199222
318466648889304527152405511746120223016036619219393661579378673696
158162597300582128112825576467959494028146674576045577470673902200
197698318259700293819541492759081133733236058858777871610067258359
623260496001605899148893422047361613271007545230048439431098999163
722188623262572247230711979182304944435140335744766397083610698607
144570069276396639734920292183462976443818601893766880534512777038
481569085406140328036150280386094903353489323035792511739353041584
113326547129056739888443593082228303303222116592985419196559797184
885423887158089140369301617172570015570614836906812742295502793463
522645006869343077482074666368747614762002275018155179697826673741
459504387058872387338963291213957303994663054340289132774681687546
695021614124655037009126591798302903887348417613972343394556935660
083801609435561378553746892071454423377646719631384646526315701017
132358397487466544236302779285419045015666457881859979478971251481
140502377690261728979301308065657163121212079142907054215088898379
545365916435512334174598794809276941751149031174605522455785458135
586702153090077031955655899599746805741613338361641691140099233415
564386836225866442807940336267010522666936192467472371364090542898
520518835100369268187997465647052545068268393626406994422311791299
733364106678173591597162898327417728872302052609804248757771006988
196240372912716284558358478403404924364878183372432037161878814931
836632132424242420147187986601290829544902098739959542872139066776
982756308916794217401688235876539750420302448986418896369096316270
120557681969929154992775142543788129467665083250351267168466448444
547240410124528064217832732227760436916102880783588718437100518084
017958014108352816351636038053463076389194761501869867367060501475
565451912556348547440616202739383503562785615295889468170169994014
332311095287212448270472060546025850066704075791114136827906978686
587117792043561148929968719888032590349546258685078645156073721715
399533910705457420844700489981128289942160212220926244947274540561
035820926425126780398190545265944373751942813217137033612591057551
698992847294695342429807232562902588836268426784470298363132949660
544162538614748828347981673228810978487694132343671883348297513277
555209811183566129984856870217344971594558142051676013631681044749
870916364943156667001634124731526314664694470222860280711813992815
888751637214266832121415092317205731891117328825980525201560900415
547775952404089135009403651970848700749678332743233588694631268790
098502313172066141132108607604861957063566245230487204929701679457
814582009036192806782139458937433777693126987686811712481640849105
253884239333690894635409235802310817255763499799694364597544489485
664773280449886762357891730215026987964549842771233602523960136788
902639127631673348690099465888102863102237495535995017161877940595
427203256807509917230406040592447593475587819231150708603864364001
669769358844410773687028457703794092834941402822129586407527063935

3993044472385084396885275777983552083175810709482686545514923467711
1645118856722380760062998781844878270052720312938847998209719432022
2757136352039888007560979354968507222173819096427575684664407843844
9762359541643789860716673486049953642921576909269615170952825421088
6026688812876213228288701239411121356084998485602261674350348830522
1151995222130947223118824547392608085441215344210434543110428353366
1072322446109504754903082384976233787723979857646707148507270115500
3350791768894285532565557856891411105393768123007641727332335555555
6958197951616787651611271588023817370581258433764454963932090336300
8884238313464741325415758340853287016214784675273660353298142198999
9001039965166398578162708358962481375811285205027468314346218654211
0028735798453064197217331119032520073461929812872295178982451117700
3283234759864039562706190854550735807916589710077764022903519770555
1651463156954288414243755797570968942232887312501544659132356822344
8563230881862148691752544420425031155171125209326672093524453853288
5759307857205196311267715965633535956460663812156991761342710503577
9689346925609775922911355755004495468093959419880769193752888650244
8971124685916195119118057366233650749218367328395749066939668994866
3881254658855888383033086427922354597161408639132801696867060967477
7934970251369670949211851826806837103932976818279349090488099268522
9794978557333715456812291190828899996496173672758296772254271826422
2328664001327243273092429509230566221346977560274971311377496402166
0451869335895994334517013147431671669925355352625191822960685511022
5521066176939130589930470440130555394785866316843769182864723435322
4859388779733700023743444052230578338504233697486700501600286637166
3548072142572742363471659825920005995273502863429413906679266972377
9873043735393795775874670438709507356712455449660309789611819455411
7024559219300964059380552294276921735098819503385424390196223556566
6509598118950849558347583267944137194334777064417430687607287323866
0319093764674529189218392734056524491250586956517611562069812500399
3153884581844064908193055138220680810239336308565359538328508151855
2860249073808881939719741926645634161448426511541316956283525195911
1244283826288101030847554548973690254035882364831424405950604333633
7217231136979737662537789832914814676854754118971023646587793294522
4553660846298717097667145215393593629565084168679388747451776846477
0197056012062911196593927169428782001047384226912084203747363388388
6274792663438170746008618165177012473800268910283248614546728946433
7703394342046484241967025618791648972518386746222304165160001856433
1299654117598208500563323952416320466755350013353729684917467646311
9413499223744247224633022021859547406463788211882345939408996899588
6677663701142952853127079355663237832561966782136657092206028310255
6539135401190662142923938164112080696172160438099387993003279135199
3961605459067259657242443886673098839494804050019958699540877610666
9138906842799356469502459908786561048152626194880291622037728544044
3107619152330967613456578986649276023103467080783909926275476450000
2311159881515250166756373957401905770341261342041635904480839076533
7485827775259666285431629883314207477826120950407760433388366358022
4308924484034883541028706147339632834646578357966974592587401134633
4576232160810423976222516895973476817428512737721348884312642986899
1670696316238734200146948985214230208355107107105025571884627785644
4037605341548737134057035304716047710677529320079908300857963350588
9136486977471509376122562844835142793387526367457072220025676912744
8342237943660613198626760944062105152371984859747379297406177243333
0777353802543022189439576676950956672798124850084862642588484579677
7193561466464626014964951463471490061886726013021674810746605411199
2668918406378835311056304417080835780199922329564347430329597913588
8943793800972704425821506927979988831446725329768967209243310967877
7870437254704049693782685353327799678176291518671277541365726683699

349108729256606562281591526335044669497567922949764583960403124782
609680807632457291796313570638055301817950615589346192005525020421
276892047265235195908441637059762275805273353990572737729245898431
134662089469356846280770879593423614342618357397284121665260195438
481774502442968737870447818458084566985918167574593630071250992994
559021579797126797928681418361794529381147438345911304949490625457
775739657448250418936610501567223911406337904426932767178357282347
840242922904037674700397134683438554640642707102611753009130847612
735756388934449578014367197801389826534243776067204873056592069332
816973770772050673214000573675534498089554053868878671591124076O2
402876493610914648563243513922828961692053847422089604660805903823
099659189334258907900622370408006699792029798194409277173502701273
368468208673831027079479355302208227752154460927356207151719553874
896681908468028606626805266261730739559289324327665608205589264922
811457207893258778236808279305050030741774353514258764320918185432
669406906760079190821342039636895309452563340221307302098645862976
896554724865262428461104736657509041771732052323741407565848993239
270868216794264326875694735191217476911115775407999719992668288850
793903934061031042132964682504077064770521769095572432685966471769
863829141153779769760002581927239446920104966004285085470014809180
810817266504567967186688064620584788093007116714190784971339391499
399525524545209494650784349719810361428778184033220570694639515476
946972767747706424864607939235195654366350830702520798246537427425
699694577564626119873862943453280541508276209906622774358444862703
767092488431396731265635680597853428581984460900825022815051063672
691418876039788319773186265729314218073290550935385624444888087051
585120556194413737032854054757221463407137369326552108693222709423
907542994089944254445906867574114322524261672343521912782585438845
595167978299328323642737457452544546052939896806263513733587214850
808820205518659958034081488329701253781235056793050818818568505731
233257555424196054273583194479764324992288226604355585233496066680
905502905216337784746351934749713022329493965510415987839740167516
618593605179338950392466205245511268837311120785257244245799623294
450168341713595140252095179264681156829820313618827396426623321676
441524695487558164084358212585044247670699693803758573003905790110
515414779557179169312729095998221364115981595201458613678920666663
532183944579112942949372746424648239215477975615733670895761840575
322098504748583570891766352727809495354274482525113739382912378335
184147182784881809377625946725543342069023837559765846744498857297
100153365702593838609837888370559661656612261881245463878074036437
755829259340136451738584462455407649029616220229244791778901424327
249245624610572832994427679678314481934670551757083502942567326335
264906514141210237861093296718863103717170461762893116167259029067
712239858836596414924553081207285708410066076168543516663530341382
801133819677912289974126655244951348338934636181282256499053411503
179141167093830767768774232569803429140799802919107611396530776180
407622194451519402604063470356799353883274378588152011080406490885
175270082056238020512864218424823002632432055997998346926232665644
701956357300679539057244150398164239082136235132717714586191210328
112357269933087662553440894151205179902731473868182626644475280406
727464085723801550389418912595893739926501687752743769741533748172
422037707128664490771162603154417119414108348606899529507444772203
362744266818471196563615713772424615456070479650878312900133434911
136292975583609060175949453796861506817908507607566212738100117918
293076118629911635574502602021275654360951138569094815424476722607
340061037334261273608044855312147578890237559057711317455009411859
748652962705885639173897159515988987014175869648654185324863779433
780506989345553880505233124949841887573046444733144459850552473986

5399707346233819398008577304356954761698282658938100306024112186656
85980207253371656135335099218859560107881521955992984830737114161
764839950330037898824790345410532502054955693588016154598918936886
572124748963613671862818854644786179243581710112551851317871774504
307353645029761507292301108330802551534951869294849716900991730394
7697333789565022956148777878048366582834827540230192303690385819788
534303828558273006721561304247679650997367389863963084595330994467
366005278953510077510623540518095062072959121477879266263385428792
589775958630580646504484526239133538342627050430867009467036220406
339767252991365187842306583966702262580562122210733541161850293635
641616655779237766395860494693244550805903617986442755741294983021
046969861644931370103702775084860153961665864512853545304815598296
382985981545562592486591863288176301101499737206920153869877418621
655782087885028970856782970192695827695239408257958934666666883918
358815549069436830703532763207934945109365399450972042836730670351
441963155288753214822189325967173707812714051334747386080963694563
512019018439160557338408051663829148862479351379403713197966875856
259482942074632416148196268288849800968875641317790265769105550802
543228031258589984582872083257358894763134926062496271832200731813
542439536437705648192953995700144554383910878449144193680471065163
474031170374482458505185788180686628844170793566042698003163236349
120302919753700996010666193896217318762267071826314852284417279433
406818103101838417534997349697901352604608389864938417085293469279
158345594247787414758186260672246624811772249856862298974404384392
184024560360919123698959782488064463195555559308328167346023120406
670072487747599806326845273202557015621687662840583268894930505193
900504950495870154003485427760246248588466667342385974454567161141
984303835706397426666703855560964523903570201073652328352769206772
213666358574608076159948257589026155644286649673725692080468511746
267024678766860322879651197857616442650025536622079972039998656146
915511996591892609987569195721982755095064759786156264742355786450
113897041993509976406676557120850295842115591494729075235534992741
008512949193855962594032638202524988224921444475588270029003679518
705235762764423558418333071204601246299399154841958135512551467709
344714433092476373215011861279838185602557163141744264421039231841
248615613047098148024733881256960519677269438321490104652409981501
183394145060084222913194160995009964496196330766171680279966145964
908485717408237805713129439661036877272697904349031896749323216657
233190372154146103647188424635680197125709771242045599277189401630
807555791531803886385226329349122868944587124407187398513109807299
600005402969139086326671417923649756297192502128839909708484680439
071763198298386258976031273818102754934261012824458351039724617260
027124726441028393060367775439840384623746557117766042747940447110
253227526070881915259623881035944912100259215675509990359849028736
639465333622278560198785244807812000092267255630431187021878325473
868804409188331048255150339506237035345911575694871584408122253546
614612133683291417713871207911325632996961058630638814550382930706
507642500409597837720091354284328731106694070419993253056831695331
854406218096083461319779933817165917065487955211443993463691039132
585349777380538014249409345036276165813689500309512610570841234456
296013280703948714675890101664115170393932146989030266726605846735
059647527480561780786795393551032684912986766565426312732988529192
700824708774002213743115658696907606589908547798087756486559413089
027045689772974195596550109221935693238497816225875176465524209255
740925717695468860519010003160801289728987052861085422973909396815
077500965971737146008611522092622608527082988364373624387798127745
117082236808061077077413663347955743533547250663440979289899184082
181502006262900581367815452848577595273359535974840872450053882741

039998701952126233169862828034388497269141695862950362027229748868
984900397414716167457511413346027344974235505878072186655258735064
125308324573880356085157662659100847907204770453688975071997435665
063066316758761134751644189050994953044117199851499167397662294269
445166214080877491355367345306518299977582014657530815794081675035
725631308268975276869491317516603141962741227162095782997451259507
368949976478651309830445539167618793163664040969778731171580041226
555288637091406258178846923903643987679438994419596332277331510624
171111117589582042138226824715855862315936615312894321916548928211
959762276658143596743190469318970709546254984802349550186923112936
640292909966700863878400428904420862483661779064302063305933920322
434365160794325702465868466897715343280772170987980118148551579281
644492135430015252996137723601077292108595131459952461659422716415
747632365702571880611706348762926273236008312525699654343218937450
779674452915427894712722894704464813147441242211665900810057217233
044387008737360533164683029287005557200190699431998706454465506242
821727117124592068124294810550504047059241052883574006564845472456
074875624763472596201955416308086991308656786967875539700812791176
866919496838135150988085209582767929487854818158433903895764802898
509257246086253006148886286503065719865793656157955982572991894328
947716189620569354672805441856350184626344267485715560888443376776
775181119587963168418536391233749766123771258705575367714255354528
010236191288246608468567360849341333119579933540423335773588963780
531839093444280492270352162230871494436067300423117979682863905171
951575052097655902730996709989020051300226332647381845202399769112
952460615572933669965418267875614644743693887290887894259922714756
326206666732908094698629295343111076243281643273608630864133864864
668368334034117417243361379086047880568004597543289332721406080344
475032843441146117190967017625398428226686468388170610025364990074
317384700086148176164319642146091993738188776548270699793984153938
974909461030806089521056237233733955299064854565477711132351150583
518723974869707635229334354972561003011215891267832849264645292657
116115146530034496144130407078693714179233116662476964087635487399
017477537102018211428142144824621320489013665523144244134042877529
811835667348556593691796255853153675107980671452796637458994210311
881547454807524651853170218249967058200928170347143305649061103029
660077886218643958620309126219537459319155011169133155954733941172
086135358840520458592736046321982702240715420614033109614862990759
081033133065914757495394365387018430653038342790401430598298810968
662879396068421340105866813687700862550410669655362243076097486920
666744068427555947082594037595454393281264615197860109409221003946
623938100024885780821530539641236263035680445023302479433432134418
808431469281618392368201869189398393330782579391518765988615852658
830313054820647419923861166216919045975656253336318446768950752992
586777308978113220554526893234119637741580704297917296184933765166
937562151464881384172662171532362271080278418137745960978655724521
653492678760880091880707514451795591893207464840761990517355858488
913303280635078797052313167676931577373187959490721237263799259715
349422416504918609591639298051537541530560108354141244266350884117
088954264409770274228232128737818584813773935509749335551140624466
436894204535523793022955699025688892472476482856987927771770439584
369247240062220941325554943292326806265100656067112487799788039988
221458634529591716662481653228741155327526411248966562365362717905
170820153100267353958824702235281639972401534641220320257977808273
135512050193684281552081855499751491511016991412711608445400907620
830051416461882552963462036087371679019058251894683895404682662971
748668008393029512607766929924069435225777438631812759679506943700
156062505627855914341512413394030327712953253107118617480257722349

```
2910595279610183431069077020322287162520836119564948117529971327 49
7305978355285212857854784428618168525710739979164485073794630194 79
4860109379383640540030350892499489138010893132270306436604092136 15
2251750336475912552993362345087462062522116152134533464059073152 73
2407955939560032748790973869426066361431450934795793642528207605 76
7366822455612777885798509050746557599952332576801978516473222357 3
4446612494779990642933510320292417061814769571050772801917271665 42
2720280245480655682926565244457107484434380924735583240595727928 137
0093179495842802006781667030234830107405474219268605401978802767 06
1773311698549010053252658070039193322183255176221950044956023295 431
8807248709892249330737590455348878518957734282512509676519718567 99
6529101719951017464781430278133357169564223193407571376783460869 67
1224381217307989693831217042049112414515862212057381989260281325 33
6165063327096126811273544576450343862718373919938943796958561167 12
6683833937598558264615427978133179120578291237899892276277256159 51
2584275400014463204457910654686673241405336586191842804226258216 88
2737103215382122900160538955574580481497079514288287427566570758 14
8260548242022106120376883410734370446169531356584731584649952332 88
9740861389260374365455710313573097870305157976741864883330833346 83
0617761996495333234340591686783886452470412755314395794027884216 13
7596849182823286006692891160507611815980980572296761164235609054 78
2755309990228360118255687572387881258582934211212064353513623423 33
5454800037637353922844133746644754648997271532487062343247393949 40
7436784904172725426574267589518279602033436226061843406548292910 96
9473277581063150058025056894921339837057106195538103699251006160 45
0062319589568527763384145338709215687875803212746031128492488714 69
7592389661216541007845166518759992660729902045834562796342043971 56
5244565003933269758416152741868905281033962277286801457027003189 62
7877077513728951374928538916013451181479091212455428351145074766 20
6145020787405527198310649131950843193937940513935608624487120632 82
3309725631065680671593587120399214096663322511910445083216535436 21
9937775858432812272309717649727002825335202360334694516082287284 72
7522818468777375072298833913187683690263824493488856434606147021 41
0159335537083375926119354381437532583680506869265160213196385900 42
4945026077793289829297493102574748519154758212368427556373977810 15
1502777188467396734397418256427158653009213673388007912311266160 41
8917228468906382678717224697734147003033770948629424678362291721 79
1257397858895723049380035859123639968963121613858310464837079637 66
2679929761566821198465934155933916744468862003556896518406189650 20
9957879494750342134510646829418913576240994955771883376474844936 14
8903373387364084487665128579060056918035580217574328223720982964 0
5413984917686142541135780192328432235662201253375699710382103714 50
5361135215808754432588751773149812341597900774841548524687471869 82
8437164277567966121882258983635864612337270873161639578782993815 527
3415802880622289603227447919731513419589488384195292905675291358 28
4702889972904682421781158812544500275773489756610693699383060028 44
2488304085568975649116156938287828620459017159206618355597055735 02
1830926911960506871136379219891638826470030323985599825852973720 67
5968502122325947960921370133154369004734758005266973163662808767 54
6868431544120054451810963963317799632707332700784242615943287198 36
7100185305221100049935858980934727278261324522255474466336523469 02
6079952018829848657929356433410586192063576580213494971238154233 32
6330818249633020386361806074300789362848049457274765559689769047 96
3077258435896097235562688527717695095754854674156318936544434526 82
5222687733165858336717410453518601689739003705114038721607492565 728
6694144634342819342201078799444793152890807044167837208591403807 87
```

```
7600520523215382114088732406797255883699722977039781747786554451 33
936952804673097919875464405405015355984214901764939708993368236817
978618263713774776189924213964754681521802356570084650624621258009
338239375893985352553247370307268761318693261257733337290274919695
015084048179987728373665255064027193614777325988089081494639422730
754621137974252544785730656206162327584566336871604105456555821963
228444258001613092292561169521705856174292971169937298798552686573
679816223076859491733218637615077351715337805336399472531737904670
385755272237382781358856453237660838981202294975179584990141689663
452187860835838411893138472832576864873474621953538997800875424150
586749780156015931136540552070950803525500481212312377181521072980
032310175918378625405659625399485447107620238523408341501421890183
896302766908646062889973158305000605416610521126183324563088749423
761321117383235991026715443333980903010767519215606860915099297579
489847091340484776037253316486633273997745741707870588584989036478
250500607565276677666730181427983462997863115472471904638130827026
950271552434583777132888840113322856123276424758054914145334004307
351368200167103048967407913220417329365588638081990240250424758979
906199739449424061393859002043745081712616036278391241147268209085
690526837422506891099193767722077768737126770152907129682261584375
714966534629615403528980698498199023815881324900728284203166454586
451867787181771772779283212526968322976641245496739715278796804347
658957612653385245739151343813784500518738591532963414053689484439
722550801196079269028116229367043437115837195386577860034194671309
665344253552356135039263743335590248778009316758556650202614245175
520231051803797924160186816532721349074474187926304637935701957254
656870769649025628311394908306598139258771657534329051829883074422
093153945626718913650932778527585614188691505843112818062116453383
461456499861027990887831599923320834970349900964482897361997260841
303050161383437573503350269679199100394576485031398899804053476207
979951035562800942711980771413862537468942006711292903794021105099
312881767863557121288220584525902329882784488972855767643376551320
983720845365197273566294540752078683774825993769508547458537785440
151866870321270325108378857553525327422467456165530175294697049286
034935237663193775815312691121571250545649366284046131575493234361
611438689415519179552116040327941387040597365968287723555493695367
249260335274498928882204488684436527515895468955885890718317292891
292345774428441927250527684755038702706328297997585388259938790678
996367663472636799709113700050045191515070502085744705320311342837
530396450683734947465152543161640695883965696047762481007698125762
324027656324714558678116653563357384133203756328577111457947736117
758910978449597487134549945405008974943123702669160022779621516016
443144632155674657986969134302091737537932953736310293482594184851
531345770064943739097642085895731774231457672882967906750299223152
573286983302634123352631634902064904297082100632648815676763242544
468703921333767894896001251362652354725651702225595569986284251088
668968471078726001673324221562512429272130805593262213072140936864
354996898787430352676884922123183449241907637471574744625215974576
466357242752795222891504064277678656611519193319181783056716465304
813810106667342459156864174457688390624192018654102252669706153890
999072549984285484195668192454519747093061422753151298445309182757
715136118167303580932146032258472352811825504706062154262243245514
468964572693823166555250959895041093425374308599979713700425885834
030449726710962996976323360777674373479878835673010286471384545928
791637490145406647519394899352212362474366131747830486884631516036
592243576762762344666539589796468790552923902702010757218919138214
831626852490495848675432931241826413466272282093653277283976675572
```

642546802512289430533112509853860120123607044969978521095859875329
303277162267982305510767692680002207418849030165005038534475971018
301673782681943612416569639252294741035743185176583656034123276433
900956511863260791733899126277207213516175222255241829612433962825
182328696862544411862381233064034533155601640695747232038365145663
557498734411685994161655182496042597983926781613148318090253450716
466644267026276118597649132476829527278057032238343515063672177066
376374024903046590962859602719797255378001418201998101813981259504
234866248344043921136487236662920206393962884531448374890102608403
614840731200674156229159669636694083603264334149637120985454752501
773669601971461784645155559941672637397085864959877953242158328218
410939164052835679070686421078034660757197989148155400542005107300
962796234727249970112217781656798449194332226334150338567530824467
734104550327428561157455387421400071928430177447314230098365760751
551277796281014722053066817420350596794105098046656313637782517247
091409925552471036812670513824675211720052849429521974886284898527
787835621060048781271144063499088164592445189801044293570832904722
016072696604619842607722478310717143909349289737950750564710538029
161874918869946353013572935018732066873115017315310912967948625497
958151216822075712318919091383383534471023659794480804712338827444
035053467999529133546094139272844651391190836076226579839815642463
829159990441628452768189353279135674740322735150689098775472181557
499848834669462227119434351395727560933187767215742857830330183042
225170496329716122968367527489832947231506974978874140021926700672
065677292131084935729407635928956182812900110978472432830384651974
375857368591230981783603031405282300813026631330413992533992179415
764798534708178636113772001408570838639437703529183497403741835116
223700401731882693932630875054875664529093265665030243944536227279
170800381578513252902365105680962589179424001638714961212169469925
442398674726206005713115387838833883078016537883877521159311944935
956949179394057884886062395944418497289287230849557926072113297712
137238869698636023682916222546471088062109447913239901540667816028
934694282155062721260541798291781738248919973382953016826679061778
013533650471886339784273533585623279185357977922662703802445696829
686254911874868530549785796598918486218623748556393532156304899283
486556154154064951221046610376548180602506765491340332738629416911
776263813834785118364105699966094920204489502629446168466855106066
242088314537401268779478139859577769903987707399417032565319355613
000528143599456061261608322358990324895659567527576985324574403560
761288607968185761977178876556198523752744722359927202060238916718
790814708867068302793989768378023337579683748471679204205611884614
435084238369737859482588497825952141431676898495921319389412875069
596149193271147035874533660814943754697142919529031019389435637183
593749148307423044934029596281116652389958110600093862221622655252
976610650745271358949733447407281673926348622262971347155532936244
457994650823197908769074458525370345709007744081534167863870148099
674124003837808523427397788174691080535330509119443313873040838043
067506030686269532124522901667503856318585859317437769494157410810
572749444384001399152295292401680667458424696605569107697596987310
950784518251857689800093942863712191016698078851710571144669507031
273706962004730035675368235205815249186823908397408809264852704581
680063915340134937352547150923527044161926572110042345848085322398
093081970121586417329130532589871715885516842060650340556996859371
591562193954595558557009347711681179835995842798195564356365309389
050941964641889243417661217711754573714429402729377177659183107443
058151531596094826350633655723861413920813075414610740512741348138
890687520896517547286443489020150187201836613841728079882729582018
977486126338360371109414086804414638189975514419051152014024187628

978688233866528874956474011072459905537992175155647819809184955876
752778280803822629818043941563979561725694090929518577447883651594
478720682678596369454763706238206696202396620665921081278183219127
468081453031421779867353368489380826681896912999835199422321272638
771597576428521321515883717146485428812423122468402839056157796819
989785562510271070628379399431907357979753622873719947845210383168
668521420822019266723155810117372442375609151489386343666526579426
037168289281580693159057152379480256619126870887647506950850111370
257880233381801903021002975975592681821635953507064188571900594974
446797417420252130947246191950277213237247025702961631681464762184
643644651799535877759048091724695567396794553734971032219369455962
778937791938340697253788415502062958387483096195420461546990222684
347461767711319737486600087935443607302433663280865368473350668707
408900184703067698214753133731542862215155131814095414979724670676
343697696458309286795212019941406654043266683440819686918622917654
410364920807857292423388755061809836591222653797288411120130691018
576030498329532694214188425942866214695276880632082571964867134224
698526419419022236241186339130284171844724822755723379969707482002
437580371792180734202080536935740618765664169607739120909813494702
120725197213699642344209305478465069237446490420888732630226156357
919606309236991602782364930003449747123779455951240858239709946570
275366759813304777505050536634574715516558372773100785781787153031
613276848925357607846211478860351804029765696058486717567636659308
748016099927950787178913104203849478943286084797051504283326524571
886423198399932856342268607883443745309272893146092544299060787111
736766959849633062177514884899337787867859785265280570548661217379
213552124702395325608190678852803832422968075544717437748950143023
150146961225494895338362756944869304674198022922555065087429772758
076095106879827109193837142290968268728596321942836727242477443909
060036804852784543854819955828743344189095523099265929588482897771
967505439205771668938552397736092582090693430578986742357295312051
485090384652493140068996173731735816222294455416149357871477506270
376192498036384400160913611713729557661808926386467940279365670385
305779912988573944783757639092679443336505496770742285963808721870
399582714758000440222404214003303590360960548004718847304678286807
740989832225262453168032034084435109374319499380299081241792110895
423927096542582195848586679924115788447815219557498322258335674226
798960098032009354865108549467676713405310343499864349758000215286
835835721365978208435732604661260570464409200520643748808684041999
585408697477316017505390253064903620449458476440882040053860571525
182217793518019414711660086532948281060021915944692783460379829268
818677784883782713148160668128480874790430234200377130896464781785
599461837510620688441358628450630346441913942893762354742777586769
014678228907006092683252250324639953337566728997660254246597951963
260902742615157481865278192977983681101331339651625793318419407026
964988951386923961261275369592060229690087420834720840833188415826
838019383358973224335136412244321174979404766824167809635203566415
433254150964501977910541460943749815990445792838028880133562481814
072611423275972894824141887025957454934254722746989976877162316099
322885042028070838100814091887352633318358420740748446573397838429
805347106002374219987211762683334909209073865337959074928089283030
107207550472450851183334676304759820661789998004462744803370196555
021320441396423674506953708781697379969379061637848201169796270721
270358480479488580658308963231288673402963848241128765952185362411
256969749199057478280326299861231724793050323637705845698785774531
610386670675558440682408910511818429025803298514096157331538756311
438547792152983663838215871358824082012778384097362326475844352630
281664756079993221483927156321249990837098930946329559859928728433

521252427433494379023824949445785164936127032642339094544808620028
353526261752981835525297880465028135399112847161281153414460389703
165467739525876538384445746110351561641809273346254142217903310714
720310599294953895958436885773489495225982103831596420623273071483
716691798967444541841890372511272835300592982739374737571099277652
356370360647348724784839684203742309758998874387876542841593565973
588345060936129924492587467691542804598132815825872999110300780631
592481722052213206010771492336601003182710066727266488949550942336
897935481055796423771544954137177407995177501466695465574100801557
934179598301318715461713838220333287263136997808093756281698575352
925390236568114358655398284283241700051641990051764383512005756933
430421802931523685411424059868057389737717209032816486238395498050
084360235358582554618855942442612928921434147898267094176760452225
134929872997435333820627622406331004883774527381188722581819982219
428293676666000040379914870018568674554412319573518712379305995148
215954867701054047820258539083335640618262225208028648666809763156
107137641890902360603954124553570380667535676524726680367517673845
564696693596022634258001555720896238403647713214296692193472420798
729629861675796746082959712748570679034670391578658581112738257432
519039782995445767430575299283028634186242545022496241979163982734
904359411395898404348957578332464822165249725331811930555856140503
105076548589915525542652886287528895545773678742029770378468475636
642494767084854307354132840361634913471074468329858980951110201242
544884643071227174886586964367223751257407663807577968593851823215
803790138851324567042252785387661013519568286523394604020035673386
025205513475307900746894526143616381246602094339688182998572534654
355285404636104131214993769216302614831518234694209627911549417194
660720665528440044356575326641438934277220905575184236912080347379
886707969228398693750888161460738382464200081539367400188625730736
953499730836725281014943043645634975213545319519500350764823703618
453849756361633974429430988638719898808188086747495831760222984672
501959183717870015464719437744024587964419343305273778617450245249
707149907000518726929283458717863091748438785506397547781397976147
102974805258930692216662225235373449011353986266028141926476293709
768013187204140666876255459542229249384946271177550175862021378876
760029805157411237809551927818159082066363654035686833244556620095
160463375225688255854582920019306381533873656517945743702588756264
732107732276466231522699379582538162507411935992575434703207518963
927929216230912990259044551217209318966179934694954150218683377015
220759113008886890238579915282639867824654608874627852622681424733
188588572416651261959000329224404728408961960264923773072793032869
835071995091733622206904266211379357378789633982192711117924375186
838175762134729222730484110905289312739756654644015991089205639539
550622684903481778340163888074775859106047358664560729940094200063
120435623081164981455149655513005854611735524052167155666041333475
875987920445592175677563283767227690115416492112246422360395403368
455011342654247448989545996792036442429665248273506879964950157404
621482511167401638128823705492676679728005746329061945617997309444
487237467063062834619376926372843710250629430239838747180411277944
515182108640001558475791284640128739950977629770826226345882505207
818345760505308157127681646160126745615131039107176973845578732241
330030005534719511669012581135208015630373046908093097927353658564
913574711350904412759076490299193882008262173939592861233365729706
646410270587838551318934657962685933047956026011545035967710140057
993336889004022075384825139930863716343366007923712406457617650036
410612205435688688177406253057006023018982911091534071177517124423
703643637158902201162317102635650130243991215404270127303916604348
528921717678005443537960268144769874794055715993778356399662100669

274192714681089620407361116347202589862464744081961204033687520897
010880633542844369252180174251211967856991105833494499916830944946
984707806367546667767825383723040528489291173054802989310613282285
243013974421278401082297992256374991861619095395092292352403872656
334962447446903480575135659465046250309625011185996363024036541878
244570740245894880605074168390715058032424183755862679604489403118
420715618426638993005968351960880991550054081911609426156177996494
555738936233509560216938453029407415354220170088505934108021537744
168969765523900070011310946928000344435606360766131030272873892742
266524989909815901237651570432773192185028448811193320110357105719
444387121835232255486772644086673404544135367403990104641792881141
327732957052332339987800916026700289290467003455063211355182259645
456365580270462153147060321476780387345442039887757315364197294374
658678276336231119864674608317162495938051631791016021743160036372
135135506555681162767164832287962390037143316348095868924384711690
483078965100591104965015992831438312018932525166768955897310518020
709156128212794785768231503099654870137801420342350862188944511309
174155201212503779765726305117588445579181661243191479349987937188
974667677782724332922702482645480284999856755494526946870327503783
940036651442685682081309020949057899622100814077366965566279789587
599381603739294081898326023119790605145978038449412185507347234440
464136333171482978197669866965514005181845419763310556350448849713
422360339130058979717346782373472329230517388505004636025681998062
728258112455591586015018439090409864180971710075461884773934911273
571127107533095079036197946170873344664805241788806067731106455884
142874312055368645075413123789205016418245598529170285529823491756
815198174953565040453735880040973693100210161974099408857233681398
906852305802152257830798584444988490026722154928888612925028852813
527173780318207628086658198702133918612113360246187362649128598385
704246054788599442082401809197362711751540474656341180486288643987
511052601860076320866403208005880981246682872769158288851453555992
972145134318817716645564502666336275157142261212702829023587031467
862427302335998951338331069080367912289759223209005353398361052808
487974347050510512429799469695877329008120707972879653583923242657
673392144380470361706529595672993234416869309201866257158203504592
227460113349178476867831063630236724355370932562694982307261863131
091050164320612674246086791670377930940669607135447772041240171387
152541478713374566022914274536828100929205588900795084837232678718
659556212837654930431227464459773811156396674092749919903096783157
044379273964166675109789264093117468241878846539287943914280719137
228194506211199604942014167567514155226569328596939900541011164776
752925649440428795835710036845090703458019087499993092734233237906
647410746289811710104027788338214509831606137185058427903895394961
345986945534332173388380442292218684824710117148515834710609975786
976196816012437330230684469271055789326166001295993498597491718450
334461056240840010952490311291513102073536606699142509744167108918
044279263850255766220625664347056888812091343129654781619845396751
548210810244160624449318587351214286010858155871519419397655261062
478092540814247596466270191943785507186983496876926575171350176402
003599383530178302781767102202449288655654620105595674157711590472
858301654225614200548268513719162768982527266000770336835926768927
117466145886443256295441705121686083735716597610278238848606701446
329636821363730331746487176320142788006742493485684457268867825525
550925006154697582885492108122224766822902775116822369502543987324
561861209996738050145752145346770108025915298160421223116328760264
578489208814442541782351787729463684916863787103355988029352879751
316600965034502135008786148165275693425491575825447858789779004210
159280113548097158154932538649021151389857756639270582004783308103

193586172095928503098371977956384664987334554901336566062958993312
667035425517958589534255685222167057206373166820932241554656528706
208202685332600866580058396609069504970302254534936941843479918148
540317521615318893601698982971238272732961881513540418704927348526
265666408136486378871680299743419921840452670036155802038750040963
721886553766105646252585967623112091455580614923744622486559052594
146783412301336488120864513178145054641794164567238577509045217705
499758332360916182468663731199597425637392431936836066334687888366
489399770870992397517694293270431571634050583519899477212598612465
956758031364020077933287978651130119476790122849334559372745446777
306994245626020238875493090223357398303966428565992346239434307543
557661485851861284466173143979975977684470929792773827647093562794
945093757497580940229719554370143859221216058081004239743853304543
467119143871226627091401261538446277366108865182715566402048997387
185384279740871780398587857487216892636293407937055160183714050877
149628160787383362335559788371360809666315218932287510522740371018
412548297128568954164194927943850639454838617154528632987007434474
646146503414460256193649389255719342320962385728409362207205517646
982530400643228756038069773146999660101861018409083474528089280983
391290914925830365117302996765473925151845027724484495376804763886
401906348729677479902124856127316639984427361862308855173182399678
817158183206309699648514729573723694647944254825014483727864303542
669964431539815277168679844685777773176724214993063597651813595392
768068710323045802519156036464184552722886148251459740929971994529
105998334724104185420272085136054307357487622738407920016763466151
090614719108133008769243989050542838285871745960020088457644825190
313755480860179403410944189883726523194071831370537998352344375954
898132153424084287482442809898880471971054529233998476551717751441
096350331443841574283608079013413016396157944559087366278909144275
984522976305439340866678264314016375717056188134506536372888736845
773001897543538641536393817376290182296333049441891940659730575385
121339862756462498470327918415111491211352501046851190089611707902
188891880624882538422836411906558748088381207312323141344233353144
433609656271921082476403927206088886262852588519928301333058905765
272829571426194979164995894363177324749580959841491639960872405594
058974095185184537010842391107823544795389772207975226175997379931
801766025841678345852154531357858420969913069952099187860988612444
010607411986374471530993510334286163756809485035927570474426589679
566193382876884746673876270357798755596549401466289989209986971648
540723033988839367611013303784045113078379970433116053326219954425
770307103968439752796919730812802511262236007775400051308597498304
645404951309704803426138354091344540564134101462193716056552804448
400880453039649492973826865022745282299484577467343378675502800997
560510091528866646587902625776895712418793158394872973887714835384
248129311916831606013543029978483686352773120290302971077830277473
895813465194275616066742843607020400238768610459207769656762678781
970656061203397304722965481373446191321988589232186743912322415257
741929078225709141401815695728457383362291885079486832949330533593
193572091676364595581367992386963556749298651132482713946073162855
012413231173726487739829651492342674132247288632846021041366966644
267728104149594302767238763428606644807904842677191598564512608618
704025727442774514307901736151561773151575005988399640141880497306
997550669101292407530374958155784627683114837351610082642105686878
635684085892011926825243703903525176669009238408264675261709260269
710407047148153102057397997681579182981289235304146491987593615632
212451682746172277968157330253255735223022968339827799416034826498
569382639736059056232139294855074276485329426710589699458926421441
196000845353331145040686537313195714843485415041517234706871596658

893468794776160506525205325518877942762000677929174286295148036393
715562492149219289945067840972054346001956298474409674862465363711
302087381417548333816616561518511191134684732365538248531987858181
814501053869413158042894105310850262582815712311114555123885490445
347986700257077621741380291892762345238939140280529309686455602087
074750296305685666872397749985911356208348594264702238540331396655
512294052067722982107716988749068312321865667889253484373928939824
303927063104601678559287530601787022213306811299142564872664971680
328549439089540115982149377017032767620929876361529476102296386400
390994665174286052716065117214013250959270559294839736129981798102
567138533177106675033131782787325121501327837748650207033135506227
558130481080029460517179886418646938301427222915794350397991778016
490528277130295624570910278494445900502500126475623251401612039820
325502752696951967074235168421191098209014734553452473851605450203
448865261194845794673910324946017546065949133473564878481268188018
707352591838083903679072721987136512691128737953771599527413426674
052986058267277760841997469706419659029959562956055600022176378828
180964658424294431160434101540324161837112411833413369086040738182
186785929660060152609792043026905143222568143657469655420071610492
607105516213629879300555912132665254334472375154831796125640786777
424307070087662202918140655021360191663843859998861232751529035298
703495321075289690611404016598021882803768053487149020830871917804
775313608584141065967519604324017985891535324434233629910033903677
261899140466810276148578721543032757524359320503117165170343024273
760823271009896214950638496910029025774167136585044898207551353446
944194185199121456681506843573075876541271316654235668122273722233
387587767393622874603410636815651866493228134230424204001730539139
580450340596804482575134055549046416335784381605886860227991791515
677699184838574581978981362012970538786726606889518850722007442610
297073712835694266977933820878091270526740228190344887810682804959
591079430888100469561835872093432323304409697612377196896298211991
687870983396113452620176959400345867383397823414732191382499749914
065896773447434028358031537479848996761924969985182240176819305210
022455108578603846905687663641289089751554366506561650619221818558
606395256352043478945916167981232360574968374823438905559643350429
275477260719321983082538071553885217730924299413144190263558021105
357988665362641510146460711985598292895490755148136944093607176494
262910340381718216143960417698528173282088210369061259770468314059
876589814268531700310662742574082809102331168159759586548561781614
606434476519302871734300921800100587395483454403762764601382427634
053294655808884773746683625616908343097270069974817824245741942626
077709798914229003500843321192397735909559457468156646947811010103
696356946867890309337571102762086607087821065555377926088164133752
939155961539106238153813148131766276032319888049792167796110491023
883227506396471007696524744146098226259447672975844810138808410145
215932988735385180730698100946012616786893086024378498607208280266
924451098153916959736018213287944079223067584129848490365630343690
870142583141989910541398522693075466950199902772011993438099809657
194828591987672414559171595955750006024391473464999909496228073208
901853177416621570733389283886391644137593974177979961906452774096
579769282536534878288646972253655754525228031688471039269429944017
744150565410839248185380972246265514689030002078212754945052791543
698175496661875134783191865741255835539740773733416015611445281501
716175117999639940611911008630470479571340953158279114969755061425
259661879019404745287573343889299080029769877893086987339902732472
293618765193297280946392158801058120917330356069608825523217970057
600041590448817939229844535803797460712947076082006515168336564512
412811294002079114608274243109520336002850578510317812969011196732

0860994900367426065788332367599890317467841882737621128226150437357826128239235832602350621502538812050380876107577923411020066388359646169318157504286066212124025308127579700257872538449057874024076775176118282802205700768033173143733222497208953222317699148309229185252467490518771658719283391451272224391076880381467647768256051991624289946664158685700335897100581727461170013131727201526453957506701723887331443852719496997537245850451851121245323760092430472399543989332763258474199660212612529805661849768230504120572683027889590129847903701012363477732672039081071163930328926896985898027604285309812579195732408053145359995068028164763767861620499087220505717926326447013802103274475785095981537692794373539955990692011086845727615873747414978132199221009794636168368837698006801932672463563313936198022844660290825749708876116126193917988989141472036055993690883893058053593693391145031666583767906825338101549463368505270216052865898969422570963534549240879532449834501523023103683349308340823516829151896416671575047629019534676550504543318915726570514987763841490791267283803179053794039065513432425793133041324948076088104697312495453454578562643292457539754436311066043652894034438429341310299218563861969039536229361901016399352853501057299327718394468786490277192411969477667967432169166174018371906560463900076521196114835072075559291017853787705695420746007253475463298759180083020271502977497891528398945332554071951666575323092649513942114255404511537786456966234680050010557665686222565597532000694853643862230379848569368223874903195490049166578339743669860918339199872371947258845288725401284645056630547236271099264278570245829223730422001039892514376074181197679980049611584889031365744048147277693497933519690791241286804950501774453583056740426732857897572640251681129144017289388939060078622033980666196507858085348249079437151059318692320640496738656353128130407910722221357665482187805198585300198832071946026351214279937006940708565595872468136554341671216007026774829236204014529850560212244185483378259554164191001106984416061119361341572843855737682243702736802105490498596516582972944555191824151604065511839707202720208464020439307298630013905543486080572720871181258779384498490437052921037497010016639981519494762949998642849373675253631752188331308710888079788392417704627889360773769147013802057889504947811588756399045026857550561741605589946250346009210210935213094767593435082242287365273888374232113471060109204939561731748853780227314662884160388678815342375391560037740786683286939848348080670719236001585719202923111341735102217455941199598354445613756191796311070401804744803809438397548267445519775059366593295007869513983479298733887810177945608175447381355918082998124982315003735066265433776452183166172959235665505036298871193560120416793837252007715931419035192724580169449393893968861290001191170558851515798078321975863643962234115591247845187082904022020705526888567677675720843301962157900852947127982339707670466783431019043137939095674184931794875599199055140961968939225573319387182240165404389424297616591282596064556576789626950067545766105749703494720985496417221922641518102798911059033065391546659670220214954542925225680119973223318629930128897726005488830518019073656178487244896155732164825745475381614346708401825711363753394201684005114429600303082324276272440249343956105593933073782790939544010805851085381144126655161542809528681170509607828910789971952998934216779462002016998984965144055336949093144156637478982789278074171170977983171522522769101762906375367829868692728058988150048230697907349216189955367079070334793754336049453072079648770163350386612477167988940461720890124337095817160070941224936215496495754923913389052137928481325600654070872192952041174511465783621081106242854180781661949418458010948482260635786400953180380553175259088246

419440837169751263837722189990351660818503784010369641914891107 16
279202497884078503577014056163478742264000063558174689957745981617
176542473582133634390557463367004182631314361141816033296676296760
016799420553340364035181666054990897216378910193314935297881620939
541968206581906428364266241323705903925680464546366588202705763264
918291587136360468736384505449748883936255634469025847997339724378
268667920048942544022238694891720046655825732880233249943539810894
646629386210678261585278995737113648254919496669990046514833047867
362138961073979915734993372656791738205067948903357851703480156848
711905733150307323648157543719277078267888498830184864653982118322
884774599068233974621561258538266237228319698602520433786277355751
555072010160597871217465069736999774885058236861823974059628238990
617184581682396699394042340380271521493416458066506094255166330604
931071671973308360331180912226728261653649177815454713336139513693
578970790381291008172086749705669639627505283727439791628764864557
681076979043877788536898437300360143129486649262310772749037059562
142587514293266878284788078762847818624595678168662583026820364459
788509829508925259441721135355859411423995153512414886100581148133
714476205748937224169221906391313782811620178787608643888605065870
832408698639465194511688837952774635359778005996422511881275601601
791590224752139769106798632092833840600861021846713981186612051037
717967858647158891191978082065110498720927329367444664555227833315
561279846519883483619760801583131790674551241002086775238220655461
455939748978692163232155531192906025885276633670028108500403677602
024625577852652669770340069592157761564764259634743356471602185512
767526489221675998015472591183530177901102148470226732507958525275
484231626158929674912801498005754149289437240746443811161096891279
255334866504394777647016689704662497021733473368079477321086193443
422642160858013035024637111089416681026503673752140328243429336919
373726378916898338751375557910265452952313728785719567272503523272
501149088524011122123152239535668141756360827968894201993232452749
115000568066190671007305621131256401829504993342817836112004413870
830454117779908282985239103165217553221739083863377240702608516018
650975472284511975391203976275960199053838294949822684160694237074
685285118596668768797229864602757184502991246920648994694897748305
050135197459222778942804945888936266175575585016684911380345406553
384045092547795648368531413220672805295322177576799732975000077209
030258510443246399340674510943312435873235986285879361322862400827
815076560815539481580746263592719106228645291219548991273889934768
984063069350153058083956513944024323250666224298765923968269410311
308341511967555361378398542211792194114495619165388491859795707643
267369745935938106687751104593905615875963496617956775816312907314
395200211716241602363878096329901332470378593865529140518948540202
174774349411807469378036532613139946208945557490603976970949550080
264968790833922207730630331527199447948997819818289566397872623599
650538450840226160712887069191795534834127291556345078392909554962
176378094144760392676202991209797852610126161587127075355867886070
133229374148009805349019787651172350139260320282831938137530461743
861854870731522878960438016260215193217278125010153660955395939482
662978500964721476963041420928560836755369714576744658251377926893
003498187673095366561540940181812138145157189348521141376323974759
124935970455118111351262638521822684142229057616723362217416385969
594285558415266032581735696873478715282538546866170831715678979869
207966857938520386732793631311097017509091536397157747854669238984
188000859849849311591372836081341437393598408772681388157631645299
040033170061435515871321048490860408630995545829324985126941508965
889662019644103033569857578797191371874747941989023985576445427531
759127821820679194009597944889802165519947491076702194965961007134

306836058843980625029333392918050437874958457839559752191849939915
665684207644401577026373497360798043894373233710357794629495956975
286424169076671625974311702801304335272139209895910162931503144061
676033044871310888930329252095324604243871580713741133334967521894
678420435201200510171558881014072790337090139060829623257365434902
251585715973438396432549682824253927771242235747636514746268630421
600367373905728097415180263325364778806297783650316962478876630494
658904139814536834964837676379724030131546539880841739693614258671
793819140481431262211849010477107002065136340198655824595491518936
088602077543374425879392350378338502501167727052720609919380261179
501593818710043945478642611784251461725602723804763286997966113116
147174598788554203789603048511936947192108774950855386289958503777
799044798797681917449414290009803597502443144572163879873259333474
843304573940812551247937720838298860465524871445206812031443287202
182491713332412389413032203595572780514404956058136514098616007905
100746445574326539394539164730244554090449359189950093007018061723
337167982022317326195486552606537659624069393912796408926084161148
803306464994054074645973148240396568634321593129943937077821600270
955871259263938062653090637227159030214454082789035367016709815238
983270423840016348101274986996593833187719507936973083569436728856
506110250945352047041905954246820198982796106997322621366281673631
229890562473777629237557610116540065593852372739150669064576794639
205897165228649774345022841403508880830934461816836667371572514163
826056268400920108841378388920760123000864027233105874037792368077
723633760999633454914229514026340740695000357192016355410390524035
290727326982389146458549806071219716833951774684572732705783001650
434385226732890660418884113825083413613799619810169705273677247197
959326117762234453981953204724407339455212937548499646212828779894
759263556471124809844788635485768440400796454086534311784469866699
631550715355346753318278942174905295408470023651559371913007076532
061016462428460575445913627408024944264302474203872311368140350116
491673880339796281288237086263147773710950925211669667728400056696
523553312357273447122505849334210006543804671533515249182317846165
102580881801644604999405884890852354086151838940436071934067129236
726069448064599978077724922090290386244074458697014205902219888068
460965151709482306000964657014364407668066372796875694071351567827
990649079063277209044524503248575792807082252032623968335485158606
931459783852836169456336186314956895751252054275834084337654387735
755811332332244587416913044623332885304034168185327650796295332567
196803046626222693929242338176147406217694381301816446836250602887
868826388486228440768649870505438229258065848704040355625732567495
201114319375477540770919620795371814882506166306925483188887162368
525155484810807574535347566506910899890257900674711653610446354848
625845329601116605524812366581334844340230338387669473085311468229
529009285203397072972478459503263973047096928772755667410787921270
676969147191629646778484264375541857569851081658718271514749550361
565003713207380545212738607371349328329950679483810466721696131667
455649838641066553851957114779798978040531315313046953559560125012
230730135122823590308974131531872460657676506927312075405356228053
969566709404554331001699200934016030151096700708703308962865869548
111711644722242956459262470228438373631682761826381454527381901157
150895220166955589525546220679207427677769230811522647511068244339
134165002524040693302585989456339270364194407539781200821821358504
554735215748043870509653774461478113468716565558883519727921318504
550683553943076050369009362962387836408667012944398938544441878508
815883750876290001144469012888129558556775237521659868493432422064
333126591574887295399533862251753806821534505112447440946911045116
323995851505755123125829401236747715157902678546634378329976843751

816713246639546430207887683845141331311200438362720947289667797439
488890340393339347714916556098784406261235812235129549256259585831
916362448449112395365765871050788073967088109766912105133672021088
490924983481410808514719793119832560149134071181692653771766948863
256773850811309929392427973112696774583490608959014679711219754532
868794910038496940607005645672377105349189064450652802955744757018
578593533302534373141442053035818947250256597419922390085503338479
296623707672334636290375995008754307502579639741502039884959037585
768291001734410030163901748485633064220117520177885798427457959125
026072781470357311880310825407223370561840398146430866701326413991
073500244188772556351318020125185808148975409736973081487305408871
737347442840919291643424402102449642639287357398240381058420037345
869531070327979850229309466268069017927241787098062523829754926742
717401093133908440060317715039883971756517156642506614086351975457
345974732854603525770816379080538731580692280533069866107176170472
318941722385413267567686410850693619772850880899912059322947870172
365995791125474049023042173561164395489353844038661966783227383630
991110052858370782496250614551882569385163576399303075590707409177
917689600909421662686399459309871665187527628061216776559917910290
609779887602911312938589553501801828282284251271767414232794377249
083446842157094679010491342939738156593513336070619125183966348987
890494708726441344580810241396525389714439089177752252411780220168
874982430162373290165423895888029875750062831044553948727701249308
315249497974712676481104179236903256497908621475914072338569985684
899320722812683150370989919313076022276809177596019419663136065337
542651454170789726564216949912776720193565187129742389742007742270
600818331468689266029409808583955345298132643374294839713715782663
489388817585285996432152468492022170504236438629715317037861202578
285472396855010947264868652733936132705317091849608428679730630043
616542134626766101017003598757979069986223205488026418532486292510
961687965980769538976545361454574455400165223914248148929729381427
906255885970122387283489024057385524642344391199345027206577171521
049912790899211699242640970409416207231803949694168898542656153032
807224682554245811114270095732327190155988537895755711619245963123
390013892387272152786124203816814896467821416667587669182854585244
394137306771464037343309404136447692935783257567547224604923772545
306631226140550175638115999431970278836561469974535618662519921774
758789668022046667762597743833899566039040362829861482702138619053
606636684579151451491296624149189690080815398786558385378115703426
603443048225501319786604767627111519141329606396129679567514855605
359664271764873387754842166807326793446827374535661080150860574339
919862152957878761118559244723527131690090072760229277857204073949
284081028003889856654021555633375622291458982640581718488090352195
923229845591916946392957967530091549871090141039887383479249362893
105797115046206176901054689301366912564960764551910533627317915600
645964827476548057231889471398410986013028648665616266629576250098
178394457435203937949163186162324508410436164555398170233396828075
408160676789235051027647052040995697148193078321599322556225791336
901779370937542504178257657070596223970542412067164187424641557566
178175183211009184626487177650911903345723077873178804849377654425
394524714942409147933707351348787631457698510024967498296725718389
578378464947863985440231214546407023160932103605559461954760831841
078154975855244947322143893205233734975829477293697854244733192165
853383175552249465848759397461203136817692491287900551784037075161
061508632843344567384956658915034942405200789741383221214924671798
085284634286820447027578369882865730447370179875498833918216443632
043836027526113090044610037477990279490761245993811240516191996059
651390077909634293583119034305624356715734095056163628748278205876

154899881362284006319511195201780809674906704976589428203193245913
242559617114164316694416180155240661886331173995068796437781553809
722292974598686743643437724460225008217013149369918140245420915767
683955013168181074034283041126865254986803264579323184502950977452
013898055358194141019131984833867985554819940171661584883611814850
491864675639177286330583746651251909951762786218221778373921604284
381236585503598776831679887691766786403769600339672806404573452597
520193285403903843278475185561475310226363733593846397999451197734
568566544674283895819110350875024475420750117547265579343040641644
816400018548961723573698765002091463124400680506656182113202933685
956754722666466024686854200803926074056629829966228278866733064506
550328841629388295625887554096946807070658121980508578924056678207
301913200670603642168364916525631353776254830895948420536098722955
582752459315009433419009078154671122572710798521222737556158326103
023920531679278861411832982264597557653404542346104902957223958753
311957961950228946050308356974031984824793750389739827953899206897
189129627097026816017999488236851544102490634503793304638505980500
689368850921596092493454617018696647022532619388193016821226484368
777952396181368777350178398760142879720848366410773287642886925358
714697839726188810338450333711815171148471575357282911137713631319
778212024568460969777492196379684737496910945644214623527467527261
285301800789573543032753550589002760724171082427797722732735900626
663866696013522302560850972996315375782433792507162172074406333137
963875171443926623811455393900438867851842471758753790306636660932
688831931210323207235140907033605416575082209060337201668113885031
468464451916950436558866152125950693828434458152870871228293140727
555933699681210990351591016421256071107576563443063511705673167435
289581947549521611425193110025890413452890075077583181812267074861
693705113747405144794546197953017476006106793453483773940552135929
988183465458679825875860431734040554601182336493553306359083224391
666806121029285929309962757245031274894390964296332087307746715007
773300833934315885963701244337695776945482607716097670861548166679
423891035060904614604413039687136894886799835087804068064381621772
40634791781191620062957770139937093439443217249722182319521253794
132602753367456855860884410590851123027060655379689486119033343118
293391087619618565414570968938743695706123428019773355738962407681
631584433584877073360720706401263672416841255098300951381957688615
124648660191044190004053873335671201528782626114531444001949050115
641718014623553003346080217675891561479950371467332745815027281127
211826466922555444131883985950933419623985945561184947674786532207
149201414043587348389120710525816364490692039881387927289928539884
606794699973386287843222510037432806665926419930608469361675677484
717995553822497857465567250555674894930960003881116502599000593740
173866604706262123885284817010946710138768352200253700490944667104
755790008627486998607580100559897397752748532074183466193993789997
610753993025114426156892048551972307840758227848383123586478168286
347239705070337701551080372168639415071758912025235200309364453816
100089088130502039169341159108232754929969978413544833229671875424
182246552003796227904310697702416765482939497616404950028309838939
426022430461690484355804747722403871786693491539288578630229892431
436841730470301570109023060675037024472003326413487285601003219723
656520159094929341482621229998231733207306487960120379727647315563
630376092938373423468209183320342088037583199689240992749363529073
564984723927517964835646038113180744527184622345859794497228431805
274062500057844200483005238238751084855426648661840587880464120681
035919898396098727131150641081845490455579927609435421840067176453
548615105282475682962685981806029377282987924425294387085412073102
529404983278917912774900315217552148825260347141601819538454176711

806252183687581941540703676815615766181720477998692331464033613803
346520401842615803902641825361857224684486061288736869992720274162
680637666211206929034619695458113643447415987140189211660466226582
661590542069763943593123662045528756034216500347360119434225614914
032015794171185171542756396517256864538467095452483713059489858255
974527756437837209390373760644875780538089666661399183963055434635
153154818588677926291272536342628898525685446469814497461892414958
636636719814006506858886086022426733798812768796940649702991545245
272132575428195324917311506620858665207749095296510075340404922735
654828295702569062935888169041465106971777242095544613025854381786
304850806058990637380905430695026138424222705375354598590993266967
332165151948172534532747333602744725852724539478578704905484758633
115718366332359132347588259340641521039072871936326796375928473313
161233978154985650774595742630192501361344218177865732684594980392
574196969999876459824956709409559549064514319299753296990292901811
334684918939731673374047376102153497902801317223379127998639147101
057364580882496403779366914426022522432918220359694796522963241504
625930376366432840865616023121610990271779794048244237437724217545
327436903074926261725888065223326141060338165320932320266991087084
758681985639904985750117619963690596992541043687532918190720041575
982623454667270157369711333570414032093793451266060707990655868796
161579984934109540903212436544310873161586375727374817450178665573
793984866922911759920434224760064859760054978280629418739147496645
660197689263616598289656557445804099142689094724970673522047011619
153600945273632530666440201002320187322781976148686634898913273470
144482032429311784100915283333037691119705125251189029708294297488
398137149977805278164934370560432600535269810691898658868981610889
926992043454781553574046529382255475792396512578169849867342182185
312407343115296082141120919999406160101582191273730165017695811861
903668977927569046785710181059393731438811929147494435652218962602
863258663651901745365921218638768107774209158364690909165182739807
531030664998062448492774756188452973294727139148997268407785897786
865604872330575242285717247344366674181812327084159179147861681978
003287524219464801193159379351523941740904190984125780990903889407
794204672704915345000248604274553073068364722207589308219944523472
145218428256209291437681438780213969363997860221263221098220773571
441294126406537652642854348297069365511680672830614815535006779927
342874671740836666020537292204848408702523012585717914569665795239
635968627064590372072680587943981340066767014117658125223348104838
686771740587973689615599621784654973073934095466043145860154049561
577645616734472126427468746140830206387935980428496622472232550466
095231768172458636261128483387407650758288956774589588736109519742
072321262333556151933824717602181868389530067975021207690438838128
356258145012500711902715651443462706481495059019699613904056079067
372607241129244719947702838483531863329817619994731283314491596877
750429032797703476193806295512384013842291003576769296985959054439
828926666087801034405967055905207668370210159521951394544773118741
072352794560844354946676079288268193567646665891613621404247856427
926815625448506316685268327524656002747765241279427053419348280546
267028159924539124873937558094012591977834653365573562593766808769
975257460270216696492529983775381969384625470851886151047522647513
648918833573818916971219583268100419576537702119211827256650889668
246750486669965988504204119161533208988567238092601814281176234479
558294288085813698860737275497491213068743742194301486676625969169
519536885691174938231969755955794021793887372251515997140544728970
839551036548666281965036848684661039274556841436359017774403875651
208077145636487772844383315635724968367563148103943896809211928233
741450761863056577773779414533302578207887933713552328193066778103

474487881453302276711598246301467739131806469901714532318195729640
171382993445866528104239029204404205030108503722889707226673204164
075838352116255472935420977067186526207629467204363364327350566320
411252128722589414997628046891485550759736146505117641257930283697
453203604650615692381197213251056180613452134381768173276284141810
216441341702491755843656811454797775195828066284424975037917477236
204204245057363060911154840242699564336003530143666597511828032803
953421054049191030567314210754130235793487723423907395938930043689
132814854688219705348268064061334476035746590906564609800971717699
575204375563545216850442243908278083480721568003121380513744755738
663358066128982569525355723266307395195406369794691392739195092970
992913488014867607271497851682702690507107678965765304404633626268
887226274293120287517384975542143150499890745646121569554253570152
014746265873588291544869367168210972638770366929876270333269267640
511235659178773632477046116118752837864088038280136349041450851318
295587380336571592375540437403660094312662493744735018369408835012
318057025510894184693666883038062360436168998681815046281454399370
843946723795281953032886060996315478054110539798317391048191798793
376309918240071636953592567835866999085256828346179992044849215828
682554266066694806590558837547674778900630307639773201191626441931
231373328223641811934383050586585544982998689911466813117119421891
601793736025759632185321048687499204734702994267871276133342376683
428225657565015748972028034318032062448495723097390057150931453891
843449438283973734515799805156259164322627014186206236944759302141
050850281203604910993905368478056021662704636855257274132906042289
956351025228475190703082532738499555533044957803309025927531635221
809889882629115980337112570172176766904545680649223051574746891557
171017567240354189350611288873024043144319869586521866760733038549
036027746096354501952529653407030159703240985115025293058865671901
125083284714968064894381300783718623924468179021621735691228872394
802164846473775176818942122207103565559650797948449900827193552814
791434040371387172081760903198856458746108109910593691774379471287
936895032477418648580648196799561464367082487089936839513150720305
653030078868598203607206699167376164147566542877193535910445211691
676928163963645629783407370647882740618340543570774121613832028737
879037858589327455361495644612550505547826687454455090886889469271
859889494234495074838218501843413032362004668070019175045926284083
850536431267686980340268115807098103458986041308417355099569441517
991754323348063307326133913997978800038210913276601454961115704285
802816067516233813538630524295638330950320192878164132492230047611
795358249560591453000424644788062130246897865596926292576357402879
940135692116755390400269966455602568297236950495909960271709738166
616867800483272929905942810296816157469630606061010622671702125396
674795138938137485389535175783294513642600693613777744000566993017
402011366773178770446945060129226074969110615762076378933450413733
651195600329078352359646532657474974352867211162980756758510838501
606998969358671522596463057009139087628741649254170816909684730954
886023329828880010165296280897698772319690742102700934888093445551
218818833651978431853559606747635072216117872873563946565543427614
068559412425912141707811630308010192587621993809895894305093968251
827713033033492664885329561879526644631906349389649279747796305818
237760658035402279669108180932267251426142876850855023403639597056
214125151376204672244428979978554541319013720560029456706340382429
917807512572448446357715258934722336845268000504905794076592611834
046276469983532198933127117327105112938774627200813172632086712472
478310369532505711668466933919993838341653013309247729429358470743
882634240070132171320972827494796116356678261507156628250212520627
638975156658513406045529010926112638166224236689927106490418862700

144211287359239399825005658006432506075094589358500721807293223082
461082581858746713340673344326805154756527611229009421546556183103
412687017228654162269075744666374025785058198390337266853912834251
138897761037015547059697842843812110416674964236193689840635613876
227959545215146892881328239439636612945013878167659092925089753247
166912368348275154607827280445182434537596550492568064853230999281
451475345509275552779627941424557440525521882532395562508572117996
353019567581177443676255097319751155651484513728542407490483808199
558809151611793921910426190928485781613905148231474455310151615917
841459994712239051659694410397272957411583397459390620500775807095
967658989249614373434878156377442083749991461834513411785721356576
510028821653437883890697229988629462268685195628832320570537894217
895936784489997283808225129585737429565314870430403219543301647220
458139790574528802074728588103114085879874721186328648984473405311
460440048001118964742165719993115889037205900314664181261792091194
088785006677801099918546190934892669185091189985328175801561496344
252727420213230832626253782975374780849648620574737729805950490214
555010896934806451097433300599798555320331318269175191591950065584
884365516563352350794974414869955459346826667617498419319232391985
849314290703397249074104335315507161857712844527045561514872082178
790107580994599078190340768484559080348525612112446388323887760077
256405950585761450617292910176342106201632276313357086984161133384
335191595521993423910456556597310082316994799784563195983411430563
705888413585723243945999371608451544322473332574743269029321113405
536052609072416883753354361412870234237825270895382357658516596417
328110042367022479426589489542002462779779309893450760213624171370
310651327597596134899242700470471725333264227742623475663202251798
424952498212765595113945681412700766231548515825735963353826717922
669936333918066855094802896639701633580306357779206151292995868181
602332782682875613869534898338580784048677565029162328102232126286
237036471167259091116220744994447033715467171935579603406495721839
913243157175868905635782011935622777763421623672475667687617220610
152133894288442755463803914298293409706376846651169968454977492484
216234549879019005901223071102488058804992082026734152133452619482
271804045246592288442955419081544921458089045704939272983216883413
429953682586746410836728468331328743219812300745130314440076835740
244627809770130021421871235118844390781428645714867130316274381405
338857603885294690507769112439640284427539211601124684855885908711
120433136983355129831770447543952735118094562627365072134238675487
816235096639875898907810584333722039754420498618992145343945700107
669930233340450706235582283219793912107163390447845740649596837368
359996102705598109343275457182265819162737640832649194463392541412
570472789300121814099502881776632184985453085666790070376265770583
827317495612091752324760127284885442229898679988266283950867894577
143881580967550580114816329510134178454949532484743502461668260490
860543498626070769522068457693088810199363886020569081016645051872
181500574347274692400456628013524424290432379684768326499597859954
187679609888240649742665229584406972977619146264897831510287400669
792362033206040445354794419819989475253195717052085531617779164107
727646139215711774029913555501516709796619652406222291416099722702
986540871469079193291110174604012065208119033479335074093396733543
810667764425162109106409782774039347240922697139986419324018336762
578795166166921148570844034731109857718604143806570305811408965227
990337070679277771732401960636690599023966006174759660274364772334
171321125640695730430073871896976082625377840761689045650369641210
172605511050631805878307177104571678917209315551005131262488507497
125708827760818462973515666064138131857756923219422161698198183386
159109111212964068347645414987489259676939155208899974348348251097

717173347748490424157044765736574577528870313708739190721825057729
931772042125966177890946278107374893967275333669497759775761401339
096400599495991247742405822602276743479140436597750071174906872762
693456375287627915386103348098692957428984900870413755216037594678
764398436269591372899723772007314533667335269968366292608565705839
038823593831189614763613317043665297644360749401649697173956600232
484753127826135116274516934985914972842578011566354011257488383449
578205475651348675253533930949298777856682473832247136341278156638
245905023073398325360830038730248396399541840286629808766899600543
606746374781759738593200109703840094329084825214860785580072030283
926248148421073567689436508478291723943135377307833828620452560793
714347989024775448000157389116577894911436600367935436393206776263
021105215210135921546945449970588876283657933406061323911023381234
789116513396064823273430276115853432570782259674566926399445065492
819990540278815070629903621350504523173250136706243941061807667613
786614145637857660442074924229269770349845012651492167176963310516
767267848837295490056786572397844276311347324977198906007618759140
896073066582151384491345615555711510845132159738291100111428738959
946162393850583603232344208288145043391507378081863193720783803113
641758373487519765079335205354261064839646800228318032347662678243
897034382828567674099388012245584182825061658871841917739713484249
557535536154283510624483282114107566096796995104825227942158706931
518686822890659905445428976768340784284686635169507359290059446525
171865184120454432711637452459124940510317749743271643304786104252
035794032712103848084464363330137152901498842752377949834574799875
281511965159434402988721491706555591849396373620235346368882181314
372081349365898041885261814616685646389742839150595583703947123832
173452734821361569612832630851650382152956508420824710834855636841
528927797777756366598286215650221250746116759032116542096541470122
942838545756721417389929799800524641816847334818021732521988211950
351846705828084292711592599970153750974790079509303318805655801013
980812298456546816187153857992812861037660068440830849839727976370
041660306524614826604283117793289355874205593451881326476050299179
833827163739598602358878513460926732319515887297292466770976350989
467701402822922459093649319251931742859694152366564659951119254685
886480010377793163073879444378711516343580868385509803616734117746
795525054156963255517565930011098710343833359200752031771871321169
429737366650481616228139743691943602197933189113714775792846048510
574542832874817932872278710961589682604151910321603588677947479259
298850999499049216484971385441255339167379832698374244874127454709
620633713904740483607941575293633639044817954141117349362026048512
012305360554368887938629144807879100525559893282527733543592470673
999726922739927756533972566967152409677307351761299221420255506531
429549087426170533585033317774602589152893788654307666139718694743
502046969668750567663782500133665443412014978039533409484305880408
087138695315898917544323055761484459929712872562076470023808833082
557863754480673545385987638085847636506952673628098611990040267981
130204610101944598035927435305744929624227377339552016915800533206
697551301973113370391256119133054329794169919294829390557267957952
817726968793291142577732050021476019969815220206152451794238984581
852682797897974405416564805621008330286054730103160503201457405188
876198453502969999264657956500190448278241738609540179103702546381
436244525088702922663923366404351967800356654544364250362507952964
892139065648414018273697145021452643164662100373809950418558874896
308246574073342630025309949781559142128751970527100115710171637749
883378795656865393442661195440774144399248742320604226615977147800
654896433496235839622113531257289820135598481565702045748210917373
786628025393070446554368897474906584779494095981224478743221866015

300973454802469617267129477879519441513440173011883236744872044780
623663700352425861765683808485736885690237092290882122272083417008
979129256543541940782689930557315191699011807050189588596474048139
046260970734195449422706775403353710296483606203275561731021591434
684415539095659097465499279153763329473599602008980264079468292536
127789832058536058561861189451406113222427128611166867250257033634
886315857173971160328821238663749187456626042953189852087917692798
532059687082732060772049087562497623985063918737880636107372115907
573362989705665746883773515985230620783485667267399450197245706160
417028656143126685079755716895080673861961335761307793385665294261
178995576128220396278616303669765729274631760052262488130026201593
446959933230130444390851414406961294909514351132404372818037803052
701614745966697318233961655755091009447720112313016992708211043728
483536465802279843531764876043952080736225958640787345819153105945
582524031075377215743214571344778184309844749991891988572348815392
190452015661719255542998753345400238110974852027005371147123897974
701015854753862680281323170286201290386585144236448649286552358171
372029388017529410102252028569118718607964501087652984253297163117
441865075201876929274642106131317620308506790665882560616162451232
983338560212251996132072858164070209062312718344848223007894040924
391632745240874566781367355700397551427807392003435331321282287328
689542061881883291418904239295829349293059481391941921015430324356
744769924306848954952023954564550306504709712589530864100115696873
272805753462888209289499803769427671523724690745276874941173761124
890701919879296423247494837186391329220403340476272852904805702005
887722635976788174715798214217641766434900548515322233513050200474
266084260275811143081158773578536814136568366713391740316070565037
908528432267821646435930151920709912343384447619748970136375189516
849863063347124393571993453325119243402486722685096712244422809548
620967064207146038923593401068440047536963864975135973583310937832
600190925157443192037612293770890557454362847738453351375649811641
233689229522888668519991591629961786720819183717347307006822812701
860650275302983390146699957479446107026716111602787170650023445485
265531804615279803013588943109664382224754397716230467646353180281
996449375662371151160519787587083429214674980001529714109167257057
969361548756157178235831117710135901253539556871274579972017592606
546190059379608978461902722181450723587954842714991331503962030512
411091650441966975582513218186178356942755406145597270535238267357
110231808197208539486018220548268986636028695668183486685245446144
084069521826332804876044469190082669676296454907545722369233274416
649131955648643994588993387758700988054331863998554049323067761591
828592874389605780464098420897405506832961139723922222697903646977
678755173039666447415747265846547806596395648953581955700357971668
912269469927151284486477227398117418148866331928994659470600892113
189894296771965704861852768613423688150000417238002829767005277922
765540848554333486168898497387186788618987323238004240096386406798
435171625112697259246586787211070538015319495771649485062981579894
694171428204216416558665990728619849384917548026958461964229477931
498122383641538557038089789007613901032349717969632547196564912274
558263541323414243643574594749297927856960776359148472801212182057
123722912544332455660534074849518144676905895980695200349230012498
661937621085005123644254782643573382132966096697316535354256247308
090288177611337397206298364305054086194062218385024498547566872126
006763397437315325783835487482440978097393361487310202390453380947
415977664560313768110629892140901661232700390050502294761351885912
410647065603129801460888994927862354781233707563735243212718006105
308551717034053603307376301871136693532176984282601761121860063584
896534143606709141997792464097211427449546989146355548286434011040

147223008474005897193942555677557844039936570126377009233040177201
570971402261897254902499639625668908485897750415713042927159289338
014627628178042451243346115672917087218116986695871312610665581097
155155569633481984422493772789984900169133406501392255837452536445
871131537749642845154005364042185976909329790072008362962024673223
965883741317529058662626944673521044260379193152110356061561327177
958332423894101137830862454629514095817189416538188260985813625507
028147207441010322838569770121266793214647280115968243771101640588
292038122298228255056488502000903159908409669801425015995974256102
202631836172557111491392144361101853384556828607031576644860576825
091946215850695082849408530180664499120714286759898668841279064653
948357153197916296821916428762691256721694722877436722690746310853
419544061133608487163008322078153715481543546458370259485036167470
707302758499672313280960162936123357408505167586704703228598724739
808647326794039491393791308497363241413902943928457663835198421867
634664301808689606921433831960422100947209843976665228225443683042
220547014325651194268703971992934248063911831147159679288276590160
658481161372294419943326376901682224343559260799088203400234350990
585913929771576049472707702830957584270709136977043713575902026722
712135534497330430506741037054407734595843092276937484039563820488
592647043863621119935422550000256114497085052705009162892960498473
983059770893141204198377018700063858078442617736127875809955159503
846662074848157250182123540825433798202752568074574177955302589468
194577646065374993269932888205395151847408514744363568210924250171
052586349534579159028715221299536595915807697373406864938135614834
686259397425293493723922991126095352789014741520946191693752339607
918005888183785066885788221740893022376707892649055901783573302904
986104756624088404630174420916460792765924033500521369750536665803
340501312807124279897000627372915259070166674643724566787432560019
555424209445336334393919567680459087133098344796212966325811637654
664808900711247402815156434346310814453273397154323345449268619307
448373532903424290227063234450908155379512377571610492712179818535
358509941553871821944942708611496926208222831105450526017646944214
984796916249383350864389979794372572585034947332112364201564544595
832321025867338212082720110236171813291628134769361336231630005551
854169945123703087471769347530638094914882823181146014847359176501
968305714704713971680541712076085415277906148408154352770547111386
619355191825554967376875375655901891589207674352614885293763210787
312387572064840823737530328846402348829287586745751741196425955473
254712998878441337697174478289514806062975015981041096517924873725
424158606070926343451122180515157680503952320798390703844559080248
976751242811186831122353594495362332805156428450911638253284682769
037798940560973049659821677479429060522906427115409075963049075004
578669480644240791416042498973639622383553426568824892490309401353
227598292226761802139944680189960320675803429397947950058112356339 8
697279023670197762898407331986142970899455340903726057682234474886
920102976488719461875862934213170932727776916518710024989966658550
838113356789617481139240080692044146256653829452233915516045948248
702775240265603080241608406335831049915931308539474269073234720884
119182024847075732911507212454468985268553115998511179393810802097
734738174933999888973856536994038759525336217482394715478348005894
603936659188928996217521047163046603844441237345091038293248389456
001298364934173204224321656427582862696462987054943708647463427711
852938248043935821601986070062171196591832591807574493655660319057
421033069753707322301404429391360729974231482186205712797240843229
742225306478470222877289889172044541364168301862059266947810650015
772730346839982955110599642751983442090994298239413615583265388406
852983750196100267432796083226766089824373992304583819359944552036

471095908251899166291593196398638609984425245591944423740980765055
611750578961464359199255642675963119627783751287465379652386652568
169570032696428785072018471660573792272520952321099076112702419491
396280748169651494320431930648996967523523778013360171535579415272
267440438354774783461541713610807197104730886023745031213890081325
624317204976886802282925502433493959917827467695941155443098503616
431316394425732596746141758066342449250604023220280294698737518293
127065541377829886195848109350266364520750931961525082015023951281
069618830208378779175314247378006661366107125561380956875490976302
248182547336957774425013633346505272962419515030700162982343399091
000615550249467371288432197235339945293806722341962170161634329247
937574459146657715626682851107920980138236027790852490861406132382
047029603361421239678916949923423217152888329799391129466525384726
864053187319793226777027601787151571127131902216836416945329045195
204615033553473446978714152244708770708456141208314980110666716520
746555367187872873746904987162762468053085758352280419199560732766
591756957730028792070628730635622829071093172254124410289965621943
930339359793127298249018850599820753028058182687345362620769884288
389289635517726997755270892807119838327126416498135850566092974668
194143322037636016031702386454010380419337568874559351238398276056
529937969463111573623676580351575643680207979098069735892895033363
102750947847813570006470316765317984374893859319928446797050501465
842227782696067591977743142284898346229638095752245329234359223513
044120345909610127448589140505827476775911036197187481126259602724
586676557147275493941205551336228916904263820835739952061572394645
174449899129702110657095944985564756710539101364046559061659117790
624364595739346071857771170611184517001545450809988500405931558750
951206165545631462007387394433274391565408222265516671149813613507
373995489339174808637419664809327817100263955025914047775193472052
693611721552919289489112851231256310527709723493867708930988562479
735893275980846342321851624985305036273164555086002448011287948708
902187528734539294131614662088148268086141620159155491220419860259
848860991001008935521986004274340573101214273402947594356726977628
542777276759795406783215499987080260583813286902818386210006100337
642379198001944237044206333199893214651697433457699128223182611409
070898614415408199147374743368164498243252660816669669733619695332
127776912977263579843015097108158562779524101403121972539950098548
370069915726381749334231984170867848596330912936697738356288707840
800239223578231036122923133431387087137560726479553068785676787614
086797853884183975085046835092841447196834356692453914539697267038
657031729624183897923545368707062910535840958625288172928169247104
639591376543397751033038618690541627854067967188564523146253468346
430020306636243007728041839150504839977463906523527004766822033701
569158523770139905412638347648466384117910763416394509626576134524
834091389875379348887108440822515024794471987688839920035737926073
657685493015534302684384838893140272196682038727684904065078641498
395483862399144327100354841428571466367581410868575684974964249205
885698437459468657448018493422802795823563775643882682326241877622
162260709045198459326773477350182854360693935241658960117450737611
406406895988292994464538186860664747288899190982296017892978047357
912479632186915368703655952944433499542516058090830492735940881012
512804591065050476649626758222413393315802709420923435482413754573
059087156567567670109209505467111783773210475976697943635702499917
247764099099618422342259389668469915441887720946530070344371831157
287057320673987595782140794073633423608496383233359179022771268263
032769487753200468018475770539434300795119666775243961591663082780
839590538322951127233820755074415307889777628677166251881091187535
19733009863771747861881547641160222390303195596789815373333258298

3600451889734131385796009297774588061424010459150147820679739436362919935582276023675103478275564866271218825552853178253586010351082258145026112047470924017186025646900684617317679905734901100727287689261945275883358758221947384320834578633052807557454938289523900598456827291413464323488171485846067830594826014535968762159670812495532305763781564934456525782552293824962657511707494988668765441038270533740898921040368677365604542858451695031402232363375702641796577852081754489246557099240813236655786985264535388801118891932862519259222135566768155760927630615575930662647392608983278347680214605557131593915751341973623814379449788885811963343728292321966320335780126113077010857215989820280124527014194405510821109126282616707080627427106208073756237473911790188063039151918982517186661375757275971038981146281320941182432120115782881787555919083128416589810129599397245828288503470905302829222517897243132948789737432150734138953199236450940330089444177994978550611469561529485655311222946235263060515601499246416469416535817179065975346477474751894490338863769454910133847537957012434353283214492982757320649460057035879741372847308556850024140578069949444209656454542067040671269177034205589354592147513994658656379532449893948072963453595989773250352242536697552026259661902239113744647015156377494681734403794532885953949267776387559086479724780886800681723558531314356329754317660439253838540575689974298509517127782488540660920326559828127019571356428139254066130983872519138828743230385070066019921570688871565313698645866717928365735525658627188144014431714367937481059691370932121680624247162372966171539343995336805414569867938971784889101598603095412935298746169109192523809742944551488558318499498594795902426366425548655563314934689356150341488374884631294653005659807467697736019455981528630627612626766355773597675811384216123733145970872986047138417401248888918797133273636262511311334676537629299840890377895200008539399763584744281979857678072104896345990778015426697174567542617238402203277336476397551754916653384472733685352496691562976924834362650474619882335945593854238873901364177046945395798118752712159776844251179958071694546686174982003891413675742529519923536302864099584773808066759416971558083354262399667913719181197456509501421574144502465594282308668548345475504175328090492439697577338340692003659626982380632105842116831153608063960003029834869862501418951961597709853098934159069182644746212429836075434644624346418375891269938235380143849573643323589088034003650356294509895717318211938753606040337502573311825257046693447027575665724602024343580517617980430500157661220685910711916447836787755287555765149553864629003147784335242018223228186288983604995567100947130736251175046568288749334511080043705700857207322750831967368410921087264603086269870121760440904028069794911183964366052184314923073516315889044454805679783225559628577325030639073862370333133937948649073559546851796504296661825531052645982451735375632502837394355101805927205309010514290924335273127869001515130558740539553595264903028131275892668785026838027753980878419695424447075982652771476368013445122704646965861601405000598135542660045553041256453856489931603128055964697097277176477513623793186953564285143044563012761042826494036797480293301901344399860928405868810026888328778606907005752229163788459542951127123641629041724925928703351812177724572063474441407104579870116834865389408608793464288581405323722052203338827824916065507514284988950267304733868941466577151917067007154628493341305459532617825762474557786052890705568120939213636020794840019045348227175019919985035172181982915553739405244748816084011428682985421981541739946151944675665399110862570266172891572116167086128078644225968780654055240840769809262297989089743886487188124121528586210714417131468273941512454023152718464160210458755

```
3986776730560640185899053575123002306608370871168691475791336384080
6232538434926718606613904668433787965167900709509607700445453556310
3624051358161617384399700872078005727973747669733000195181571665140
4821563406218186569555695230599732096135279604360849164654544009853
4029608616745333438096899823943693824949550947169541670641283942451
7141260191624738270866976931180696097101557258147853194625746145350
0397626055465125435021452414343298249241381217713203403237186715200
6273592816337441031547104639631117708345350154482594576086659775710
7391457660958775366724984051724570082595821245485219616437893131090
8824817003842233206328850043254208492159442373470693012469333391880
8763956241425406832442961396568624031641284030578256465234281833600
5665189585503699094325953094250608082468414255314370373906650989600
7653980873587499377033458953019495195170217522645207320360275463940
1311371161162289900457080284754036140638147708904136189639814679300
6090582948205418580676537456845215201141451358474004249171418349790
1085267557869236460840510296405685471629114796051660512986011642180
9235847067744783697916343692203021988896384901524172648351087367310
3458395344215022462615533663361483478643233561380530074966762084280
7575307448476761221295204640268425615740219898496340903610921847600
4312888296926638081824774162432149180517457605165276714859567236050
8544814854402132907532311933789705421376692598941462443758062516150
3217997346963717964554733535492490014056074366310647417667998572560
9863025283994443463799305182425984125798358649590627890695365185490
5918160282523112964886245466247414987802348961990493472819666583070
1475772532015868355470778367282575623992435546393774544779733829600
2922395391013639248424579908952046563980451217891118683684611173670
4956595237321588319222476966429594969400736850502840377689062658080
5002467172958976399152885528712794692079212206720132052809396452260
3227086822123147580066031785861841069345045525790977739566180123340
1874179413622157860500783390345912525245404857543632770893637348130
7625065838802048457015329172877536585102843043073380946458279446310
7034769613448682880549387474207836072091819669560779780786595574070
0960443029786093090797207307496070101815586850159480975343530527293
4116171483174733966100517521700230147990108690345229770487759840570
2862989418102152969007455565972433483618503777150550860330111505310
7616416961877236613812722787109865173452038573787302166127222580620
3945634238951182710638999382819394680890891714268678748870323697420
3698254355888832460845820284023483626235883643493266480175590460340
2821517219563963495304300848416621782715144945954090117944885259500
4947265569119457920367936540376113867493761913528838976548861213180
1153030471606196867043644288434988225523312968750291635458654268660
0889460469293705964951284489740485678003585270403993562604898282210
5255571996993525794545261740743270859891130014290671400225943274690
0218982995179553442748718111642117429343466185812595775017541353410
1801559140012523994839390176617521611923006320392693503074408005560
4853217364811681583302031705897646782329202381824767644909239971570
4846690066268487079269797454361050266597917187256545818722599567180
4718953968963563982739116945400828677220839735648520196059606726450
5519342952306818637594677165747163985103758010266451314905894653200
0117025939029809267215532618811216870595841647293227196915373323150
5149878813034739889489625427170463108520020308934976074748096952310
4381448595233628759639315703476429000525190908752531657407924492940
6317641112816050060433236781492170244718104090002523572907450940080
7541909448501323427737790282037687775988388910224289465863071885780
3632584011400402958201515727775055220497671741806522968128144535960
3077468399197436150775560849014830488152662261688754968034628330400
```

535622738086930362906434961063890273423981644437129896752462213499
807179098325385375182451431822981701498047147440520820677173482930
757424460717784725245946185748904930509650979539054225306921236070
170383791085714698302257724852517384591456079105588537047066293052
686115149625701677756605098715221893119931908608040958932727146520
031599118043637406495544985222116311079240253084120858743275083025
736046735590502428762009605821782540707253588194242748229061261156
506709906900009646228666919350261335698044849903060697708791796420
344947066473435831304985932397059589076521205938976976179995460990
255750129252950517564633281937784817982728921626883979150390284154
892848405010183293430169393085976918820760983272888921135516982344
564447333253072962398579235645767684446557407881847532800320620409
124850379079033696967998575698548117548118386688492826248933731346
365620962364360176047562884825574687983523166892032758120831192672
738707762830879194416406020746280318221576402945658339747608798691
752555031704962919619171215072124527733136375472863049900375024593
485960032115144992840662158257436742274475501063912224218890391206
885714990281250332229301019625987938312748207951457466369086901110
213105305738750610287625824804729782975970378866527021744112460837
370072764091503713333614971740090516021354287018659906055371259092
989698875726878000679158690910845740780273990101872583402502706752
349279084556458472338387936948393212193705663102735811096309442346
293573358743954610171509748417603259483536217516712490048287878693
443178634077789561343147653044727101587308359186544227533506094500
454270943829595234500617950815499122260677053695403470872316703773
580038588820185360607740785920309100130736686153320514309483297128
610836602524555926973266001032976141119174374276782789747510302954
650308104060484212662927492258713195804378325583814279728206104671
644545396678275066337611956154718081141066372890445086071121650660
339892385555376753205387993468505034921858653621561162560773785078
336813948450925056970346054311689014345656230772431780451284414990
211879930904828898966619644774295261486897574573204681701313930905
117056129681336246565767032752997884256367261046845139557876174554
261407949927885159419323455730645855363767666277904565194687523590
507070290264235937689217411258335714398554727179693347166990452457
385765734636323402095801122354476444172330199688759484111588591938
802652082412625415775923953557139009940619257885762438343967082535
985086771745203064771259716871629271981108722640716731620311995057
495353350785579058055228056768709400358862145084193945110212966418
030102507190041435180262583918416963342871083924470112172842730327
747013437984117330124469137759748817280837808632835848060410924220
865767728752209963240080429944929304986884989845824998371385891669
131411594805379770420015970689347111831573389010474647987808156521
926441124175366266821681770769432381466336419486790863825847134143
907867852662542025507987500598344208643353203403385407169700485895
423819416463202364499218696935197625148758953644751634449406416198
941671134104435014824843798746391600097858007148865413513572346046
623479297272831424155920800251034678954542752194241325704026306976
946540161354854687985714429486803039101844108638904144811237544371
285330823993732836681962313029569185695856627411377033885853664622
747193167250336110407331565700765207124240797569995015171682190064
511788702874635229298088187710072903397299225664211305601375775977
190139941236326728084538940031909615421499319261336411225553601183
662732783852674019875478187635393573394928471029582528710380997565
439732567129487558224783626807452739034903745390658115194195726455
858788269618859947491839526549635447571365041228605931178327747043
171702175554273381131644612205777914607365779146307623015698777942
799470800066693339080866312852037258042871394552756934418643828321

630754249357674340668984292481754076245634843585997894799507358408
97211272601098018591318726985820436021544935373422820998321512799 6
754771510867255688821989769067943231991850034564659754689420908586
518688541565005301770434777447943867270393095250807174811438806676
944040880337002762289229403949546456869467365627651215744427275561
854727297092316077100833032761204644001950108825543666118384017506
433079878960184957256409227026453638384878282644377846764678894521
861373535543656377606476678170898450543551146912314274141648367976
459749610075175159580739164799319511126936601648584829390733187973
916908788195618674358831373515539061314098627551521295444877104580
897721910582763398947484910278339169952254377681439407418668237567
124423323514823465867649964519452476330870536440687141456832606639
769454801930943710086795751239891190608598079561897702861704671402
026390040755211696790396797200717133559771467791845713611149407966
712462922999331477634216541227783574825862753499006792111978060307
885749546973284196464487248154923884504168748844032656277470952960
677481195278592514810702849070915018652287534294183631406112370858
732629402433909819838680897918601274620818939502098874883195920209
202041991431102432886184043867214698447218058827477611885531433454
775949915708112154724304881109826753085018792712236067265424722554
951167783499513760704930313675989221645767417615608894882995093114
276568187950445739072603866015581213816955577135842043549347895364
202338974649549276685353620131752865705505880944097716682618778503
565693248837006191686881325769898892153770164294154770352800561194
842247298518748777495354659986473128376681023168478386420803571506
610437608027590099241762684102073191041168647523492506036456386777
296180495366105618145387474536473562955760076828385002653903382204
235925539829328419394494205000899887248918062112870490391300294855
651474954344575244822871583406545423074678494274930963470015143263
124161298211097657797808646208963643472088059155364232264831264515
021605196502658472066706130120493386969922060721205507484698313250
445679036197937511451056109940597227061024553164341842315931353905
307277364315636726458013226763772668618633479296099412432700177449
539823043204002544985446412582181492056014921887888485004281841829
538377471703679191289376308701042720721052793407615909556056902878
841543593704129448706773764271263821528379114630861459968813851549
589693494777530091090895645062887987494998791897733005539554996967
223113032996223735743856778002884728964213358326697472583460615280
371262743221724325293392575924924474115450585976031425395401902732
717953482453447118183267533772568831319357002083161783185793469555
062509874148085073837262014483584640035336912705562119589062989355
562779178399088576495624037971430883909711102642258974629317668967
223404012789249994400301464679220203830421619880771734664735164681
098205658445677184987499742916295695277744170956312845961026901454
223339536443247908988275294511632099275307499379381898314756794441
245959637262578848779824592170128005575802716147579216877302876772
414278390459073912739294267884385771923968052299403405334707357333
451836725155426582639619999309836730795037244868646114937304957612
941570707066203289181107238915527538336663817080430330556707275261
677419463060608701555657445230874655839612405030148057910614581630
901314898861882107938274751304764124868278016019084967844218901110
183925596780158884405085393976838736014191230960600868884840958759
093979887545712509890277211540144019262217279654656498959921437569
114290202041111579248731265080755959728472786996827891626827869491
152476745851922811086526770098192794353379135535005146898579382371
388273535727171789383014221648571714102900699728235323288484621928
112894081707974024421905443693038017492997032084340110873321114536
842359999320908951566908596491522776672296396942242423418831803521

0991164028064348473544379832000036280819097685584925611752392977416582041844810908951268364708462474117396127148810193294558673819432508612653588373685599202590781539608212879073060653335449059883474701016808696373177373258291394020149923292096927067831675968147669667520807748268071111678492935846198418626172110925322160685253881591855988062776872164381835160495538527936421090313103607816243891471254331653700492954357313119180428419650321615780904252013312485605320360859144817671705497669073927648951405521520609254132916570249181716644372036661368578767332511082859038192154476652862252463592191750130342843933195491229727926976395317486223207878920268445083520412496126094785976476405051344616457799579741920939413873112767240579405410462075597015741827181915656118320966587774081467691697748376641838608175254785968515246948077528179062939927065252312930894866450147904547107615155399799638954573549956164209646047474598537724051945204244695155110576014942214459850785628980052001501442172360303944283424448708887381117569548458688101588473050820293605143013701045069902681581989459515045374857234879807161323689988928620499341347784459648262388689344956413068869396801624022569609725905836903591447830464096918458265475815349450077651607581956641240074336117286660578064097906850684383908655013660341715661217741365352466629240385273163158429247510718207376199742570052663628734248527697091241463260439116468257193844749765274127912883327647534004587978756721972850802515877654905655892204714962052521040120166987644538174938761539621762099818669564349377520407867045502757495420828343826527279906640404630855560157935415801838870887714052300577774791013360758346370816037414033581662171052377954778031082878931138989447958611693922813772482097662231537416555187072430037844683873444610185876496624830235355290685620889367050129145033747129770829859205524051878310562165213224612288003475473255932571175091617659416648239497291335237054388627240848370552817002962980905865080479495462458224013571415849006733084457388181196744514291152143427939269965562257198643919606775303837468667222170355689084250348329476678415036783681989747562636076056173959495904537872887088011965418789247653078202554123421527555465375278485116421930021040069601544428550071743703540070335935650539865178570700658209980419574495123587644781248610414755659045005537787454257327663137388250201726489696161291010481279326374567313473871166387709984542067027002960111722637627272674521018111865591052586145338819701289911963847350290174000706806314623013080539754577602872474099097506917882619283197164721284958737918035495590854500617988132119821142982583753056350978991235940544860179024862607671948351844234905871907533729443395924091853528029572010383942096233427756208747723170117527568724101224530281538795845096348579839104519213186349569030391256411390596412540194362929494919213530717153448409684207106422822898118640267635071607969350470210023187810985613567910659306600789776255319789825006631515629833544391644492580570078010616280321022019570475798656581168093324230005606648967369487624261724101199075962069434685940406164109057115534545376809232712872353999074435919539839737211013349491656927142993028020314774560505446618457532305901707861564430381970179149567653940459435606066050701739389313213462585038107319503867448355601949182280516773057685026884325495890895608141995075256518935540226366100848161462038654464771071218795163069833620214074612432738560733007417290853413714676466208233265300757784578012755078726278301618250870084281877453320787977894296485859758192842049964829620427354073385310054699395412546194734721703995270227035779385851296836644067889837465656361354447806668384669765176614999618543989623502371767110274089091939958314514687236128713849724206908095044223678551028762868042518816358402616298518072111

528332237580970905745356917840193816002439654325782504544519694034
884092434085789998277191512303094738055403082315560818607161583433
955617281063733110606015264964148156840431946235604363017431750920
771304908868560472738551730953808475014486517604975677360783315447
229253433596384563021521710498738592531020397895814333663841559299
255089446027807017962274521055506711913163262652799366963598923830
060969816001670881344300203911771916307016377800383003711264182414
601498704171480566260338497751756038495491915791417929536849597062
863297174522149452643640004011685127587387943486668837572288996134
299386640236405894482452912428201658004781669418478063660478483817
618565155840174603728978215856590831489306511779198572317164764724
189304315319990881549971377420721011833196861968494405184713480510
376244887581817277233442721570874000852493919493398103083199952288
542626308518145514104967489642576817203145774197605540116651437193
376372188186506524854450399937660922677770747939980142258086621498
719124701387469895676580981634240795135703733486839946074001483818
809109122785049874225639471028589036178924869455199904917116231709
297408195172163625062371420825261190713179187638003738209989415516
736290310549402905372537989577370008817865887049043920910261127811
112935519422778192147007436373735190083294733687322396549476329928
275318351812624007118920345658855468095030263042519219879777374432
637260928911109005219868755372281053055764147614824639858637189772
797761937308767042649704411511412890062126113885306037367229595816
117021395030074141766112848332886016736716732035880470158024784864
639807999576764707923310644566260337307361589922617875267426013536
007295278511447312992791452450623340290963971759232197958011191466
992839606609054620079823745212245036016910411563221953297269544555
151828409453955078674040937185323660387796280514312676627403643982
398318817605978528134663946956055581184589398070501135751682068069
489438552913508828544734828151760971542385952213273086132204129233
077656558692790957004784303955681994015964223359594550228105654377
995470862448559080037069501560801930721464897267316393892553815995
861702205913973027090063858841495346162764260442837389125191847819
850025498263035851006341034437633407334699031270355830801124354815
866116467990470409954792381287897186780109815204978806050476668203
662967250936939607313669337472036724300318966620499750652520175471
357865734643836467637926850195846151069676536390244748995432199723
190611693296228778946122656683064351895616389957450772210945127991
885324449376487789040544224233470979587265217693777637079604073278
970245582953869616412632465754921044812238089336380764949967196348
615540609091415949767808774619122770828684460532572718019232463199
946676789260303597957811827900047139409217193998740760009997069581
634479233605106166460077875593954578581356191179734847242756439544
469001853703038899472199830805883405836720473010593371645853777333
738261881997860740674509062922554889908344358447071868346283907929
470801169686394851185081164381346028123477126650093708643498081061
192511699839069112411278849225010740467001002498755498035240563673
715644048348352246102196750403342268090196091918367697539184488898
103076931360368464907518519208998079674849520642528150398821992945
016312244419290790582175500212464156438780114736353486032318176976
992568862342243651161413783584860345729184564597310145801442339103
436619091211751414322400433814714649570280368174781573399255278767
581363359753605595393840176867674304746201122791642138790162717829
363320257354064982943159500554714114406637280225906499077297215318
483539621059207509481870223099482546729701848378246112123226391943
500731777620066954059960374037654410628719241786662420882146482782
794381136197446758843595640054338403532921278595748814000737497083
250483278575562787738665283977888843093724219594555354368165329370

198863634368179075180196127350934580410944655172588496802612101534
669727381161498560713593318421251782086976441366280313195926629980
080049993801799153449183914186001491765200955505742943903063235405
129132722852895331838612800900242018592068301245576885159986036670
882144036179949757693010158168404896369505707194161819229869985461
811066431624215434388574483433888071994766742309255414252204646132
633302192541233265542251790039623700118772248801218997398674540232
783101115581388876253275234665649792845609117583939724769569229581
647917057753460630170075274314017131409347082620758055636513836577
977785668667231433899715854051283817539517288672679243351932427944
640427484557220427137038338428243358563142551037569101814745835641
467895345086856290184867609167706199880065923053443639571728250888
425996639289712477574013093554392473324079182045553817663723533222
093512540132481590298902642974259278789454750065499469212124666206
626082143493200208055224057223984383044936328281922845488932895874
022579320820777499212636085911890296466098389588810029238820492462
297133229297037751993136100980352670524486853226374478046384325646
301000814486291219945715644790099446084293682847965274461276533324
488110332420251747386222344906972367553788033995966640307214375360
233103696573323152377229874426905668961779628232189871890962510009
667381607994856169911256164664511017559716379949948745173586586770
840471684474770849909769979447071072216462801394627545923362533618
584457602386869037340402342997958662188971415880457537565078504261
021206974369709692144094113468994263087437856931812699404387145948
959983500248233904352960214104347987103068353429770829833207751807
152848088283370145995157366923408952207467531109020004087766773452
083041072554159390052576034609120814028417742037713070084243715867
293605934806649256089060062140073524622034669434043776297371417295
657738443594005475357783364787173858831995725380639809960264540532
720562873151122978157160862957374684566542360082900430576533154127
168989746228513385107670664275111061263511316160924243466569700709
329952178094757112910514811469846816362099491211915737590934654666
176085925249964296490499530084958760781963600471026739400740732084
362442003461449470324239517678532424534809937752802805259240486576
284671412186729974077030839307426174014500322104972620715170167184
912813438981955969766521920190813700036727820825480844407705488949
773529007409396411521592813098524327606824759522533080282483140911
455034964805588336314363788086926850984022754356109530387770701212
000289803251790104923250778850259083263414354851687722009982947719
960230524388558044873833160358617226920300671018787426069417860407
069990250750049323362692993214094658099646531428589402950507599383
742467138926990433088969689317122152250932961528322314040152247200
696225193121876948328674200291736644356806635821053328041767261515
623021242888554325019347747162804383419392847933486122205735240128
775008941153319230976508281811189635502495846710684529264484555742
771213474285886128655316929049767845662596418247266927783440026607
279474144408370664575851237670920004831219403483729208560022902779
773086485414746185881115702402873149042337471022051225421059966070
353839506482334592609712442501847780809771726739247319401655039607
867012304739576533280658508348412526008894658467311915729174574224
177949335497989190078206213086108038655162237581246483554065072432
929291154509266713185929676930903243805005047057980295747163465461
868596663348164737563758253029528629237343678949466010883529136131
466638328071198371749145500064298208172745061751153316431785068605
599580414470505145134434661354349132377733690004775758504046993549
528650821231068855209618997124393434111680987953962481708529338557
609002702279497412417974047571783589151301019483114208868324943314
817802337650953324299552859443683435519727318517804946558350991943

093702256819318024684488318677087309347278038059155250231204064223
609470988530160758370543677837414644116053421760045932648901251010
943688839493750108078235517849509082358995407264746204374288810508
396455884266548692765048403283121820919225254911472664063622030793
455066039883857248600510852807898750922954152053685929261987791689
221592205220551985320147926147108591884318686574115963789931121609
989761109796335654449027343590983289478678839749661090310579656789
884814633779378097206339058401025151480018804012308861312975542072
153696492705801455275021614678086913660749812251621634354851006334
434910353420533934694524974769550862656936276212610915726891851277
303763839795715936693067949298648386633844544863127608551005181894
581104146470845401536291458227457290153410579881693540552831867360
382508835230473920214607034933191067272651277179914138372267050300
428895154913480292363239288242207957730865773544300787149940796063
032576958992623920596601385554180341966989324367028510494283621790
242444777381925654599726442141645880183324265738561878678959852363
478423536809059927119995597854014483596606215632951638907300427568
254001923588832030019113497523286130065109787692945531139668955583
476437080703315692464770566831708735818394804538875307517147833711
950768797640772884648664979182318450231538105517587340718962538798
537621437172436820710468814205249239365508704489135541655626666713
154192176366666445461341970658844048039586284011613603368548134550
509359031416572525760552586987870305119319075536363454402515452339
582336587160381608528218500714103549936084662784075061236838764462
145891921779142593006497312418546365599567512958502030822784103261
371511351983906124159646333982390087897387993371672887203567894869
612784691802988400077024046141684357096461162379484580024369153446
231929702763704581054668618648920064818578844704613762942820614210
703480363084411167800614815023913690139665758030939659044159125120
969529735812730979032583292747939969282438314920625902933092860103
323917403838342745711351398457747832024488386021968076716331653498
673034745401399921598221116711151254425853760916734327699904005844
974347590556420630769469347283565064991077767958926507983234810840
182244891170749665104061875918474857697210367432681514842019569722
074734482208792869960836293587631653890478390048742747935151528419
727078614321965828417069390496815772568280501220207510923634655176
931515723238105018022585517518247822341364317881652068506613942262
653446972893597805755792607832166738898066125192388000169534325226
200528899561086132879950759573554749552474243083287019180305647708
273670143445393759376315501750935185158798505069463905231958292523
097514074052545956314007436631365318598857577378878419026141186895
445650421603370647862526272470854341759779500692649026951120275026
309543674058016472131592679453794369497526101844660858000783127191
316021235091382970425870233641604644846128470918634315198653654660
902676182247471801225355996813235223866795597369258068959753037126
480244820458137018203432869086371659833775710858330103687434657741
106218143176556339621006385831674674809188717077376535589866220294
998738274442554792411634654679406036502474602456189525909349966735
951227127320524011111391852302706761427723092229472881013319296333
997818550318293443263220646980101295170455704300632501562504315083
657628326741230020495063971736784188910051425253052496886256738738
477796628662164973624028836930961735437337313667758358357017944714
574708124906980229776815688257225708949890899120660554025206080132
245048251208137567437661986796426412986059431128300054088818812331
334729747312126744443118675943209106681841781026557335326213877374
737553457827904151159313218518285887774417722966204691386602093496
942560536450662133317325961987682588594298794267093801141054153993
918960836361924874187916930646357844807215658399312037940118324480

2015567141428598637278598469707404755436182348158796011413662352454
4806720345614756939780122895658521622561696712983690085390043834451
036299591188765554858501038242000728257383191539907527071272412720
1395819955356637355128909069762510703048920279452591556500485530046
678249437412341710845111272716022109419314554015799975257597062357
119650920779653514558284969464124712726275202638568898807683516345
882147443822121862964661262708267422635057195047392386350754121166
483026200971276710048982671613245191035278362070128544446701113052
325226930150870306338451974189270634069245458809137680372957533468
067793504686478745877073762192530528748136198511385757845429291076
654292834071343502250882818892915244527783987472190081314380720430
207245467931758778466626656846528861507688403122321218733296072442
684943391394823440048794731810015672345261770257670886790779488435
759761351817722330324699911665759663561548880404973201297560095265
988234960223329496042355117872785529240802858776678063370228800022
181033547073381937382622675719292593356370954061572778057244148311
164042802001173781045377356971226702822063493126905454981671849950
281156890107968802900112565891790555923454722921378528723282181212
503451820595480967722776607131017786034388295733182064235872369934
100981061723496382468683009133256534492346840680693901747353201504
440611383475519042887754060775819161781154130039925740378644359130
190784611315598122874333833098537839728020727286892320298943973394
819817757139247491694646666789612342910937033879325123377397138802
549835070664655016435659685314582650756107057247029983740904860172
437641981422704314803738452936874360053639126726507615680781314200
412241195949403929770163428124078720808521666306459230567032070981
617225290269602306308129260227978177435716138102919298062250972902
342140272119169380327132983200837285066796162815923582866865760999
044159915872039418368224730903669496907177457012177714467268351119
457992357643789469356535420860629730104673071982276607743930230946
886154828109515202159705255023615647835579419688175556091385778521
922259647769941023057003836674762355069788231896556982414650286786
318921131224060618093860488326451730840169501420821573260642922288
691115250254993602180051041932609690738474883239140241558533260115
285070430660932244424825241441746074448844534185528241403762246430
116084929486645829985520542671517405454420206280703433294106933727
264297066891081535848401490945693824621684799297070687378774062892
832545134226040883718721990383555267472209412312676596338155725024
484611269592746975238144932164044468989516224006928370958627380688
833629163724004187595864552046374627569583050798453815347152306641
100725692362934927639367109131351270403054576969966845535847145591
819283771246258352141244581202749660784771810569945704336050816850
958945374630795183950034717251948443599494854468526812973590190690
653598903356956031006447968263120457319101154449655511226841732515
227738630308135975821722905976613762764176616347690912507016199955
375964321155753439662168827108403675119299960008862463219998754180
943600530874133962692998207629668015488090523558718680761698556060
434175392524049679996465176000269269165516987526525067739508513040
805388445060926010964368810894202685615907389985099082501549463918
220970064845536636689968589228638215971107671797678975011542808723
039128878869961867893071982867549919743262709341000268510325929632
688185241489579124432764872180145298218574977142929436668937836438
222965859111417940518494207357064528325988459122593949274445571466
776515114569290313952537580463375857964158116362187227661518982412
205351322422488256167742459676541254033178449883841113760735007720
085697277626487365278338916361424964210630883493089033208487794286
440497222682861859946689343794851334855905828206964612394649757412
236283139773381908745580331675172722386647360336322942216531277635

847730631534297372774923149739427339168139875171650178103261317455
063776577097886959767574221493716352884568741975606953600393373320
206516493697081492295834331163103727409098662574705051470227230962
962287676893693773871239020465167295592539637570422968270137215978
496180362348043379063970358558242210882352987239496885357705045182
895099193763474732035193828111442092546739336782292584965322008001
526318029140864688824845131277306267980815595476482850417271139264
031553850405835488218720632498225683083781144749672788833320028717
670034369690112633377140681492485609922237095518547384457179054176
873927917178516593198682887564181867768768410191491807939961128907
075158854552469947650821767567721164663642768199852241009652130504
650834072757448496619761614608405906074102214449762236679944848377
775461200772349048346804799553637492002228711271851356066168229123
239218005582781418581599044062405262546367710924438317339847632851
470653500079783184915824011759565900535027064038408707123341254904
300589255268336910947101885680630974863894076607300957659146215650
501976073448899217067459599010087108436183443327787653682587723082
233443454329275264503508275103688527876967792475290543516348471778
306004342306802510416927098404294232615492831376121879256159805549
617993488223404322283765653188612575026407804515295419719419547797
740615546242676714336470510868155403216644525162310712260123089371
047550682508603240463908671443690773244134398492841467882847147933
299621527265676834010400353802720252754184352253722834489706581156
085710599336667745198566047220084160190860000788059526818266913138
417341819905590105823592902013540514610372925984089244405799629261
209829819631751214535332103905565185435755015506330230393983217424
163887658141828375495738248135795353764156486001991020976353392075
484934838869281611240904286051688972415625937579583189895007292459
604679885544190931123329852718446233567773599333480407470372053280
347069763700271572820230730576961414871966420425572750491950366323
004651395418140725108930709443978312110733276687592803776507172513
608755719376485636744855888825788766133918449302119188229270901950
014684234363699275319954611567138685704507441432163390407802999234
905814465652044962335154357201055418206044026298913181811835652148
013865548465039391744227689983209599473453261485153380908818440673
726935784293194085081800541112501203457445303331397380446948852770
989980569600051914114142279477629690392241524461598136812557513984
9403015890891283903969068237159676222826543755905616030020457274561
527945965191825007133487156545101752772344850870016671622088134714
559577331251876805648529105407392802221120032862781056402439800380
701905697739652978779098800316200668984037215519624334297992810095
608616020149108819115097937929003694499120621765377482371955324580
636150130098086504424252934783569851494288463384019650862826192681
818133145352271022161786885639796180425572709401969041952782218972
371157604600163990908887257182510468842834618913181202969641334893
176005480461049509279897517631114740226421967576451984887245324003
253302518248307749123252566884756169542393488977846191575794422770
855820578541006163487982613985550919249337634419693864702415434328
232827684568414269943837235402540939373073804634343828207196553140
326715847016941483313914996741090812530998431631207310850501690632
931910182110539558556703982419629169520765719992723071763730813642
389401482326265470550903841961663957458368747626735252558119907191
836237358436871834305909221799544996922233003428708226933056544676
836776755877960837571832810682555956854316804574768968447920124443
974874700573757245740874921782756424733258593382718360118550637271
168238142462458904590760296921428081805778560188655269099927922154
71270895924017947650784451441645517172724548476941607662478326065
735413894461988583756749847678053698392957632660226472399204136521

623736361466640323155185415154811185810844339855150436734807098016
306252475101971504669545974914141011308661681636860421033880745651
994932458611604871116486279983802481253941899013637727311133834266
778532592324533344853759664716208338537412263555308937443901951541
575679942453238033408307216841994789968124268884616597060833397337
177144364986798767187972512615063197288554063191512610386496203137
400515441345721044904467051945156922739366732497685739160310543114
816892225159257668780953462048818191351201284162581472102780965979
991944160344384182326231448081673218912379946597469447371994734475
461447290698588542010162523064196805972372284233732451140654028375
526303970784269804597321019493457002525055959469814542210702836906
765111338008271970430451924780925117856272910352627916974258068402
627307584190214628440917898442561629867390910053700297885506985823
190928855440993457717507878492547917137854322631465566615358702116
916043172098426523231396606548893085383019683401924136867142069727
633671975814778723614330172612555205583002475666557171155769559720
731113661647649212410007432536721117181902698647349658130101367113
732229307682146982330958626017552721672584247759443215834483251667
717953813744964440656738138326645751744905748004650568402111858982
706460025498422293207274982206974769806226082660110961848556935211
806622929940380157261010384282960889874606563046798502292990209291
932461771606160384801422508869123734248586441731267184746634512149
080855321275811894748251785940175844722914259381996647903390875103
666661541646865470072022022545975348098423501364848574947885547459
213375096376782165427914135474670062812767442222991188279981357269
073785469091101556810429717305778842476385374026797415237094624639
434605121639245148210507976032685986609400873234156423535667666375
312276348010107704548905103240579565966741308311579279748096695434
726072474106920153392090933458277744733961650093575247112417075063
503208867717412193495146667638712637372846116040877063015837600111
533649212190203318068500324666379221737979268046636376195583620471
957455881725511200080419911461363395552011300394239159753566741136
702232551901934176455731469482202225229548333297745560513730774251
770977444659807607594546606593103127348715553264389083383702773822
074314568944586437175412106604933774542490470014903959465745943771
237897280058429396210552675039566321492376729814066350061202360793
595072500261188229881702440932306066540097229824335376524377994159
185149339041722501469974253353780504362141093577235490088186907390
269514016697214836261672332385111763897273755722811689919980399572
937460322172819284534093325844116963040623830035426887429021182429
856451245795207347984627346195104552425613736299433014983937291110
870529969519183533650534844242848046310476522083420859365173096584
496130285833654819543598186756220294161384321163257685982562967201
828906781124186868784597313387140418729700711919867701293106293096
949899354813921875928804839845138740758643027656157147335183544073
506255312343664960053109148813032384462652043513629026265933306812
051158885104358569945649938869570706111923572757110294196928994187
655436884256928063861435130310638444334313396196979733561523242666
417066390240362365593574944251038356824938543066369565628381349757
831909024277049909445141111241230206043422328892597491438231575907
984423517784541128724259493533330985710978755386839817548252121751
810308218176505155563452429948453904577562674465785517591489162251
178070532881170459740597598645830080056103223325384300750981134310
120450833190422865468587799440122965616563890815959223940343922260
010197221765736217116359902575752900033866022749259421935596129475
518513699393249692786635065238826768941167638908982314619095968338
612311858671495757270575686399745427113036591818384717808180841752
010065566213047632675249466820599746393688064999076413425808930820

409023632383517830632217617206433632011040344099409158405146878326
260477568040712615484260346833880268094044793089193734239036386440
254924481157390763168026646799679017556718706413633240288705087457
165871395916429253614402597840290871343774441758956655811307376296
889347527173711313779000310082729387122487986914124280084502725122
546352721992018762423508027844796784337023680736143992859048111112
541931101350959312727661124888954689194535563621879329974884586783
481448626980470399009162895288368911374616731561702410045157757001
603778370339572539261354053250114657419224801218243169853035363945
988575041132622240054109705194916292574379281916876204926756847774
860345138396946799094353055861500427944483112063090593443247000937
536710056044926556034727477807907836395045187624483926979873135352
843484638881332304397927748022015296192355486000473427309507867806
253076736433322824196132427056557794522660656950161892625626948238
005695033649150471162428579689682908969058363758278283892895520363
223930960420462287810637658111782742762532340505564585343287393682
881055582302015544340256660561199160833124527637674938321952334956
317862584221341665748262887744717933870329083625039959451591113255
050506004055193310678540221401392893077471031783341055948291837130
769245697952206414022795846705868857757846872528947007508518500453
068757078397466638450736019535737073785356700362585582066737243338
623835066357323527255029874737157104957827619393563501675928367423
085383857351276128826173200948780873129781099401372087532187962176
750723431042040983005430754543089347560999967053006260089395875398
030047816817127195093934191068331792429451595438511210909172267321
832118162279570595447822536126829509548622172312363166507257218634
531741072397650382647261206586272339002819509088574096005768787459
135373146151589768102894467346086010997367803156129155718923796941
378351231627407516945694908680544178759792003655469263444618177373
227167003444266776046094924812058797064658497352880792415315639324
344125789177935725901786375825693806342504430545882795813741075602
287584447810965427651767062223706972419791802152291454833956256228
460783866292668106052637667735843824873378258642885012123329247207
096075960682756058028935698068009042477941448052246140801929827445
359142613006731539974220398080445375011953972619448484495199114028
190660032628026970954487165779231879915526503112323553639686684530
243015961973492262352274323053604223375606417777316238560718050405
591823508941421612604389373200349753263396931768396397408340876148
966053467650020180180950179015247573815905144668404233204625195859
647960289531559077547204187516804221048827397952482737496742588221
229084184267327354050629747395625186097845807013227661151733251592
801250066103078456552279339066119554676984670693114454341653858729
999117725534075836267307205981923186415819683344077561735876011585
410628859025148597066302566272781881934320443544059426417472874446
384970887538924365482695425677051955005030485739671925936918313224
995238290395190787500747741928834128339315143605456554992740034000
514752860117649136881975405835181301064281918542772397988891155654
420613295214834039746559537469317679712605932622735788936943952814
037155498125622918360289709884483519995909272476142927384761134273
942707646596586099675123050324376092528374535447598558719279745135
456040928446312383889192976387224509486946643316458586117087884072
595663446488772838981448069759334634892092084747933664114769169483
043759599830298448510466971167611803157567043074868319135015108662
681481108066267644488187637341191046301486433951333169438648671912
530511977122226051292720662888283651460221944708196954631978248180
623064295919343803191070756728911303416669390683766514373340098291
769177833502351137723780700801344937512480505007321973812385162506
320810496669536517392868858139774907719078245297941956168869722316

752104506671379192086538936749986314866410170044457629958945321955
986639580917941851056008686137283520554464203327542845960428433290
425439861323590988961610491104037053812955077468163875576813162991
902508157027190764907757843602169772279041728321524831115467377986
753486330097098407188529177293638378968670454493802201715742430599
092277663302429744170450074589944751500765742782902593896377634935
315947832002254243267251842300506882020825497212921405633725581581
127104741570185389317829398787712798274886567150463224664385555348
488786876641355612761048582420589186519723435336395723609948081681
739740311903093045100043961806912344770855514924729667850895219861
090165524898357700496192037386165583736313265234598245319951059160
583790003979969517770697952469522983676074314990085485641433272200
349890338404285312421123402100498162429185892422243495753608082603
067624015469578806472961926684527511160095905378548316367124548486
778820172520546398558650035823500202150851642366350078776071519856
145256264951591055507477278537981540632716819511175410228892983782
225453367566519025121682301691819839425989997594117695875215092794
637661963360871925094446859022136218571040722049966386358712990530
555270284104677262021274466745452154577237082364445335706452251497
567870517194800500802567824607818185487583892258881641762049036112
904312816748320766934852285776067934846458657669095206610087879064
301681675339162115205681084467894159123374146368211388337577326140
274624666451509892104406381808305608040137010089074171937662958444
216912279089328319826772333212792007680916610268387238432454420686
797560394826736582067345155042676308818158558039319161732754429557
643651620151664466855759372594097115187836921732999384157882766022
670925291474618234186636721931122698423608685452041883128019978033
487875658827105870237096130274799132348225813769612170462274224961
016733754788010634085514866228699569152716263350178980339827364487
742823153814829001343256788832287607165110115958718714501057834762
975288743392745818276860004527757177247880209896564385294238817387
999187471297587411654133239155651089087752576862866168071921090343
265030146354274920447493124652014732319757703612045011747939862821
525354972026717908950519508590979562153891218056487899678573068849
963903277813229401520730505202096076546883252517526410135506514502
980213435336587555916168367494289441214804865998225611198797208669
302128499501664795238170649564273515525846846289200141881385743510
670806803934006493980278204084501559743111297779680235822043997189
182740819520245230161759947970828058600328892886741062471328690946
946684390420120577410669946215723851109156923759278558684493977360
688001922934914717580163165254747272854041117825003154049416495321
147450754966548023101855184474204933682149867867387892495751470690
246623904739663020066157810935792046362771350394697843435917414264
377803119021954752269208954796887825485645707135566698502327546123
988230278689013175732191716784930163650138955291315901174224896073
749460202948512798559245077379356908133386974703741573965582588573
741349035827814340746255423551643968904607795838664996649445074756
579699351493513597320013303372081166289167650986948072944032991761
290723011140465891138242727263296851760942841879398026052395654066
459330078965741576697086719209294525393155274373528280048234575091
448533542404918205830211736693760497384227215913483552040425761699
280642854714019368556141102765828085704034232587586569465009202308
498308267843272774869277955040661110043597647686786538495502722558
854224986136573631739614809989360847409717649547154334062370732658
697527645921337429714124207051122564494990791441924448328393791380
420636238002970028206239318075619226485397394933083251970873138458
119857017100249136525227198684083184047846936429025111292751766993
988620343878917517267572217442145385761100283487432190524333293992

809488390615294064110187937539411343031719168526355316735298536398
677616350661196822472348640340968103858504606729601476131074024007
426534024368037921068446153419708080881220768071106438905886342157
852825822877436754701408310031138429853493528249358504036234592392
668831144807754671059106940654723734658778941924873236437024020240
175175970468295402550796773098444317986354134645953902276574320351
884311155975463523304523550552928506363115240811767846454714132798
134065927467320835452822766343059530547343093817503778723176464890
964910442558796940411705277662970232569524636717758458788220284605
305305466181722096550981839675444011935670278272108335560644665752
280980586287033307578738210821129210210389616523122848792032803766
538720195820518561825014424096283660984633225284110157417361488234
146028737749849149845904629889032849256083957399140591024120684557
624803459047235953688392746190063530060252868443454660057857456872
500288797407812699501435969649476414931850496255521694445285574529
480723243302773655574561042966088356974253073611130310823800031585
387362868849210917491803390548285054627864100013169471998948894209
530864465026796100715048439262063311711860861208187157327345513700
141867686921967048859010362423315167760172717422330504953061585247
919628896759444213462356519312219428440714309789550559858862442017
362826073505116384635711187846834715883490623342588343614274793829
548341263417416882602888735547816653238321540768359096812054884548
171892692062036709653605640898759284508157427478621941367169943229
458693304953537229679978004871910203478247394917971719767262538849
528173051360089227052430855648745041216033551514931039641863808926
595190343501979503662299371070610980328347286934541513596718150629
766374430994932369570495349591419352727151418737658694512236367107
938011524216624494119800446557212597535804639991023932797708387184
753479985436030661774685481674302535091141422536356338388354354468
600110863848021479276315847935107106227221232890003040085551034607
680665195521686567189181334850680693770039068701376254055246684976
710144599349618422468980417018602351019649901337786508323436869443
782163920218850859965813111265889482917734897361101747392459963978
823770846435932564640554466354564506278165078910000603578958471411
788114001470455657924684793389859044356095193955983048411907085968
390826969646301033129195405483566334707548255971622409572456078995
227513743172920432944214571757021119664927329504589243511542249892
841829309680162790999014391000651606646547657233038464624395113830
677701241619910309399289403981160616742076033829175930656604400871
622422948641209276800548402635460460212864312968617604867685458733
245904904458154468976001291767593717828778157463447497443174722457
141782495933265895990123263165431809785578257753550495717798199024
612577728494463535252170334339313436451383042589815147736236791994
181688328599187157217228765199470314248248787596411631008191276660
288566473109611646667671189367634721290963008692396234207088629338
631255612873406534679097419938288166385060278732800485054491820199
337871130829493390254440845445283110790671755781580093867536003860
355768063981064235751099733184028068507709966620139938302157057490
439491154383090561583001130141099139583290325869583767619707448919
113263121640981843472560087662111596951333890462589050402103971688
752976341156530163761401724174127541567757338515633409689244301406
179749753717442566473493454579246967993011026194337636448667537561
087691613043613748330479441225952612415148981699807681018481895780
038325763387804353118569719951740840404292157320717893877089593155
702274272826581614974080133206343384008689023889391833314200451596
140089861121562924364169472543135617018117920055959285439183334369
451919451431715413500742006390506071676581905824216202897696844090
552454470012818868581435495454205566898387862448320752121741094873

647410915786040991025855586394013091551756002797658653840756823639
735834723364825533520524772296088193054082105328131130488092625504
406239557590391944317141526527540808952986572630716340732243950379
987148815126244606312813852410741435979408576657852516943899274655
621805886920662325044937029212885127459270213767483429277774127189
221554062683300828600901792187538294862915016245407247778866998890
070980505692506985818373901350763943654187232930311836581686333644
779329007078574510611399148091109971279294384939136701584241896910
938623802725764521667814022361322904680977248778268641881228779934
329767071638705695119901896323092458999902179757131345174259360325
136902088395860854675047773229290628464817203195072814911800959505
125813447571570554678484436237769148191799840529066771547929117426
693348858567088139473448486214381115954754498528040574225069514664
720084246244722993215792264672178814520648208302769361192173852367
856740962791085234229560906326667104802881922637139090737684448093
184449896992931060368700953087272410262136788706972698569734002280
363504755236880049558697359039020691931283070710617913556767421587
725759822191294318645745477205132894587827375775210307119425545175
288583633445935800552400893120991882570655486639231448086237441459
530559003065809529244042617675254206771033553684955246546435770101
789796416313038674952396108289032577244899186229965198423278274995
958934083103129326766102724108073795462818807634269861761703310093
468008306518347713287054254996232603250116192827302129934934139799
634261512648506347348327591363178682472894449321466061847965057304
067187482841519147416841690897229561606700728197766108311410618692
743573355356462725414379409813463832286243364133478978890912641531
873040605317247738103537082669410654440622661293766233632590613838
197631940130398985582094820953918602481990989664862713566498757014
979534006764478586820491076270969775705491508796197691944139015941
256832679426395790182782545568969968032231781589778225657613871677
013214802115527276731117341653118922038451473699023164776475938806
016075537483382328545825950629102393600165447353942982876682073712
192752999073697440811342372020131423404505259316972775590585657003
132022465479086657583451266545682630554619314754718739127923972118
469217783762006326246681456322496877656028010455006968281528657542
598234834009264710128328430643025697652644694135027062145831911049
179319414349301770348702633346016871029183608374125327587763154022
303956288776844723280302142987294946474939503146491804133514202393
881524875937371592838326600354064289535416985061992235303829218610
486918255269025574988931977847206199198050179722622476371814459796
413713846207982090839588211870480155631852081343051685237415842174
781458011563058311767877089770942569153607878091136284354824022828
394565804195220130311977227965959838840936358355901185742704482665
056343842162536049834085438904028543049268993066135302956244002028
267343672871926207940297350617969254101291753762660825396822180621
641812462173118967334743094220887670606305467169963141821559029243
513784364350632262093120680143169163849781012271071502167924720562
634687067390887567539442448270838250878826535655819744166357784924
176318814836216441222232363549938942990784092165099611235325120110
314233835506549388909275961105369398083023861037130475667626055583
092784943578519785639053474326745056049094077085918865106554530081
982644525281740877465478991615116348753930674193023342758365049641
568635388344929702698830212838507501173072253194741218860306066086
656547714244902618109150955830742760829485811035201107033030587092
579512353389221572474247856716767883589737676177211751305869343355
367649036743738817044405436987385939266494471535038152596325055300
204906764624458522605213931219629970578255032704577980448589560702
099021534466953706842845217821445238667104694081075061673174785697

842165818603669376409894336493814267999197179815233510841570647616
502760453863299952244303389384486986948796492541509969476646546897
692164861423923568275318509654088135332360315018454309181115897953
009608823801822954836466507308346158753739094675311255426108040965
927527145627225527350121765782432201282217368454018834091125636960
586741735441883030291042295409312139163251665271628121402003934852
462926770253355678013212110007468751049272921506773328246340543305
146411016945801523928106487165119374849628471452059228602216430922
707329113148505514626087654910369689708578112513994774893832884265
980561212183161153030419155186977220696407051706558307454295568020
558606479198561137208320213973371589718010093419529924086318041862
299694415900148445348986206796450530773611839913611440073059304205
749678639537317291277835848215629148003084298917636403425821977585
909988615422915064881854893919464924424427073460653662824110438068
746346424452489217270080026988968400977049906879061899434905998791
232993931686542977873405363741086041168212232803214430184507966819
142023711524639482201770931427298845543362729864739836067619735520
646544148606206582731163150571772420887655624872226829266602037148
511214624414964999515202973569634689579349118486138732356737272362
867229569307022272826189362420914468343426468002660771995022165093
651912429912989094967348869160997557084153856326697999287231066968
120038735043570013902925124556125259003322197307426285577090519268
291420781601592068468518949602680324164502321797845493500368685311
200743998786753531880447347851395431773960600553611144063536421618
126021263865730946905932559063972194920380562876788287430919096154
068691521231893627892498701245388290743083816074862738967877453782
363709467889116034175404550891697540592273940492666239120406468558
191112089876130221444569725686254529501304112171998091066911541574
687485849599866199083983781712633106815336413320408361685895059251
501682768475240163473441553578291894992138591091122483181871721948
688386571304048294777424440601099690156383523782761290959674815107
225829181329982874270549873596774842208562067520671599421809118851
021464388153607962345415092134034300612997868682579602938497659187
127068051527071418619605739031888512824338726837664102076717571266
109274582233522905129948528281613592546990669160185027594016816244
982303860880138242498830699639176230939613485459945178006710782255
193242050873901267954978440156989887507912316527747212894791531731
284579690612883700197518951091669085067985676051658730747998414456
849249212797902881242410088408847783731591936132045482630356747492
775654138559987303215905407629514363788522298669818514967083486052
744652644981763753211029230625903686358567994891492488089604126511
073389096634435933844124958326982682427601020989659454604773652398
711801751865136733933944944267677542321191082309423728968680220347
439593248623769437410866602820174765563194951087225889202215249803
245839271724279586677378380658016613729297771184466502501152581240
730709630180406940749655684993087263042183001570412013792245678731
940258213109454610936997492226185837465197323220368127821679785482
540383535799586852096711956323580355674470587232686279282060364871 7
936660559441175390180898377902808434422479687088565952716127883634
043276080005804509481613762417573420339477986303223673828425719406
058354743788904464154065790999931856081243084635339697991722881263
788799160033833788588455471982316793889336283777320641146617025954
208508851805129008431630715043310753954554199079646509023164428834
474068719718934667356494568123543049612988845376092977089319942 14
630482207660235398243215761625487612842541210616113313364882245413
244897545356626534914240822491340206617500721126943061249663413321
878554680294591249172174641038186713621514571649507321984896957276

872688364311053604427157154289136681571274428332133397633430005081
273956718748263504621039314324319975491958090961365625700243201665
807095188730589919787067936829524404853420238652758845077629278714
634221930150056380457251317594012285611355436854201081823715869013
394106131783292979204109913274541054713071745330047233839565461871
634457034018556047181381578089815947036849425786821926363634477428
221860596424745794721701500258173333022732047386573215346948938772
327005556946006475062041487331611568622852690741688093499017469127
606927198938505564840024337032655975293686023874269601591645095650
888549020898484988762500950696676359381231697790345117805540667349
228798064769733991389207356808640524705632218673824785071644611430
980954948809247373180071966058926964767438019702682241032886561113
658492748669631268073768413367434448435491569475817285335944010063
322752305783993875303255072167800795954106798161051287012353251890
481593561721703239862529923914117857155601316674495414313623225134
309093811713897244019512308210327892658362850240295904940492941532
776864235979783872458059395139056795560489022429600734317792826184
751933955564593715046875619975797431690153572144066875808686554524
850048273571308531731244135792320993581940084780355302666178999034
001645004709491013833401772294629926564233454788104605647714030170
786181447811161624796255323924406800969011790922851352731479550364
501613013003828067884533563391135173141024267843977071244855993926
430774456327190903208839351852366033052369976131317334834526825772
849605743916715539515852956333008194184226563499761732066839099307
221655802854054586971189012219660240669460829422781432154324365761
017502059129601843671261833291453458847496239270976581693940402863
874677684859859221382070348649830522504843916857753454819551137494
718952124618141523990215336585113352535414111656440333991014817160
305596064373428680323910031407094111326232632399839699537619227136
973500148398583884971448168151714974590795901774492774451113062728
420131635843764179209332924316744005546123114291616263976070663560
386006560837140438303004134347463989548459451126970333775832905533
641222535927524973455348333723258625709633843342035966408972856443
178958761351810094275452037400888010727884818895951672312621189125
001920873733861336501373434886404252252001770462072668820070446561
847389682355647147107826826300452149043241055363554525591842135419
686511373978866630975008142564319847403775914587895906098457426078
857863202314402576460524637248392024315470427111900318904204033381
700098642286434180747777779834425559893089229069745701872020468182
941675249134855996061980098948447489166287604198006025970012736569
393629754093208594546675623408046150135458215508632072266038934013
767305762534065551698152777855992998824194642665167687761191736222
702092278336052507704807075907180343633570756382836596813995390760
727068181365657591986683751054611521808378119196475540967095824956
017828245672736856312185020980470362464176198682717748478222463490
327810885463141517371814329792883256249937115629715737390115836310
870448602510300496946914258386937065120377046630824216489443358000
596868730214852492879538242286100073642036496791486942425477306447
281042550872919341960667052564506409608790024404064247311413566099
006514678880932791384938464806546101789056276456355644526787973176
600856459859045759450452936327322914034062409343851631402526002102
085325002803141809837523389639583076237367334254811893427718926930
339828412036495177176010034675192081583382936321282066313108914560
201482252304552882944291740051438913118279809819848432290298386962
825148739445820391094065328018875407720949074786117915770017190387
912806376236617440144045207022924523204540576280696579308502039812
183784020672025012026675295531308349435347193634177273406360262579
603136511978554856693728464042046848927715778043458677610085289607

3693144133464873773525015924521197659754590876950206056175781935910774036258357653600808937653281370843694390227229865322218288437400138825811162971553457567403214986097554286886579874369009497050979860937702783572233883314539804939892101714335826189674003122527997303364571061607284968264026682347704558301545855748271713724358470994861372658713025494024495738558899660535370903389251145405558124569294137888271651990004376107967257280599874820479895678559388584994834696519493089781499727763473305857071790270935682275763063930497022966339552876337991307858593142078113351114320121026019873042167062601435758411797707904580838088498088166626185358835592420063053024643462899230820307080649410730415675977100775239855868675945731744767094556842689038531128494988018144774566505096148989915176299241642878000474138508045203295305391840976899463199695591278676949319592733662054309181205566924621527407866514323526592070708678795586416860452775357502074876714333770601191294031585743107677777952135902613080828983248839483209499884568307672417592994303402094399322708275483573885074199171369400498798586194234462796084144473566520379282953170163351181530293127230254356291055458639577778022116588666112693357407294436145574905637200712825448113557834029016048517605243296981355027471470526354293526481366238869584898195167904761247474468008477258871394552736710887847508425688259839636830667647664513308234299538406371493965512602596412691663955329422216277976078749552917485688421824863746324747783244929832354402571567607928674259528494338989676434365754823075754784033503696537687365498022398780119203544049128826835941953971843647255409053142105566632073204638848382768379261055003805739537940215136413662496749353732410440434862382336249204953544285790530654527726507220346592904432022017163242358313783512521095764152741244657762616754360947097433564007690414362218068299355151091385565737341194890321845622044387715270048211012761208140782452649886361038326508480852529514952263554264606718445430426533826668610066557716951714429565559054236819339387175320386411552242884740879638726559965035453160178728429959062489756943146572532979956564427538102595666725587611303086354595086848420817023090377601073137106234293378074547508237856054947987690213905665585892860091990456026032063782729076155397038311018008449011214811927779674839102728820575597820535088346150021903483765764631105684014250421063783316509790934725949942661704520723269101718680689315989500806239975869483897052416122301717289403904669984942721339295681261610046509028456212675739414392795031958650235048110471685635783540426485721275402638812871946209203813254648116170313586767106436587660551655133113317022718232156877362195848216856465284606970661905439540140651063097333651381196333165949030392164270853542280497980267149118956364251748913441214263615547808921452836708221694025987112632114388529939169630480481789296298820112380749013052942492948016114353302390080670657213781679719856861302903012993994451249846901001989193605982791697305147594346496028833289696608150563450566093781292361334905857805509456421035309073601958446371216507319820156424220132684566877418323310247319218685156434120327170305730660785175385097069171707917252855117436278713016009522089202424050305756402153727369592667997478107072793723912355777093468284756010763012791311995391762818615943038207783982432617319663133362063793496768750895240236424692319045416738623583604828374392788665477594859028920402019395937706567321194909910433528551798714035020307605578201914838828809464964820842417669924567583122624780703905576531412632602429224362037195329185547180915964431856852057882350103091076128060445704425147997589608880281259978623877435496599049296732208449724434582435036897803651849099512142294015669174534168383090352847796430676086115997636787204955057956365 16

693834521021205712467189023635837908339119080206899596896990188122
321855252869348573651888630160452941028179736080689549524036066488
944683485357371170607994305471921648759431314126975952516610252290
957537550950933718544900072907676126346765291664645580371533060205
534741620555668380872331011456706082197136019911669601177265351241
440510936203601001758405334468987565349002447580184990285112905603
628154372796762883123816577437517662456404578370496485690904281846
741434107660754984114657421533437962825237739351775877039942552131
816901739901861642141354392779733470876597369481710103318186376892
728376366023019205919792959179148224416394031804147790028285712517
764484105931564467536330924157970212626481304280838933770672398228
654341731736481424562966180793136953250911287546949801550317994516
691228413844646308741027987820955877346176667793320063616141299836
112387852698449676224949460162224198481882844175972508965043238838
826776211538694490722314080038640966747955659603365865500834501574
668100371549812154559177082855269058782746268018954840985480647767
322593083364643266678951981323034384780554257118933244880337102766
080664261976800040145768192614123421421090837882603488039871589674
691868127595035419040689672781395132198842118325610948747352764866
436713359368373719071671361534428920725273057077805616065916154423
589107846465547369563439707372217818591230109443692313952203010113
674073457059526133029367437932120406159970890681203507862354127805
416826582353742593856966435762710973540865230333395749249771995346
662569428121211926674888665256315169706607240021939626684282515447
561496357933365845237724099687357953227591900979741551721334845333
578681422873993851902093678274021559991420456446438381600099906505
371881484938160865503572270641774386629751678966655499987889572179
026230908454480646518569309255696453172241089451645426796761819728
832958413935133844596041672854573991415080495944661353439845014276
180542209659848671099440825081513239252136069510626733736792233221
425995230222936409047664596154505594842048813114413172046469267049
759749059935116920439027605157446677396870803247804063437778416725
021988849435409828211600072772915050759869365684722016941046189444
582618551160041549451062815887248514034519005556346661524473749607
661135778748374003886293884886101950281280781792745034958405752928
452983890915764913247310105633314781346402650462629156753779092137
247828970031963259689125133021524656120543583762268609282030777416
870045904352635817494636724551789784931750675390464041603363847240
546498075003930024576610714660605719495109140248232735266912214960
160708972207220546288100387307622968906215262971114289273463392143
785758381679957096512975121288247076229375657213489062361860141899
595000293934330117463300332972907834026382527837960530000473559275
468487189299720656136533751537477921962495517969220085573147944574
288225924228767773212885980653704654024619938729649935943563230213
110848242495018006757189398611897262182430778317833445857036118160
941397634465162725658288616878213013425589073818405734222752790944
015079633506963068315858425959758344133931666799730480514710420516
213562175409048777330227396980656495900945695698536584320835620615
934529254241892916173052220979352465712270664005413539212620953741
607025988131267956667461709323717405236296319608936529844425074302
280497664164038282925713716360306176259672499571761536958524866449
317201096085345723423625450385444144127163847672628333081895855593
647600616352498590632887445032551137768181305334664669950154774932
420985686593504901062114129914177309980459978865399855599720886527
297388216508774800198668603163056123011444933193578407633418331385
977273234527021265265772962648846204405032377509270264409159921265
248626771659965913245715413925400153811699661401449792205985286546
311988145874191873375518550958118710196924176642924238937549451631

5947724531101984145080087615562644078821720935112593426184468303521073794000418382893605854407065172644916885787285452650728104911722412941522346848448989734965331556939326855402116655944907515310397083246234459570196856432675680385445193586873351496819597696008201253799008400105463352336418912796054468763570371065141356837155124483618491925094994141446246321784596766719116487767444895994644315839584871818846627420278441899928803275124496669648679345894132986023303482928876260637136445807371340101726992400314099962898759328239973248787138226525474190348822177498195455707963780042780145879194411890770714358011030266245429362515054346165151986079342385623906645515459086899700987275783385647691033468638899428963619169533138310635144431946929978952150427343027450548912822404656751683738409173741484373181971188226411967029514001048449736868836048926288540745371246015784688794778131708392027701850083959940135078751064535614615484503534678749015340275140901834645675419760454833086921693902489806750922992294071550692377787826669912301589909380813372850555299059934716784235078673905803655389520181114771552751613837266566870550325145683158295906535700608065726990227214337914923752422195825551552739047664151524230841309327935561940500532444145395061094916327038715303701528100887540809332947909865917839654089741119198714373411365127164382405244158428876975714977114147142795082
9588
7029927924683321337051526756439423113502628776890344646636321844459217157587924113199632987541312018325222678696789964132934113176366538896832051191636223996373640065062421869198223064419813515321973198591015636256986218174854708888378220216171014912432492165323865576908527472547859682981249480686066444493519183037483665508175542257335268514038987865030070402899334381972301961473408648283476126073019822268614411798984367558389159008469991314054138319391816564308843497882991517174297486490963838967343065171217360275453757834311352172150182695929149322787473742572457213602566263841389526262793021300099661963003232522013138218844822215385253127676763048551870068314684039926818548765384056384831921002472223191661009134395076785551383704821428491510169897533907897563233992177890388047633681374846516892263571623071840641566324924108669239676012160108144560923213374291457844880612478637738826410208618024951305733883694158508782319709815158671170951738802867958015106788044933902480689099052919532844669682882545529207870809050166148536753308133690700480133882858546165406413320250693835596317424365884064726157576009934784114084062998236648235748554353359050536126274282001878480529530447698632263662782963274163701153111823408178673987661072812732577851392113807681541894440417632946304900618647807598912642832572998735287161277418336805175637941952440232128885491177415065311168183622698953190049592292508376260805003317433385637848674958223105863188940739807614496920179175139335329885885343364499791300165712868099995155763688357969034499847234260419431859912204658274956441376367770216114312700143477161201646483213292711825713287910584135786193118937459532363102391270889013912909166527192377458686417036480120329532875161201291706095927090777356167401939117441244712460141784967972824936614589907255008243499708909680963641689156962089845192562671934304717145630443239981556886935433726230261498003528371665135912169317838230979648522206285418847348693935943843252998753765119249233509919666893106839343099291774291126087972830433166387584023702201121723945611447336541276334027058454177857748524863164999917048547694843205312092927399866107526313197643376538029522141637423602372218506779113887258057677755437425357442389979633581971403227793561397447071194611416517615151238823627905648863589472686057334479728309257094391377951656305853890416816898769258083650

688250093611926107891124270988122269346853198517066371742046809637
665572893641713249386443340528873527902550868995093761512046498450
930848209464606417794077592727351875061493452818176751710845023652
044236776815132674319325109519200587674918493027769559654098122539
635771169467112606023606943945721364807646499016637843748410973573
009874973387215572695976033113712883158380306249032383304861952114
982623586673336359436081533096204352318069905867253166796719897757
396719850563320391627692961278450432509302784936557570466366500504
235380700021043379054365452676915632116230708153866793287528039918
102228796675492741413814600656548508777948994478550508894914805288
265876884445662729390819614400683983080524037256950641143899331811
663770163075193044500215661609123977876500738743398612137767631637
994991603580029425394150936118287792564890199706361151183343773228
780513781700685461893977078002754504705746574409611518650168872167
815808016185486410808986322233409912474922580811185326998795736203
036360120248633970529467124019239868886269983543109200456227916998
441698832120180955945505348853261754254953638515063118962530921766
529165824315900458349693970687658654243281945647647853735525153068
989109796668878100256983906871131921425419886641928666875453772443
176144635415665762871405535364287851801766219647126689962431948273
100939741710425905524374968733303659687214601889996692802582305000
250494743195788736177439811819412948004602950542736866923100793833
552779750038359156208164728629951618141511824304864955870476420387
157706452758723677080818059040842352377577575400884687756111665813
925119356731909402096142890110964899571503972071590504247857829641
913981864645698069003883646798367981012376946322888360915854430104
942614876035037990344177685999567599330502253234765254688995525806
289317811721823163795087784593437287864151446387303443730098248049
249554003322343437805888464426565167111725408902425165680743457602
957148128224094784605585432107534563821848048375625891575170601371
468196421689717760549530199246953023901996148262601706284818796357
991239697001594441468677531856458312725471743944500821629828304937
569597521339743912031065212616962292811787214991497537254721293068
703850875650615027522642033730412161623496397880994352708041669332
272183593224279111657363092546649671249929614399075000097635710205
091352172102674878381800459683310699995590778255464891283674339451
582478152805746103415113435638105663770354879809403265266548458248
684582906755643586324591807534607758016958399758064881377043434130
105968843929917931741747428671251433132551775956673912061135467364
831151979354208615472228327585604017733891732111418623652027644917
630825994309391671301603555506639664006376991774482383984045272776
417201122912290611855951275066506498354602961029657475437590695555
075101859350758837894692340808844242104044351784510184494697766022
432572757633721382667328318485104191372257279900302301951988145702
121716572276518919027375580323980856028541791086963303805028305825
415520792215467550998816607126379666906266962228804105219437555359
934786322393330880742944031636339297431845742947964530484812667242
960554778937241659254754257272948183035852407989706013833901921881
647311557850526432810671078304253827862550735751441780943794515208
769644180392945053371069580028806009592615542404295384939223669282
518654578715415505437681721619494162439802366970176458516822418482
349498561226122052594069468684335582880004236042671649205198300304
006390821604948418223179378001878971473265413916596941466285446072
011663385275508319000328193491650079157574254157264221073192131424
033453260766000540883242243039536428517010190719596899622216857242
058270080448845866160085246854711766440337454171697257316643571299
349388718199291759466318132864866184797194493997567376889688942187
412505181286522654368902847895239453189075978183784293738692711233

245224027048550113680713014991275390637656776195040824729457795161
472517833405829363143893747277449678965136481140700231327461989829
246746688558984335554557604277573762760912058048480271999849553952
672342626971751222462169690127800810984444255359151750836095039154
280956032294024501253444800135907132943742644076571567194141395791
922713654100901617029910319986570525840925799843414159079094047612
707383047817895510947309066257504993635142278626507467665960409187
273722547794729752121567356857260978856646457541013819301847821233
651569977226361516999711623081473538687445595345401386655999956253
624683462963168340979755045640501027726108378785182034937053327921
389434083704172860535162743180971964185158133463861786405808528993
261626747920282290077980614873787741173358302761312079602560784881
890468622896278439020499837036853688102771127764916480789399444390
409250759048332890872832414244933523646308167981840710198476936630
553895402873674490337398474907585950356060720035878450430168811021
442649884729717744959410584520038230125316607870885990663037128041
215289078551534221312154711394784386313780256937275762215857679289
152126300066971699384726344308284630682783195321247975797640301436
814373461526469198950210342181762593659784532667602506674488467124
609668661730097066252450176922788520569067359008431604124592847480
566894697818276779475588470103788134571631881749442899892987167278
685725466553677372831124446176730408752350650543917283196198305056
380900140711890771221329209083766220493049679457866788418100358637
383427091353528093942551819807934978546448024964273686684335521670
808965836820495433367410868993443622970414443439610988612701962123
616829423893580447197020973891992082302323146423505671064991136035
773195638847191819769880710058067038705725015301906153585559014641
266966919236290595038548687337612191372174427144615555267452282574
000580347074835030814855107153993891683837311092750205817019512313
117767785184487776872680421393928603421323899581851327766693655981
806455715585403801215929712677686438428537188809179099573080680102
254116516429854798597801200530325322575249974103543108838019843220
023575674082733060839664255593659517583051615349713444382636799028
717942890826604259749491147707142297025251125860603900529602402545
826248325755757561216131279581281215685384085591627805709293724661
243672058886815767907693035051789575639340031112592979725357775442
005699664022021383473566037691695441318906191606144682518131601766
098616772320663497456046509728826595127539727333686941755084872178
893886406183984600371686896850918522983446577551641562981490543467
626842114452483994131230385258051848680195708892261883252235536436
873645834791469035678722598255759755960216814522299491973792689788
065727749960206183123619125984229997445917931919070044699554058436
069326731670892612617013398426718889342124987556990243059505953676
654027533075503094906893359337655573217538335664890410728780373174
267309618140745156207212598314724574830121106609479787962949043813
324270611655269733179812220467342712122664138192914732789436609182
787888276414614697642220502911444841844138184923763527714914690697
433608081450429276576158754217052493940928386373294735778423424077
954882095312623534027550350571028903943133681481995153566109244774
270469991167237898955163904637498396603242741993113944290390576059
074155553325065548215179292254764254508718962213113516699330454312
000074718298006506387389892625648905439739686679432127054939232744
427995717336359106333484418514695342981280896868740832375408026260
594329862056291817544122229001840210059258435570500116263341389111
647224103293543067992468631553900279513923299722276621299513099409
795053020739055958119151243330404078852497109253724174743013883031
797018441085704513576815129153624429492503752616110118373210046518
961467826972444261780434649844070818194648857015566472912494001832

3157474892122721505485676173310551732867555555137275257228070158444
4430690911684207944852719275167523884694520140584365412441900688299
9574590054357430805615594652422881931272032923409249033976547146511
8111314592519057258049351511243668916540226006275545801761174352811
4314894999161467189352381443368464042767295387167526081339509875966
2077857277892498559828348232891772081177773383466334241884922791988
0529683092637567318470970487223370762475836981471774211480432136300
9414872547814926308746678478952455561005349890788889845921184102233
5978273731652128019748541341986977054395388749739014574122119048055
3900855254751776918270607097967712659724884419682338105825994352988
2382672363173424267155782964601050683100461379274890656030763632599
8102793661123570622546093038459223095699447464899594352803595728122
2073590021484674876096284970187989807161708718601131703969843543711
0968435176649247955442027422470637716000357522942761374328810277377
4243763384654818232425458586522993701908884747767400268001009673199
7266849558645446767079878771751353980883983207327701780462499327866
1888076713309254338928428954733998046782679145981967469019830683988
9226343929035718573309596628538845031122658632570149517844368139188
5358392042964337583894923848122175655021035540671058277268875751377
8435979790444914526917905927035088146718778168140149009915546216900
1425697803503595872473914976161903348045649916980438948284871605733
3097080720504665480348755712333122224862473301639986713795127886799
8643810255425604253579275162413162454955297310236459199301134961422
9522185316982957104068514803822139888376963907581425519571199359177
1118755087759625477377513592338703229940139176365803706784400859566
2468763995140147145722468543401280785856430439394706997121975940644
5921442901071293191405742653347134164128664510758564588123951140111
7795508073216367810160373433760157315563492556593937367165619355000
4588107322335593024482569699655583883053413186676168998085666828277
7132356870681226254846298210313176077180123905872553472420474152000
1617666021805882461949664874606456387899631507124429153884042324500
7560321404776242580365260920519148291010271577459241426287104455977
2926999565011286066885462748757187665066969779602823041381108469300
8787168483579092546278034944264254586120459971996080031663034795899
9289456325531254253443173998536945980583602867451850810533130476477
5285376287457097777675413077142430023253474040930306942182911681644
3903916557470032165881100062420718524757974698052765170972745153022
5094618659937285040116814957781425963612401478096837868885112514711
2762231791533148704487120579377655030401664297015076738855047738322
8881787312124600745276123541766657688170101149428992573491012356466
6763962580651127133964984228245027305669370892373581646095356116433
4375056502996315167974534093823532899970256271802165624362551179866
9721624983230956587196829602546680650000467162224023966538241855055
7306581446063590559554964198201139696528443930157393105206283089144
9218063428715535370059003504708646096354109788410609656603436535444
4906170070789958318056145339065047705274315664176097215179194892833
3648128660460835307188194805344217730423604084119157616446130913677
5931528639923396387054072054798847957988616996218064420153535317900
0447652558227276706459363505728780427573485589876829668923357242888
0868062463248979189584169445790029592863228882938280091603246063533
2023822312473773678740617940901041381533057610283008848864941592255
5754380946063025862676989173184615639367995770538251493415304834933
1224348063333168840269976702447327490618973633454332782808207744022
6670977817820783120857263445609470855485214265848910184957350318666
4217482298567340632852564208874664253405339504502310275449342901911
6000844150398495202153256001639276693664778095841265604290645256800
6170955852041871281481347434451539374647549634422053892610995494422
8914636753895787605584248419025831181728850215885837880689893059455

215392813746540887775361004235101493108460765432909460867969100188
403841622500908888374112448306461237752331765454201486843335315258
075266734685821776462554798771580037807371524124840869256907049289
006667811402797916365374333951451614111072264561762822599124788490
992848729888705298083065245993551737408241133677579727233568268033
781163575499834766699082383781455439162155128563855768188044803432
132102939421624081354802623088222419123119256612052550161349899775
372113033318398096362166247186330045413600338330530579386079021129
320328871749951452720450623958005426893173481835408084873470938873
886190102136217565025959113514421270210116480930930374941672915936
245687860034660224003395352251424491516060429976479424234983779611
349866591027178292993124652842563039824406078979869734206039517007
531875786306161264714500100320811594169604357882471847656444135133
850918620897857462180539472705749147588858463426651824146838798580
804824882080550201928877634902053551585624001893464332426863663411
740303122342371725194337451401469092871472905890150615238956228461
520314617026175667585862095154776828356254506431626437458076071417
857800275876080483325967588788531809595965814816008065212981791958
439859252951844584692724664770760915168517463856471876221491932156
230101271310528445074810700265024464178587243567799221303996521838
279141848265281625104614721163047907822420234842176235064391415298
230055436257014763699588784501440513057481801186730872375965246520
464350934313039666558562939251568103816946666203136394399173448967
139829815627411988262723510952512218217047506862533992675377978466
809995420421499113435407010220569268199940425166589339284985656673
473325794934373118520489614803989074911035031394683340777866936873
547549110776112694702987821125647656814915666919095529367320559577
279295129820907151577300917884779633743002145803603144778906200771
554904764805424270513167766893532669437359097224008315940237583883
147635421404408604212995770165367296599020483630577798545770670281
905871864830613825275278600758790702435874381211840974392908362239
891253893167318426595596595484958814557352731191074691798283457437
678588414439809369945323260657259969758824219293587914543693491043
904226378176755732249873147569570134160414988729796068385741483909
738944823408130109565830165946280719178447982633284555363597451218
726033253578179240687001068616506202782368106342928741450813730609
548908924744506622773308385268109020908132431315118495335969689890
391031708643714328426824906481266624069267042133671604705742655313
824243422085609793500650141268388679333024186546223074720253207178
938050317199178734911226776218997084782360811160079801956068213562
664092459514059419198886596052780459618987988978124604450752043911
951087425762535033742853443599668025487721129385619101220742965112
088842954685544392519090409345969976232251366844939593914310582200
549776165119323633557844773870072751670887701833608973105333061674
009407487608485728705025403965220624984387807914212329203806188310
481723210930017201914851255372112309151571901612713740653683546403
459090175169190773763122126257226586645496449591237496183719395515
196650457314903486981404538174573697589822751137745630786723562169
766825239245298159046756218528584558914530148222658405359526390985
393717661971015878387327823837558472442882536372621081866574074402
013077213982940343245984550348898469160893504606046029720752430522
250508848409813445599990758681315475222046561457662331404526326965
352792923797929373929190374869671964630718060724784747396550517097
475096606260234290376468444505822472220213238638855714513234749820
349966040662117777297860644341467366211106480533161430705465164482
946603376435620929127032409021043510275954039706547299792721089618
396731508470303622049840208056686992252465953583737496013314179003
537005246456058694776746889730944136910107442020272238575142052254

7081908963499142194431740411237485172889543080184574091031999461128
055643344226228957934051736502953133602831432970249365622011880738
379659686693856304036554244937571974210249374057019369451384825698
600195321516256897140183511652549600636011031202819953496704616097
462082917736572997856457130696516500162277788527383407259835596739
708240463152591767380422974317852831564944495450369564011093698558
179851150272119159176380048296581307898594703973120653780203227603
441629732969801205906682469014302949399525015898904771050802538351
571584280517815264264006574584027917036556283411551999177884995168
857898088140509686764825608157587168921378526143482401986700859594
918343799687274938039243990012408929267645452342219220757841378533
424878833535474932468597801974307389167814296105044449662857971986
849317044974915506755212791116158380570696819657259481529866721333
525808676789995964951596061098889306228869661326312175575758648326
279685711415335026472140795023598595852764820313639865562813744761
293814847740202250445085912182509562477907388965494526502037078510
053115696562202984993747508613619043976299599031311900804319752458
094000091455821745452662580527403998131693257001827684217223180514
068200502309296617727018544600370433014323452516468664561052172069
290603672027333539615770828488823728813588940453449022699453876878
465724857819287558201702633487954178253554555754906825006997973950
595046134205343092698190572553123033871330291285031922813569639601
283853455211594356914097746015819877741238198895866727111427295831
082708986392046218034537945562433971856249712714095792596192307818
840453555297977742210688936733885548588107223676878485938225154035
440994822529059283484780136013500915158114532921442353962987554830
680155885316596656339447818622239365806973126128512002422047411433
759617769910633391463758052035815470306229527721703514821238530493
265705844611319656225452385286064449330918167471272369727202950026
266949386760673907235744132686071478130632796252252135834514617531
970735280901135431467773945347102400608951781555830757211821507995
005588377445473192612925383049981743040230702238909816076153110992
888652872560217949886633964206182092131604317680117781542966367350
787241918340580597801941364138016285792981832086842444697476095946
754659351392719202708606667009644691264257896976005037528190966153
756618554580626435229492165790834984379257431971587706468682001818
232048689982564563340501384966557640297083448061878669689312163513
396687831362351177497941993054822898661904006471543595957922754427
777943966726337297466277977535731960843472491811902101192943925903
802602648417484447820051685684303466144125006122544118553603669682
994806572139535133407886924532705912914982801741121071884134268787
888298002107119318415476906323213303566470428019983416257261051670
413116849386770027750949884410851369316956444860759317083546767369
017773894297315455114592277011103608430557718241212234032928229874
439864464019195609230001439499345306044257996938491772397816149451
131204204868637916752530634900665239580440289843539255578484580722
003320292503465974481326140173373348415220872649858367236488056433
128304693053048735390596848977694106624899681646551018255627690892
330654374747732515748234642076182693720200111288490837408415666378
790491771579162617447253356921102796313636396193338303169096058563
478651583641040952185421892539384536519000945682188235121967853491
290747273345761908795277007145342964288577789197970051773733189425
647467787059514167095015125436325458585059092777722357441369061070
592541796579407364489401336846212597403776943629267107864806916569
414494764962755479752699750611239290659055560299806182775792321198
690451590594249076760144944330214475381107886168394173626824737953
620485786673661943401837539950788735707695697363348906096623415203
303273664416840915597267506068186919542897295549678007420888087318

9998422933180164226391830114079597049126719567266193876235342306778374503739921556049731619654537918413623760136660987343740561564616345985238478285233197307913701982509058532692942864012889661556236653366808679676269021933858700947062040850270178945051681786827703193427843070164519313139114857909616968441606620928373208333878676414883913529892584818453086699758841288965867024287556877312359003496164995760829237752268936555707635413408265577248890243575485397525790911342017983026115347451748939422823882771044974234435922820366214729739913674036710121597094308248753447698010669769903141940785020801000638451622035427489532856955258016698714012790945546584468531729766388592232722802392295725516217043953779868091887085119555014834500653542058958817281907159463277706136347609047316518417732001776274966861929830048478422225166252681241060317143651945672834889281095890446951076541036189885348326694340218479313476380613355515202360217636561827113154532531524831850160025503530023509981187456840139784132450412924899510635618839886059399851860662669837430682156089353640803722105692217062106540290334689571523900667996984398197199449488473637992656271379144085545126277376803369248790964745110630943048104744082597529027649301909961828672066800838124770828042534854515494482673351770991586513972074445355962906202978965148227996438228462410049492538096631715849474649687732429417148601177579254648092229392563484734484973447687678972551867684457804193010435883847874498471915754661252774210651983403688768217709856479897496641796375853276088948339937898038693590500388591500418224769262139163222115117073297407572999505921614195341795453956482580695755819141010547408583669763889748544356703808877762253423725236685862528686070111207376644417703475923902205402921183363592076828746819163573443621225846855184911737828149893317329432868786673412770941950614067843095596346611830093772355931550084081882042990111253625495158658798777933201606023025399639582088857852464068389306031488155118818510633928013806882947755338685768700628738187175509620230167890829577299370394812355322511773065141377479705343893796451947762336804445661037728242377464065317471912285508752570124855530349584254775511192310417412608960417645304738448618495987688244554949742237079627425226137859561508152707173552251406092471466287707657592328000063623661781851820146612996897320450670193879122339028019679284857642151753605032428049537412601702757403030534227164186394957860000645995239276691728895103478325073178136744237372764385742521775361860499615778451640562125120075711268692543912705480417413062908526548019648798111143734157554754999170822366196507152797220508969852051364905355472784482972071078145745145684608611157295659631757938864748424633163765649448116161921604628737029040409753672886130662513809093509849143091520136859403499036297931364403312590118696488012087866614120892897810507017559263159328947593335355585396153573748647327788469290051483756269645692713830108719298422825611441326287969308554351482195929480670805922268245906695987049805652022632188600045813384821338380107140119370503199986558912494606613140897994650450291669772328830601951849785565324355223479476176792576544582052660516799447607227055026040253806930549886106509217465565888873017888384128959995965915932279304573808414378982357797812216639154001074121336095716896366700574845895205479261619617366617966892473003181633077684291239817740029380690464820050593056721097047072358576276559715286685740580991153560586905724472242083498594270765145780433987934167957681379636208339039508329344955805953637604854622311362516792364381354247748419480435893321459881942136057929416741226159998361085501672440264935290269294762413425825638723031974350786160646176951012185410632203081716611241488676440338872975765844395220221961007374345414850891687133744267835955

270561315212526207386933218305935658930621030492920953553814549511
601214641984397939037189643986934878415890922090939070427578219059
357094307672370243890053453120967003508961922242988160876864329829
397148824960412446513280821124892241883314447492688036342391829666
716628224156781839675243796659674581169914928136901450227978013597
693138653945547206845777717459138536178479266953710369508963772279
061281236547915708886907667689081937493406103684106738600541100262
787008747059471065864514391969805597069505565050151233936665890571
733133644764213070656757319919039388118938253075210882859516147506
809468839245740011135470274786982168621274321497665188300319603334
562235534421417368117005957663493676223290691848803237345192434912
465665329718238417396919865871333134120710517073683417241234472371
867249415108427049615549503795501528738224860538460676720392875868
626261698438228056015875786683092512230401599897538489228360095980
752159490258074717157735374806690050015349853590369767229710437159
210984693912049054262463396085508249182328658439340262938268563715
259623894744717833699966180201154788249913323652215956653404857011
232582770818886215013071577934600274395168927524355182396240839150l
234739246686510022276735154153398174438062936818843187021953946745
788068387330450266993482047409308509529400887069518186325483548249
662607066502499564681941064704071227311004215441855491231634073401
961809074987236703389974993439243991658806512836679056416957365216
050282244885217573611331742947356357748308478330984929574305730604
504128402971148973655233370252933328593415448831374005810726242464
775155356118977342742942596840194068133338141610749091640443078052
777297700938276873668269280836283467725910032841281289357876118865
330379943915710991731308768414829814731674389241507081233304663866
652688517152287734958086507090211027130801148807052510861641816152
556748171047862194001056128001346471000489600816181363398613143759
537835734463470097380223970667333178847121742166237978339505084945
840564804300591120184173005022419629811376432555086259936672979212
123968336262543307920197457555379857786033399361139731479735858804
674826631905550317147510740826575455384526005243911716394798545438
546404774452744423208201158058940110410945383309204755983130127803
405876733811345178470423962088493582933994762948927492630761519594
758086352305003906352697550897371963214320279758088695852977316219
835944256689466634986556641828527840530710394603837055319530705781
433206165483720876373054215104981963982316223946155540162448192274
889889915481816161014129111422332742480710512027849723849044026429
680769803440316010136598808564229607540695695571335080053565951888
414562653198365459981493547035333965788049476127320900485531135908
697471453536890157189698415775481177794107604190191456527288840405
188279508490082645360132429152288889379751999559985968288936089112
710910309973067570621865972967346398430619284295504292105466916091
541828035178323684155218186556796340711124306438551541562489751763
977188126209840873609829923836373030275059418770626263573784813688
683682933396988927744468696966294996896602769160036986059689047184
171600019718412362651997930127874191392722798698022724595229401524
322084048175398124008257637136067030334477654114042743699620541251
450482700210515174118908825492609774721462055366779007552008799892
323465359252612737727695456712154449995014845268543259740875744450
235595031758880946786706242795049681223173819531934699556151004209
215360832603215306918572722599298600395392547343180646477064441544
096308598332098942811738979506827495524870662827328053647924741072
354611945944265379787498697959643221449501435206166327360833877895
746269008896094162137344263820026765753391034189247610563598881700
835396449558521922407698130525217319501323741964534289761658135407
331302630302854780224851925520783232374548326796328907125627475246

1229292964140942485029059474155759927512425085069966347353764294073839182074190916254160972844591625614472721705908293857676713045438433599082616086284671963287516028608040353472085362193749494136655199150853689237351274850742948144519654169382291793153091803782047287223894558772648584565835836888354176105647212556438640151856876957798827517422243471111552623436972117076334504665327247663342378211958791176157833972287470845127988659924518327916414459828021443727018946266803579233337517480641049931852188455068211836085689756325118138750460942419557443263971984310789277524723566722883955273552360662259878859853963287728445468369492367606484412273536829219132455470407442166820674612656707964301200553515990474170854616373362027038659522738064231262187406268909039981671086150219582650064497781735731850521577151608476313142971006824824913916249289255612638476738621823334522830285701381554374765057846569058839482531341998139629573625019198571058084667923649110592908805506833821771837094406269993826919706824470532005794540835186100366116156094508242398650419987316238383574666454521887359026121934531604858339212666377250620851905462362373372811425816643842498897985966795417592003691300453703719627536068718301376481212741056148244879598635108500548734902984477529757534939665530715055154410390938653741666521591833549973825668932645268620823596214023963409712242682687002533955826932243805998324269084530478141991498931107101571694749555763521945874576862963018389905824201278786755796934723185030744610603483914598794977948711391354629025486282224974389447438682661643606831812488598723268791076616230538008602921583381443284532240746235541879988817472381285359683828950062022524646357316108266360364314856355316050499154144130887215414433174525373210651396119733069487813913096873318613666961939402572004993080999953482194276558781132352187781245718342397646806973069793406081383018907785379227621548466408823126429816421239417456971356661539825555435294981722390145222359192574194314642182664776888667725021712329415228978446162910566285420071453883416781489134566744506929129063191356184691533158558135076487541321859455745770156486652616746620858010700235568168758105959676998901985472706417531288487494139070866420808062769445013196701970335518100417192425136442744852197815370357770961797803655471501006418180502754073583631210075848690420360453063743034388761523828983557372852602790109658113583990082025106415006848860377851451703993626938322618882301899591271000879075416628748322394452285023918486791381569205686354321704163358637035324353038603138245178894425200804853014804232640065209962960096641776937613082080687020130883472099991664558057470297265064248590310071846989531069017432283784757153675098497043483482059931223224751985485354554519808422814507464169325174171166032932667762278190834897227510030808975252050302466493512646127848908741183038587799856966390634505200221239345266865799204423861484757242890105114328883814583700148678228333016407276114036941362115737169985605841735445608033688890692524353381053971593150452592094012886826761957851131323839763615662128576482640972356689220650459488317985864101056438186988942899276314816131141197839485143896402016161447028222175648273420064615301617015673045186184377052127062573422587150628959854886176205229616886565215837784247149479866487707067324817990842497441413030772781221727570393898531076626569482761976332874465960455935018021232313616829469105165367996131496561881423079790028197020753072041377449667453062511078456579246380038273856860224589540614393614994048484286969806483262108464380743336558398688089988471514295701014512054966842896674866101866876512973139628321441026834165860359911438931807625644344609274750282537324722439635823144186618983533732369168086950292204814362372048123472392174822684291066611784943531672592385

0410537793110490815729580082189041099653740522940462019848205947l9
5654684300768220373283990614679207880313062108074288782674565222 79
5103975301854924501080667143565220340985209717127515883904861311 29
466673396409924349264716231223460568160044054572146286566256168928
699746838216339043317628637080958285081909697417621207405508511001
815330208171813671768543487272089172606743181038687549909642874280
410652857444783139948749789244343537320188375097795346044005281197
852692754424896325741629879428825506076395165108387117664673766387
526975088379398903288357402112295635432087437414162494910515768227
117417711932232945391974092136590860000476202432818881161163644928
715559908299814543584037170956530527567900610385822005025774224292
569773732296609385467336929449681543260568582340850739091504592315
081322676781063632505958627455148922828565406945222105635580286224
167640557695260322183356239637398808014246118755054609555094122720
020016673290789780007096384828312446295596545022120743326436571869
713451446896888922951080200162568079951218413160983309702178276860
244871443361314459232060147349748148691320774245973643394920073088
574820672241184792682405330459869275458970721315458415404772322220
839208036324127217631162891881643394036869317135582680006351732097
450524378305263342982051585729293729065144082252093140038334428600
943717769650829294111897120873707840935527547594667607700739582996
325838861067374507926180433067251405352116133767896036538579850322
184508372345725893974482449203287338602799204201279950628458617692
934934748645046491968635338391449569329353499139922453664188100631
030710950568773445423096458954361418630876241494447419866698347713
405300957797872079291462389626390884639299690639310163833638321319
539379304280498487596279495977933648358480037841610186331741498133
643427387980302577970218308865036418106221571029282021587181445626
643551912892390746449491042735969276652311469566584458181144763773
752350128593126592576075055650198433567754320917506901135878671619
549365520291390943771733201558080722733319100177931651968154228088
370576507598837597021754117738087137985338962896830262981628055301
107557758978534823854812982810501496288588718232218048882601827461
644879029172841840002936746970665260086282843883088133563358024351
057763528593351544438010102449942174781896553188470873781552236440
714542829548321311895375408182160748659797776227910230823364843617
233660291719339926167882868980057891191156961286113730404222996244
445645228811797471536635791461504943812179699911719553392915467985
888569082819110314718381129967865651453059679813190800422182709005
693044807483008345642515023609363752556383193517427034362926840234
776413899039043362351851350263930095966444898776041743786038172342
060790714371194447539522843551677888570909340590984262800755500248
725830650557511259353896315171826065846001289326829742161791308116
878212721039217579698337007055600479265997432364952544755986234844
740409617147566288275325091737591546032508096011443270411987315969
680079689360407597750180374094171469981885046628982630503406281730
282319799125530918362807799251133065557772589580549721937547685450
346607218411557053096747424723286992072949541768648905189668750773
621842979151157585159066981543346997356951876553230061572819957408
798433268786548387703704838867969634471044881036662552958628343438
480462678122147642516601976422702036294508798790923068997762622023
363411771182015399050173071493996115548403715228268577998385109384
004528263675761260563984391352940873945574277648499241414685824291
385791775623295402016824786676002102151382666411013004217800623443
217919523686177696138805006592509070252900155749948752601372319426
962837691739012328002602992780683106302734901011003140670779004579
276982793136849451630895091829048302965001908928336498837214396180
793208742328768068649099204449507359860528957211699519040408338984

0835633066438541288900714754386656707085018469537900325716436271571357347225105929586511098406845689715658254557256333555945765777476180201568378523640314978538098315662531858094257851058139046154652480727198329095772958055300883576439354657145791359276746643254483845455827309139861671777995943499455031747714365979668345645423457823495302203379392308962273442866717466798655682537325145377430092237212027531693856659776107823598018863075075604083677127418714589669834570275384870360304258428021277883107135113277277749920464830373404864279022583676007739008365615912219662842690363666029854323536317994514703396394961386871671776305654609188565518443445168903346245846829124143175358657498912475955608960402921553939774464288772795162230612464779308826542357480107178091527740165487104682655357030989813108310430948667539415116316766582814037008668655476438339770427571250447492214229690806227876085444048588139957127475613019026375150738783118851441005371031418316347433675117492320531768258391503423545058602263058687027792432296916175459534404574473997117121192077899158121806247094415506947273514012345502046290677833781762141918901469088070798181006734859396797266934876965238322019108208731369657998137760621433456083913079919949161884261517662011516330133652402686506039217254034345528675180825801046290582083646811073518337297519614561225253713567711887842519940235439622424775173733453533891507262882321321961008440631515479985995215888847115997837510010625972439544429056478372428271127688613087297176674503877844706050498179263263011217932857796302134172905845069384231278574470982819903689333232252456587848649909717503202216728521499882775465830655901595843516185552054897826736135796279574933052201119161084208023699613890043993283506634518164076838998792241323064158867250047986219148727609138365553456717578454745567604129133224460955148312535361128714521074304363554789821369772009279301372617048908280393909998355740828980056130326328048784070258801908538773031881871273074613664205091087496124190176913449970774575283633978323156110356482124709035953681610393841375688337715782793829413546239162705760560227177983532110818251344199026772861049687035060733720893553086400814265991650872235972694624860483666589113039585075104744947269939264526460552266518924976453080721310935209623157608488311382603647916682468008414223588777410461155693227078879326874361283795757685381840746778209846776273672446911189165465543048479710421523977928920676389663589638356641805894421474539622623175326557972526814663117697801299371109824534052292709330039962951371443831951350573178805666396104212257787252801325283848330829972881825074988518314717684205642585248475129554534338676312935464407773405000721996838014052269594709199118955188405370077231658195210530473114914491375858791465623887194647984526528480268964713172715150104012602307121222879224688811363287433466723782055811179420272164792256537042244877632969701292769131500425202259810800999798855902356890432902314785387387240209474483853941778367943233256130938177205017343097962453249510339055541937311556781096005736046904087935062129874882405284974393964694658467774362370280141430030440166179011657976026979051136930909194130040439925056649562424683802034050894802900026376220057864055215622740155388028003468955917543127621861314760525568998950949893238347287908253998423463975791200470117076564797800856626474611921493201815268567633485843782147331176775382064248945809658200413472695716237038372077935659762005227802251822529632003253692765319844596583810375507928342527858259174564350062608434416311297019553754748518566666109082130176889791760811867452945298815710766128602140319019653862395629151313915405878496630720194218580207054805464477547695567872089601597179073617241406114197727650252901385992813531379709256036120684983888215192251813217998

933359782131751070483806021942235771089945917470518247086038195757
880028161556161166585331628472637285890244233722087246293327028353
174928410044347263041799008694169287372227838080288266378013920532
928677292804236565912564514894549289320084006096841831783905934902
376205000224339124261918283291801084770695741558673641192801283496
458206833115754626233916641813906198852245734368306773983610104642
316464133135532352273752664453382030040955103219288976104028788257
165434018178388741015541294894921097507774609559771524786912881124
482928425719746882048856549095615768512861059352678026561439383358
592178867356600596530098536546206425434637910629888573898318354369
679219264775780379760991519977631569730191138816967770383136174316
445378545700731936052197001517354100766869601305437275881607049987
789365017422246617438193269192862429301061984254163460485330613122
444410053811904217052950210203949284993280229187520880003182408337
953638642237831826693545467637857036064039922133069977838159195259
555588878191552733115500131708794901667727284263034494974713565876
728143942654920526300619080005910944956218545217579084841829846693
841949558812074700106037683563441033205450016151657240720125298604
958894707972183425970164384759848658280980869054903677261462021539
863629416377179304762721831365968096850029180311410797043113811864
184914158697584986602245464927651310287275862938670400213000595178
163586440778748117806522568506530713507342458944602683462083458031
339214535422338754448630386070872785027777775086776299202224027107
432459134004028928720902658654473562108425271922286631314180112202
868765811959272712835132421508816450191208996690219737367720181554
670989399273542199116273165262450694992002848523663471619210135317
944601819327396072124887398157503934618170208223027956393354316765
552788502935163273459472657951040114997344823774206587573034636025
051615613927268982066048321085542322084072024003091375158254174604
210574559615339198458900273971574375380224264155677787593079348042
453601445500804634412405440897265939966330079979725921292231094194
707838420418307596625539432445609635895414225661367379692854237137
647495263375569771626021992267349139427236091784676196587205759703
367639965915756363993425058788943766830142891641757385033888100786
002186250388008817542820476784272595671960484060571210079656164926
916996939208557744994249775621141413332432379515093640966413408088
447528614157971242508505925805081064620512457221688482019793441153
229915524820485989751301007559884796958676495248802794232312419807
388735906666496491511396292448871057045643419978174371093169216647
620690940566563479122208766735316201439579444576216638684660165500
533947931580877471076476445478396109992723489002851457979744344660
440511176045586177008433876053148008853673570211565036104548799284
037532175914820188299399708652690645538159173954184083082629694214
229603619265688038545999145315531953051939183455018588454934955788
237465098560821298424007951804176372673129537115995329871379480968
186645546885144362583503666244085255859306244147002018821220464219
259172345933989619257901631775292386555197656249237162846568538934
472706908824411028132584191077575505481596675431195699439784151633
240688940704392660811215618619025303220713906467697732683681506611
237692800248968355757137231128842582768928782310956781189966969774
094893487241466149739727949412661076369402958236283034824223717633
199718432399903981284939398586827244646054071699259408451818357188
910943512641768469147790922528378375323956019474534909055101642338
078115996634482620714158723813542032404931757575337079350969461348
285274651377146834205626236321971729619991708666383043815049988726
067172676572483899667394934781864059973119900529660035329941393264
852632809300822108367705019712870766990024781300851305272966844060
610164378230719136307049422049383419600953509993987135158792521973

942806151356564313408161568884901747824857836606800403926759962908
630937903111506727671941876882462830751887950689739054897988939998
836266317622380542165500973438436075892142420468068102248730738778
094787723527390164557430678984175586097805980115965586660818087835
319300271259737913358957897334008763443118861486976837881121510877
126677572310183325111391985166650796809644853255663841758311669493
885921614166745675555382378604424455712633396677214622446741586901
295620545652768104736368260978986496300568279073769198631560129161
425693064781006989197020226538674162603049633997127366767957428955
566314014881184615445820075931678835668267089919455698580240906776
542744450798568453549054758781082742772021979660038202099786595196
994492933543169491649170554412944820929676092514799032258890049539
549957229930180621792621124071101622095785527048886053819284236085
295701406576742213483133260263421770543730953570108907807302213549
826362852887826558691568563466259473132585195659138468924462445834
402277049670665349938723547520909770648817090001106391255718740682
470471367225709217122286721119172532204691459692215612297524374555
068820561736954940666273325260413704462908654461510442560076329179
486145106922588794437481577891250885911134172233031567497381920863
972206513520158280086982468765373753233446669783773281011145261959
958866650108216881597594956660290352758332149372864358745996764765
258208605552447339121171297976729814037989097286506976492032209927
924040206629297956083333483709734371103831364074116484760050698520
156744722778358999608425578842724735306117051602021864568180148423
771285876616591199011868814095160940245806826199051129611212754574
119275346382293994753499107485956141079594010471612965889159165679
925544442250779692987057697058479143040118254545356111724273371399
999432672127719323193391300068902626785230042044775981367284743790
128671709651764782559460090760126770386665955450974046848589127133
239909729337983586535080972670948531616454530790871573529955075169
242502752012548206339883390344782848948625326384370394044505434363
120240495681948503862902856933719155477929628988453250576597682076
434462946898145244334205580449946585823750536595705362244038184299
359354415068135070920577444979990516676905380209621766365606835781
587320379231026760528625950776545290579527210420413852461568605765
628218747553890252057850800740693372942987737463057925699377504026
856615866168040761542216193889794638536338311259582441879709719549
522407479595395194229768582513900769548365478993386356880618290567
934076519758024674217910250332216964377600441282077990490323308123
176112379972572146928331218471752129580119166422739351443766983751
998453064578736640777380699652918703123194595004164514120225826972
469814614881846289049872728361923812720419779161557131344799687575
703393540193254612893602565440312289217822034095434438836935460408
954458024790180847629703419796230216029451611247691650186005981164
172743826673072895297840924401656957765725555729576130242533547909
506761981919701424151665935744603959673618532551061007924434665329
441818070121476460301248629639975333472474598343390086851604672402
252161362633437520040663234254993084214185882609820943615743240845
647108254431018830907726258570251121046551644620072039153746473932
152920633797970985170747730380605362838981511969546064420171139295
668853118832806264334317017170205170264852813184125163439247307171
335549506051586713611010580504697835880611507229261275769395214080
302982839704314172657091329508291125343176943080754686551329222427
649904174047481341076368194957397746437968651830444656683983184841
948244176059863821383782353137304522859498310101762249439405390186
494271323242724894320227208505266704016110654929802726886299503201
407835500681783266124429473379347087040369339474448836401463014307
665488869199951906540631166476781940497301003751985705416725276470

230859199227399842341493349983909786403439076922427293376689266049
349441694715697155470599042174219258825753425929995659864267684514
106123846878806544579177175040277628236063300794375840743669481319
018301570912676662079893317001720205395490686409451497740977410943
734521094285968471727023406450671007142874586355868678236996339079
944880743760883533682031211373244515710404186429258794496377751457
723511758660881749387826753877641303779519060504064763026806474269
068794144744826935195393809701784967319464528914389222386328010121
834314118098075047970243899193572725705270605470472814746474462082
224540747141694065206969516760010559177013648971322784201700335664
320555114622179331323222171193273734777157788658376969907911711317
283537401935719863182444704761544818457702523441248455096812811666
497090604322747183366809811678365157004680187681709568893241844073
623106025662747574796976374720906354840294917347274873081201950527
096375836074625454793529542341712726819813399026419025763373611656
973855087441302411046998364322210012677286821809498907623686892633
844175188562628321296599412137517969890520924751998789466845120578
373584504314890844568816138401588039698729200985408629694713225986
323131712732194289141354943233591229372423096601546164996985579627
632755545779844544032230409049346593466633088573695960232857501802
685021200071271420765556577264079830207291948065129360217761136093
911359393530812177362843189441739621355453605002461387879606760408
053169300250934418257134216136144401191522895306530077790539720033
966289417946596417779835592534222877746564000343935264773517835159
330557199471598121250380175595475777459515207139084336100115379049
015078907056708818445741145088305122997002099696113097121893728533
083594288099049270074679601280969718292839412819896052132336107052
144317601499501995895358971833690782831386342368579567965634192930
497698152099745288783121459736846075674766360600396879174354213850
348953262958205447487692413303661183286891127783175460291279129556
760074187332255429082447596730430080458966345337819875369948633505
377224451590105416566588706989890137538880243275851514118584543536
529874915727188653564082683272251101470090989862716772241528776898
622677036983969959782816557940888253718376453740221818841187448133
828794719644549375700207234696253360159221820280817972604912829242
072671298002423760022464850087988988572315884221469461474910278915
546521230594248610272604712698132414938149901767357489609411826805
047717301316458946823454090750518616610154107017955417597598280422
938627155510777799674591224661638474771460111789648586772191861044
175307319060399730981271821910672442441948147115278343039206536812
214246550226930206567588127647543438557347426328154927705626600665
467118487652072673356547316696621770698336458358268302077663377786
474895873125521947859905262932469843638246857387232842777898399257
375979040382852977873976846638096241754775914830964299155281451658
037282442319575209277265826932305975246124505643557559140534014297
675082443347466243985834376148793899641444025141130024688693719521
578304486934438407345590866909464481382353532042192534791688998014
391786232376017552145946542744591797476063616652401862061888475586
487380583089379600294714755490167949590428156184287251585293982983
632066919799567389478800199587980348283125992532975991050207350736
266025424310513286294034955417639910947632944854290444810268915259
589521457320122663991033216800068635048620073570348589610617902776
607181677068179366318045382379191259561984388576095180289432157965
212056576110939899154968840313775493935408569534499080090127038401
726261801820466737959946829543697194232579220647470975895251070020
377417194263893464967571542725040469387188357277563444876944595084
986388776168355058849700286857269280542129179585813191422838038713
728275528683064543331486880921516972378760137552798110459669882917

779624109707091370378839304506450124598867858957886370910482165298
115205833227808177991560826384334655419760315048884191986084840626
490039958781693294270619622974531271365326944415146362569734275241
608734402697978790021463626962620120377533883677549691572082415463
924241350575749432317953795546323852419930894060511341098518780558
840927355911671616701871181553505823660985691939227000646822883589
944865752267148004631656752194201966183070345216392823182511277834
283289541424721726008101478758030212294751534403712723321670865724
341810804779578414426518997788601140202712329482376233826892096411
116301103221754850809433535043407762834159300300053391341804464235
262400380739510951370111585095347324822374380350893528935216587780
976003257121283541953122550820412698760354189007713651108604718194
108088431110742536391713898813597531932334651579864775046845758698
939196311882411441543035704000826401267779538411066650964079049875
221717226415649043495962670656328990457837601136851326246834602134
845575646260777352188360216622168327326752466009077868093484338791
981564661208248486780506442969584602102930453724633487602599622427
009937073587107459968463323761050293787408079067362688314322240026
342541834946497494337876425107862940494355970889509680133762402481
909558113491329791361379976702061469804350149070673315063787624020
662390702718327904802851482956061154740695997256194929668302102119
305025062268263880896540629845561979338912841975881135820072574356
168576772366593657751853637407687010082486727325827055947358759499
152718354765991563039137857167298544040275171738826587384991774630
988127769334111333640702276667861050716885045092417768179812507691
095090914411583370303122408506723054581245437271259392013793712989
504976889641046583485074819454011857702359172469012966511482150074
342299282812227997873756699200712962845547283485924965768203078846
795064702194730980149497744309949779104146628500906670090421960298
663311030110011132782301805889845558966393460875372851907436002284
230389621517161956418065627000999956013489692855739326265725163640
020407998471412336038886746834087295761469762659715075739705145216
740388083854201531984940935620834941808448216518439071864548579354
119886526224032719177493852757822503822693597054005494545632486688
166937021367054984886527552679637592542953115273943389962168015529
068835137032902484582581507127636005483568596542941065170419630389
669909871302275853459041651175097486133027113604353170371718416809
873141819585031564870772149478854628003926277518456650430912603414
221842304210133140285691502890354177509096542281408332982473259563
282470188681318906977341224009824572047781021244257867941962179708
245494715165738807912130559425579829365910185929521458891989736410
574890894887746130638684259775642325268634634396779838391211928481
559046325726844298069038018489373878662677601466755693021431752909
926805769995789731762891660092351933832885941822434444837337458570
622675049768273613929212565903885590939178962936286792195698567220
771145167687115170363840099018030189287795668529739177054160961453
069937591596227833917112798101298430027748909358748945398117353381
576846139508418563804583298672849362869287801275472398663892413528
204505780353885364847097327544925680094275596516029103457751245135
941932152689509014544595241696113698064708110231533338827666229870
284100724886605287603754043021857842964223163531335034068040278444
147520082689003786354784156536368956731168839504335456317801016615
510364384488614982713956079572913905790086466129824415228134778644
226453349015863717448862610877742837337586277208196830696383180588
110789389572837677830118699970089433656062457189838150176746551248
089349447700083734530619482002243872523604954757117394662913618252
446057626009488669025658132931135067443779323202503802054350185299
498077810525998530448875507249312550651003885719082485478389932660

2860337393568736445574675696601191916014388077529876643945135794760
0714470988893277660530741163115301391104923282193110588097367047430
2344058908828564002679759435346427421078414906154092962408338791080
1565789909498815712690236620043280215289589615775222060621457988280
4049182338365735125431269368037987268684755236920077772138117964920
4003615902345770034941153406833557382487391313539191475815852762540
1337589984203618879551380731734622447712358536727905300835514369170
4708516100914754482240609224557576103986096635678828945500333604330
3720846556725164894623272786930989509455863098301143787825883212290
9067132918193976522025724558400814024130932133420950689694190778700
2026153575802182105123290813528319509157259925550067198796875146300
2345713152953633128866961298875104615085935633839150702622404505670
9511688694605110946627672376072375052884555332350474774899801582880
0457530050460916902159645238303826292106303225851732156752808913580
4883548548712412653274247165752201579356043381984648837766055270260
5526784708356010243568608832661178809770313606137790936091130872340
0772096181023901578532920354712719691056747947044146433121314395600
9675166261128944405036638819340286420485038072860987919829804873130
3499004845521940657346541555378014662305783921185114327966512426180
6044783706656942465109697462111066255726717641719632000060679461540
4447383009587849363123252352851899857198091600068141919186168389010
5181248046434719012840007070411738005402484159936710284924075593030
9634872403013753519911575118044416261522200880897896833332452200040
4042973126122516176556942460302075359589324020852892492435270739650
6513885729183164947634488104460321439087648326551672987621999736830
7094585493934338332588462059401572984464297862778628233290690449230
2765932924499124046133833969015247585601019682761237203430551089790
2317528677738782013266845098807180268203922028048492157434242568040
6730697177121873456085715203567062170083893698275361917197371079400
7399906036839520612792285497138600225654467398562769675644776992560
3009495709270131191929824955775763750859423691449068347634546043940
4371318679563025883826817928975790738675774381973520813126806852840
1944365315510971335681257482834697901789014318277526180883313639970
3602829599010000519724377015255864819106123617676411458323693479550
0647375250340609927897230719508277849700119603131624755510087288520
1738129185608767747856196063893512409250785806286682155371236246480
9693259203489219130671347171099770337362895959783509532523598271280
9649198711044815613010829185812739292211017926948642819683286246920
9210846438959969288165507695822898733723861578329428488075621932300
9574063083808914628364138116929229083233675690406787653576163961750
4266645938024360197471995403483650856887128855292429351867994655680
4451320866302074899531698788879839089895456546759232060581922617780
8376612225335473386370967729593023991180556622809264950044987752770
3346252173326389389494914654038212432180736158453611404602155525970
5366720223919881050284681073770437723293121597536227505157238764520
9436348387913528988787905821550111104233584518310227611334335804780
9714149771525157789848298936164491828979317790326467228749418675450
3269183640879828266619105953142528663092670701732405950706227386670
0579287338827184951879760685322806148088186465932879942779746124520
9540952671249690441046131073909192112396940683518446969930471686610
7424050519222976752985296684867348368116257663525545752244031843920
2462332556165225141042359232732018883363608901360504726337496205820
3706184846895299865267466363021958716136936786724835356109940373810
3996989004405153889590391032827945728151139758776274744531842930530
6613637951698172746052088546469232041759114350013503950753484896150
6924950900207556055754384170290146078181869924923939119438241100250
7051821388230098035458587811640045530311277179266614743774837366230
7460202073538403482309687677893623056167893806731785663137163538700

471587401728905272491252082716893043739822965883776322658705827681
347353299478638582774381654779360985613156779341800020147580424559
377378303603956313491357634013503034840739927578659534391275516020
753086523653852745037888597124866716500398774192653741450111604950
710018649301535827573069553026842287584834032104779305348522095490
792573563977027573732045507991015818436798097731303516449897806847
700924124658901717949986390152345242314464073314623970454745260504
301874212994382125706449150735469297093206380290377591188187390157
910382794684926286064935577057927535699984474878454484558005063460
310194327718272203125080977502995191976880246589639811415337135068
436100113109962982510923400768862681738142858204141106078957011495
068943086279652602887953539761168458073043529965627122092196855252
288491697344907527823409037934366431756882408319294978354314731449
399148449444634289657179655332850400946759399635949907338642665621
874810726324787243040629049467920253848591539802660863028206837106
819258636756161674883003149343012810529959988806188629972368964165
204568060779510100917746573081454691925813273129673025592058716565
880420339131539744159172871948757014262221471927891106602750761289
442732242913669946659270657251041936310238598175490798699438924388
900893934634121382813621947180798111450300020433901579204393125539
951922260908899971256309233027142912501440398187050042025960878087
081358686649017724447369521944670349065022384096930518259399680394
210385832160164007694778919242121489543593744009993796437285613036
289431167437089467447379716761820864159316835618424004845427076218
560664395978802142164316651900960728174606413768465552889186181960
348825852074845556619089693189055115991626908725698821763884637459
550647377739141580667407665009778341493484216184403838831093760153
938893631063154871345972529048370368834150168607515857990254115319
535607077038149712274469609689409530526009808174927009219332674213
887448974859582609816278112393479338277056197406663789713662110111
506206417832730909438642104434100156848794624469675999352470493084
992705731112482397592335989864528277041548277629085165037960815057
835098230275874215779597790045920608880043548743473529828147036766
578453926047834167045941769174894680825377283621606685686911351945
238338908918911109127985145874140765028525906720911115279925914212
843554397575899852207175909946241446928463268626340215826602498292
115869480107657666054916190877864527586671942368344343143419381233
500212897771314250005292249716731077710017168511888229988920847294
675210622424553471607991208322047262682159688664508167350961643933
992447751146236967030823020942646253674913470340742458766524088261
910336251980464113711612273934979644756701454581288427010677196263
293691784822862791205618498280909007323886154644578857828584097096
047744112659161889517766291662527168721718762516513631708979617646
484309698655020383274990369013956616772453347773594284105079107689
335238793554981878467812308125958197651326354256394023461702890360
708166424177501440151789871339407195406803684014377753823023274168
651273314467179277067541525937370934213197597040931481491994889872
286782267469651931566305291459571361839749552474801943392794936563
511497990843542657131020197144345952011778759458037507874625180499
505221398981478459503317862578489997751009183675879921922826055038
028507810017854415807312470071381244813199018918287915172974806315
354184027249731986676050304698921400122077849186541670628894614373
987362169714665740339540354657042071325868663336279283212156148810
545073264464398983800241706849219912974800942850816694356492957834
344123652616984823031940948633410216606988284623109833870141892807
820737483205430749541642896324813950945589933704491826186642654150
819005827072358841898280998019151035177063881643966427934704377364
802695785368103679866531939037687765826608503687181035728366643365

057717227866672879309080997221922188813671309348050105257264157381
340641178162104806787143023891668115663107287605369457537784780650
585029952691170323330371671960112352344407475268128928355868286840
263257160569745001944449517018066181081906441054552466219974605510
837665828858129814451526749038022695678115007030523761609761607516
474358393341040779805577229347759913433136721597866105507342278371
167761831994859302758572613970586584709868667953396094629188678055
717113687861007855007438818571064126920509669914928942274974254510
850292688489395447482286306749939427340324868374491455647358303787
886059847569183700454233012866275857927095106789690533359734293614
200328824671488045835314413863079762659740489308711076885123230786
079342451220171545580184169216407608957713287564519943138023272075
005128755637620339753300605391520811588010542944317855589339824635
645831046924158286611788490100463414076094998214465414603928375102
473505894407349196495403702722453457220124463687003138483398423091
832402490157951865933785992829470159857959196859838685981944905441
298097484199553689636996354717206181813585525926232995399169375917
092428613664218558467667720623054441488582625370532082454374482290
302780270657510380017108817832818651459865583258719554645852534197
577697821772923612071333641240281634970014373150088552712917589196
082897388980202176082104877545001333613191713197620856414914813796
406630406540354290848751446093045428797888157247969800565660799594
388364820283160338946679044511365530731423327605787108207926760439
535354664225367818502682922884743863930395837400973858701426510863
446204705175847605499045727257503150850585856819746842929041706716
614133188504384698079735601777104811768366410779396794362026848191
177314824089976925551288263145278012643813913756963948911084647998
499767790629954544951573178700804071248028333711170264272301544239
136020661853406307113082815207497537193146840097106225037195921638
658242822894083649272628245596051552523505912078414670760567587367
336189411174276167915553017199436756847928532673226578045593297864
425666944433892545689917122537729840428963254728255533679890350772
231031642222559368662240293289707900774962130957134962928964929387
875975644766633100100120853804913657780486947710046538421970875553
324956543022798404918623196790625317699314735845280277782058867692
232477965323935598599179926338256860793129715306595539487832777388
807136981497464050662517514314106466975480788825062108036930301495
236024766871701778304594493982135466681201891555895109323779106639
743217171528610129007333518011331114135668468007910399622453150596
207162070402855157026143358103007802778165023775172009678562401690
788151274926741016063023195786733309667419533263179154869008772125
173072357898092522530225632265419023399141038030993435384580270680
334442716519277258234253690592492756444995993189337302024156133921
728688211168862565852709987290601656571088073894875836169521706209
563268246090399618378897872018226870237853568344181001214934617230
676752863599011280889100896473779245979568825007398502380775095912
981555306692489535737276413238566846781778140576387723487604032512
946764713594366594134553106314069585626346336897310651638074355343
126100690967453765014405198118349235318871940799399909553892895779
875047727780327084660936154885596579969702595623280284617474416482
632048724128298949478102988228377461856185823233466718586883495818
421919438832220412925662182136162779449004102238014024216582366043
639132312295443465694982722206790823288070241513735241623122747757
303256124704268544338532247012797778599783518199543021487594778160
761885270838202084834997471275528202579694699665538193959379828961
377844300795301854003696616611576286812079579716153560662027357626
724712099348262911458725856610030202616546006848143871460837279792
596145993239211017039733769873495439047451665419547377441161880916

700728524973472353480092459473691441023228343634384103720771001346
205794664067075309824085550357688699700784814367551040154533652214
918883832021327917374952250859104094126462615872664616191769631733
141663374449384885148595135867901870079581736475850407096563444530
063108651340154568645460985779376822846423033101732799271214469555
313966505481727640197265790621953881325350963250947445989891087859
016969225819860487325161631487225332086811525038778390309210963973
140870146326277487361999251603690351640181322863841015778913834548
208695394911416541921224410372588235373528329662254040539599551254
188046955347016927965818084221691146779495770903181471995224200489
058855937464416153490934682109527381906924934012585401621298388823
606711299047278538251093221267600863567887299111060647443854063130
702691579329114716835748930860734176203484242255797432510010038643
688160552466827732801481669787349633209963417237792378830351660548
716717928740011191447262456712470024976224582240273977070270235032
377124169131491308448027264009944972595720923593023069927300052249
073651419777811827203052590605053664793101848708283976243597654102
454719021296462440721878685429130719911560221343598109261278099244
929883532142115168804430524223133517203668759109206121817157215013
230491503852431276016026780710277905987836302849099513323332565542
428323312697082604828410911646293553079701347159992856426889889700
767442334648965004511482489448843811905203162023956125155080114293
455938402258904452354916067705752641775197360904402044833678818915
689702778936604499831675503334609974345390667968123798331363984458
622049182698598018506280836505817585632828672397021647037926115754
368783456640465906078885859361572607337647947362839418949283576050
483364494011349865491208210048132550683756281970256869699549187621
976397204502790044667563951193761315600645448648552507497994208500
289544449953357450468366227668720824831641359948030706016118223091
561752595284900289952934287376173510267424188159371599489096979222
577214401390912722461788838743608051967515303147911414335732073658
590493042773379844712919544464543044009597518309741823376186337811
529128015178663600901647697454495895732314679554993893375139495643
626014549592646737347216021885313265446860088537207722134527510310
595262530711102355378856916499591169220838887707805173543839856786
701509637808868647575766982532354044245428400882687477770328538261
997625425819929342059121797780827787051185845230897298563876865112
750727634100359941466012227948957495592036099460378048483855259599
108171236282744204017849802110317627877887503036302263616097666758
010603955547997874569981579733423399742444758845313933453664591755
258134755046344267161094890817996895922670464402169176805100590718
447352631235416442486477743878776517385347897501402520406932991135
325561481360435329683129289912953561529040275913176773412777046365
308522132575486307933547882996383406999371515422494380882406473326
312335046138821594795169175925450984809791108923311377895396640746
083457300976511706075241436288346609180038490635692685329640055165
359978579130606640474557190848662504146276335720442508766033207641
200276837147202583957757254830817635228170657759415327083266255391
096897305850456225936898498975622702158265265280620025184416489819
196909558212078989726717646133800839566487722019320463671881723954
704930209279866104118469570486847004963864125953067376666038942176
189387542375224018745815972284425229079773722925510180886739908398
854921449138635625388637891615918884240512998193651713692591695779
199884949414977115194315755826307059948575863534749639756855970386
267805400720089745025270519397969812529688311916459045209756305283
373094860238329627213225690073767533911168294719812770574262437522
137582503200873637532046450005738917934659235577083622041452850139
079464067246673601827537985447814076921256085694481054166195675264

7507452390254517053109406626366824745757346070406527575357774320102391334113813577503322561143900976099469521398177028414610884136085659183932993931358081952707069270927607721717779098763854634460240516904994977475774828730093397894064147369457198504829905134842867070991509330445796639195355891478010944345098270173677240979049480592883846575269082140684093675224888424661256052695097801012605772822873599379849977045731761175869419846747429048640123199674462226863633260076411070293488972360126214984596841603187424531584352054891859045356941964460633379884931531169547583611157497667740703154480578978170504573192258815494311439025793450499895500372704236182645680415899700136977093646184318296660690731354763645120346808804485446394798810546809849670131795494286681388276584446505851797712368142674585475571382290726634603164381375014654291945359934360820628279072531886575179734245746612250274430190922429627769366531587168094442427862498407494665563526804502768435782113627340694356164251537923224666458237107932145904446111092210703976045651010128697605935565797997230393938683961799189869179915938694708632424160101610309887378543956773148297234896594767142721341284004376210660200550566202333951957558645128302417714282371577693287679784522757104237281724684454616224472703666186405467249250144671204565478357269214470538798054274436506654919003697852300703720061418372571130911168102172286721258959485746004353295033114695796564900624462269539125112528680378263752083462010919392052994067353165017583782632764100489944648907195082147841020836669964415554899731185444595312327881988435193646690756191744363874898626362765030272068609251880972360884793801625295039225521028318591195209521603087970922306318243049051361251785266783098964840762618711211938564523325880156358456632568153659741400635166538527833027993327618187316429284343663516156632552808774054766967000188148312992649497560617449945560484056516920660629441907471164740575561955183453874064046634465233320351037784476758856666090172875209822447115645045567243110919573466626779195035111191248491406455543525772549656639267852219176238547538197108319832284839469222949537374931725775542580964794985891026863594535761355146603751391496845122967455478424717793060749999248715039173711331059841824004577425759178667121950517428961196738988890457931260778363161297825694113684507888434449786640449589578177977537633175038686569214328456590763877035085454922881774592134644790364096719378848001565990852627617509773940164374064232153788341300502620171319759124864587923006850025338135489151319984710933979884365446048272891549710760373174451260971717062154915230935580784562839217995093664990440960915699421493871124291466312054691322386063352687404641869764975134600318428982574751202584459831039820140609484687076916183830862387891061468241972832005261503851106819905835176956323656969414236261015215538172503693919882029271368544401062943672645120333442448082952850778650223588668479872923363345267582654613647644880992776331655021300744945298954339617526611749016912130058109554244912267385285122343836959839780483975246451012058826742830639996089076557344363448014904882983571898164096451011928985398760882296922643542082445684123687523325897783852438454227592056760926079584672793866397640133375298174103800839713669875017925188890050961627253029063851224481562947348667180792167574395023354187971463703413595070688796031015016464474223923286729237172565384290147065906886729406948425427159052053409339014321081313854582597043485809314855063394187054375482019170226882317510901329413617187700580911334579616422030125980220431073829668649070375044538684428440879094580713838962095449207599615536655713068440469527129948783097134507882757248448998973497039622465110498777004179371263198665504248264504271322916123093246164135330391676894514835856927021667507452390254517053109406626366824745757346070406527575357774320102391334113813577503322561143900976099469521398177028414610884136085659183932993931358081952707069270927607721717779098763854634460240516904994977475774828730093397894064147369457198504829905134842867070991509330445796639195355891478010944345098270173677240979049480592883846575269082140684093675224888424661256052695097801012605772822873599379849977045731761175869419846747429048640123199674462226863633260076411070293488972360126214984596841603187424531584352054891859045356941964460633379884931531169547583611157497667740703154480578978170504573192258815494311439025793450499895500372704236182645680415899700136977093646184318296660690731354763645120346808804485446394798810546809849670131795494286681388276584446505851797712368142674585475571382290726634603164381375014654291945359934360820628279072531886575179734245746612250274430190922429627769366531587168094442427862498407494665563526804502768435782113627340694356164251537923224666458237107932145904446111092210703976045651010128697605935565797997230393938683961799189869179915938694708632424160101610309887378543956773148297234896594767142721341284004376210660200550566202333951957558645128302417714282371577693287679784522757104237281724684454616224472703666186405467249250144671204565478357269214470538798054274436506654919003697852300703720061418372571130911168102172286721258959485746004353295033114695796564900624462269539125112528680378263752083462010919392052994067353165017583782632764100489944648907195082147841020836669964415554899731185444595312327881988435193646690756191744363874898626362765030272068609251880972360884793801625295039225521028318591195209521603087970922306318243049051361251785266783098964840762618711211938564523325880156358456632568153659741400635166538527833027993327618187316429284343663516156632552808774054766967000188148312992649497560617449945560484056516920660629441907471164740575561955183453874064046634465233320351037784476758856666090172875209822447115645045567243110919573466626779195035111191248491406455543525772549656639267852219176238547538197108319832284839469222949537374931725775542580964794985891026863594535761355146603751391496845122967455478424717793060749999248715039173711331059841824004577425759178667121950517428961196738988890457931260778363161297825694113684507888434449786640449589578177977537633175038686569214328456590763877035085454922881774592134644790364096719378848001565990852627617509773940164374064232153788341300502620171319759124864587923006850025338135489151319984710933979884365446048272891549710760373174451260971717062154915230935580784562839217995093664990440960915699421493871124291466312054691322386063352687404641869764975134600318428982574751202584459831039820140609484687076916183830862387891061468241972832005261503851106819905835176956323656969414236261015215538172503693919882029271368544401062943672645120333442448082952850778650223588668479872923363345267582654613647644880992776331655021300744945298954339617526611749016912130058109554244912267385285122343836959839780483975246451012058826742830639996089076557344363448014904882983571898164096451011928985398760882296922643542082445684123687523325897783852438454227592056760926079584672793866397640133375298174103800839713669875017925188890050961627253029063851224481562947348667180792167574395023354187971463703413595070688796031015016464474223923286729237172565384290147065906886729406948425427159052053409339014321081313854582597043485809314855063394187054375482019170226882317510901329413617187700580911334579616422030125980220431073829668649070375044538684428440879094580713838962095449207599615536655713068440469527129948783097134507882757248448998973497039622465110498777004179371263198665504248264504271322916123093246164135330391676894514835856927021667507452390254517053109406626366824745757346070406527575357774320102391334113813577503322561143900976099469521398177028414610884136085659183932993931358081952707069270927607721717779098763854634460240516904994977475774828730093397894064147369457198504829905134842867070991509330445796639195355891478010944345098270173677240979049480592883846575269082140684093675224888424661256052695097801012605772822873599379849977045731761175869419846747429048640123199674462226863633260076411070293488972360126214984596841603187424531584352054891859045356941964460633379884931531169547583611157497667740703154480578978170504573192258815494311439025793450499895500372704236182645680415899700136977093646184318296660690731354763645120346808804485446394798810546809849670131795494286681388276584446505851797712368142674585475571382290726634603164381375014654291945359934360820628279072531886575179734245746612250274430190922429627769366531587168094442427862498407494665563526804502768435782113627340694356164251537923224666458237107932145904446111092210703976045651010128697605935565797997230393938683961799189869179915938694708632424160101610309887378543956773148297234896594767142721341284004376210660200550566202333951957558645128302417714282371577693287679784522757104237281724684454616224472703666186405467249250144671204565478357269214470538798054274436506654919003697852300703720061418372571130911168102172286721258959485746004353295033114695796564900624462269539125112528680378263752083462010919392052994067353165017583782632764100489944648907195082147841020836669964415554899731185444595312327881988435193646690756191744363874898626362765030272068609251880972360884793801625295039225521028318591195209521603087970922306318243049051361251785266783098964840762618711211938564523325880156358456632568153659741400635166538527833027993327618187316429284343663516156632552808774054766967000188148312992649497560617449945560484056516920660629441907471164740575561955183453874064046634465233320351037784476758856666090172875209822447115645045567243110919573466626779195035111191248491406455543525772549656639267852219176238547538197108319832284839469222949537374931725775542580964794985891026863594535761355146603751391496845122967455478424717793060749999248715039173711331059841824004577425759178667121950517428961196738988890457931260778363161297825694113684507888434449786640449589578177977537633175038686569214328456590763877035085454922881774592134644790364096719378848001565990852627617509773940164374064232153788341300502620171319759124864587923006850025338135489151319984710933979884365446048272891549710760373174451260971717062154915230935580784562839217995093664990440960915699421493871124291466312054691322386063352687404641869764975134600318428982574751202584459831039820140609484687076916183830862387891061468241972832005261503851106819905835176956323656969414236261015215538172503693919882029271368544401062943672645120333442448082952850778650223588668479872923363345267582654613647644880992776331655021300744945298954339617526611749016912130058109554244912267385285122343836959839780483975246451012058826742830639996089076557344363448014904882983571898164096451011928985398760882296922643542082445684123687523325897783852438454227592056760926079584672793866397640133375298174103800839713669875017925188890050961627253029063851224481562947348667180792167574395023354187971463703413595070688796031015016464474223923286729237172565384290147065906886729406948425427159052053409339014321081313854582597043485809314855063394187054375482019170226882317510901329413617187700580911334579616422030125980220431073829668649070375044538684428440879094580713838962095449207599615536655713068440469527129948783097134507882757248448998973497039622465110498777004179371263198665504248264504271322916123093246164135330391676894514835856927021667507452390254517053109406626366824745757346070406527575357774320102391334113813577503322561143900976099469521398177028414610884136085659183932993931358081952707069270927607721717779098763854634460240516904994977475774828730093397894064147369457198504829905134842867070991509330445796639195355891478010944345098270173677240979049480592883846575269082140684093675224888424661256052695097801012605772822873599379849977045731761175869419846747429048640123199674462226863633260076411070293488972360126214984596841603187424531584352054891859045356941964460633379884931531169547583611157497667740703154480578978170504573192258815494311439025793450499895500372704236182645680415899700136977093646184318296660690731354763645120346808804485446394798810546809849670131795494286681388276584446505851797712368142674585475571382290726634603164381375014654291945359934360820628279072531886575179734245746612250274430190922429627769366531587168094442427862498407494665563526804502768435782113627340694356164251537923224666458237107932145904446111092210703976045651010128697605935565797997230393938683961799189869179915938694708632424160101610309887378543956773148297234896594767142721341284004376210660200550566202333951957558645128302417714282371577693287679784522757104237281724684454616224472703666186405467249250144671204565478357269214470538798054274436506654919003697852300703720061418372571130911168102172286721258959485746004353295033114695796564900624462269539125112528680378263752083462010919392052994067353165017583782632764100489944648907195082147841020836669964415554899731185444595312327881988435193646690756191744363874898626362765030272068609251880972360884793801625295039225521028318591195209521603087970922306318243049051361251785266783098964840762618711211938564523325880156358456632568153659741400635166538527833027993327618187316429284343663516156632552808774054766967000188148312992649497560617449945560484056516920660629441907471164740575561955183453874064046634465233320351037784476758856666090172875209822447115645045567243110919573466626779195035111191248491406455543525772549656639267852219176238547538197108319832284839469222949537374931725775542580964794985891026863594535761355146603751391496845122967455478424717793060749999248715039173711331059841824004577425759178667121950517428961196738988890457931260778363161297825694113684507888434449786640449589578177977537633175038686569214328456590763877035085454922881774592134644790364096719378848001565990852627617509773940164374064232153788341300502620171319759124864587923006850025338135489151319984710933979884365446048272891549710760373174451260971717062154915230935580784562839217995093664990440960915699421493871124291466312054691322386063352687404641869764975134600318428982574751202584459831039820140609484687076916183830862387891061468241972832005261503851106819905835176956323656969414236261015215538172503693919882029271368544401062943672645120333442448082952850778650223588668479872923363345267582654613647644880992776331655021300744945298954339617526611749016912130058109554244912267385285122343836959839780483975246451012058826742830639996089076557344363448014904882983571898164096451011928985398760882296922643542082445684123687523325897783852438454227592056760926079584672793866397640133375298174103800839713669875017925188890050961627253029063851224481562947348667180792167574395023354187971463703413595070688796031015016464474223923286729237172565384290147065906886729406948425427159052053409339014321081313854582597043485809314855063394187054375482019170226882317510901329413617187700580911334579616422030125980220431073829668649070375044538684428440879094580713838962095449207599615536655713068440469527129948783097134507882757248448998973497039622465110498777004179371263198665504248264504271322916123093246164135330391676894514835856927021667507452390254517053109406626366824745757346070406527575357774320102391334113813577503322561143900976099469521398177028414610884136085659183932993931358081952707069270927607721717779098763854634460240516904994977475774828730093397894064147369457198504829905134842867070991509330445796639195355891478010944345098270173677240979049480592883846575269082140684093675224888424661256052695097801012605772822873599379849977045731761175869419846747429048640123199674462226863633260076411070293488972360126214984596841603187424531584352054891859045356941964460633379884931531169547583611157497667740703154480578978170504573192258815494311439025793450499895500372704236182645680415899700136977093646184318296660690731354763645120346808804485446394798810546809849670131795494286681388276584446505851797712368142674585475571382290726634603164381375014654291945359934360820628279072531886575179734245746612250274430190922429627769366531587168094442427862498407494665563526804502768435782113627340694356164251537923224666458237107932145904446111092210703976045651010128697605935565797997230393938683961799189869179915938694708632424160101610309887378543956773148297234896594767142721341284004376210660200550566202333951957558645128302417714282371577693287679784522757104237281724684454616224472703666186405467249250144671204565478357269214470538798054274436506654919003697852300703720061418372571130911168102172286721258959485746004353295033114695796564900624462269539125112528680378263752083462010919392052994067353165017583782632764100489944648907195082147841020836669964415554899731185444595312327881988435193646690756191744363874898626362765030272068609251880972360884793801625295039225521028318591195209521603087970922306318243049051361251785266783098964840762618711211938564523325880156358456632568153659741400635166538527833027993327618187316429284343663516156632552808774054766967000188148312992649497560617449945560484056516920660629441907471164740575561955183453874064046634465233320351037784476758856666090172875209822447115645045567243110919573466626779195035111191248491406455543525772549656639267852219176238547538197108319832284839469222949537374931725775542580964794985891026863594535761355146603751391496845122967455478424717793060749999248715039173711331059841824004577425759178667121950517428961196738988890457931260778363161297825694113684507888434449786640449589578177977537633175038686569214328456590763877035085454922881774592134644790364096719378848001565990852627617509773940164374064232153788341300502620171319759124864587923006850025338135489151319984710933979884365446048272891549710760373174451260971717062154915230935580784562839217995093664990440960915699421493871124291466312054691322386063352687404641869764975134600318428982574751202584459831039820140609484687076916183830862387891061468241972832005261503851106819905835176956323656969414236261015215538172503693919882029271368544401062943672645120333442448082952850778650223588668479872923363345267582654613647644880992776331655021300744945298954339617526611749016912130058109554244912267385285122343836959839780483975246451012058826742830639996089076557344363448014904882983571898164096451011928985398760882296922643542082445684123687523325897783852438454227592056760926079584672793866397640133375298174103800839713669875017925188890050961627253029063851224481562947348667180792167574395023354187971463703413595070688796031015016464474223923286729237172565384290147065906886729406948425427159052053409339014321081313854582597043485809314855063394187054375482019170226882317510901329413617187700580911334579616422030125980220431073829668649070375044538684428440879094580713838962095449207599615536655713068440469527129948783097134507882757248448998973497039622465110498777004179371263198665504248264504271322916123093246164135330391676894514835856927021667507452390254517053109406626366824745757346070406527575357774320102391334113813577503322561143900976099469521398177028414610884136085659183932993931358081952707069270927607721717779098763854634460240516904994977475774828730093397894064147369457198504829905134842867070991509330445796639195355891478010944345098270173677240979049480592883846575269082140684093675224888424661256052695097801012605772822873599379849977045731761175869419846747429048640123199674462226863633260076411070293488972360126214984596841603187424531584352054891859045356941964460633379884931531169547583611157497667740703154480578978170504573192258815494311439025793450499895500372704236182645680415899700136977093646184318296660690731354763645120346808804485446394798810546809849670131795494286681388276584446505851797712368142674585475571382290726634603164381375014654291945359934360820628279072531886575179734245746612250274430190922429627769366531587168094442427862498407494665563526804502768435782113627340694356164251537923224666458237107932145904446111092210703976045651010128697605935565797997230393938683961799189869179915938694708632424160101610309887378543956773148297234896594767142721341284004376210660200550566202333951957558645128302417714282371577693287679784522757104237281724684454616224472703666186405467249250144671204565478357269214470538798054274436506654919003697852300703720061418372571130911168102172286721258959485746004353295033114695796564900624462269539125112528680378263752083462010919392052994067353165017583782632764100489944648907195082147841020836669964415554899731185444595312327881988435193646690756191744363874898626362765030272068609251880972360884793801625295039225521028318591195209521603087970922306318243049051361251785266783098964840762618711211938564523325880156358456632568153659741400635166538527833027993327618187316429284343663516156632552808774054766967000188148312992649497560617449945560484056516920660629441907471164740575561955183453874064046634465233320351037784476758856666090172875209822447115645045567243110919573466626779195035111191248491406455543525772549656639267852219176238547538197108319832284839469222949537374931725775542580964794985891026863594535761355146603751391496845122967455478424717793060749999248715039173711331059841824004577425759178667121950517428961196738988890457931260778363161297825694113684507888434449786640449589578177977537633175038686569214328456590763877035085454922881774592134644790364096719378848001565990852627617509773940164374064232153788341300502620171319759124864587923006850025338135489151319984710933979884365446048272891549710760373174451260971717062154915230935580784562839217995093664990440960915699421493871124291466312054691322386063352687404641869764975134600318428982574751202584459831039820140609484687076916183830862387891061468241972832005261503851106819905835176956323656969414236261015215538172503693919882029271368544401062943672645120333442448082952850778650223588668479872923363345267582654613647644880992776331655021300744945298954339617526611749016912130058109554244912267385285122343836959839780483975246451012058826742830639996089076557344363448014904882983571898164096451011928985398760882296922643542082445684123687523325897783852438454227592056760926079584672793866397640133375298174103800839713669875017925188890050961627253029063851224481562947348667180792167574395023354187971463703413595070688796031015016464474223923286729237172565384290147065906886729406948425427159052053409339014321081313854582597043485809314855063394187054375482019170226882317510901329413617187700580911334579616422030125980220431073829668649070375044538684428440879094580713838962095449207599615536655713068440469527129948783097134507882757248448998973497039622465110498777004179371263198665504248264504271322916123093246164135330391676894514835856927021667507452390254517053109406626366824745757346070406527575357774320102391334113813577503322561143900976099469521398177028414610884136085659183932993931358081952707069270927607721717779098763854634460240516904994977475774828730093397894064147369457198504829905134842867070991509330445796639195355891478010944345098270173677240979049480592883846575269082140684093675224888424661256052695097801012605772822873599379849977045731761175869419846747429048640123199674462226863633260076411070293488972360126214984596841603187424531584352054891859045356941964460633379884931531169547583611157497667740703154480578978170504573192258815494311439025793450499895500372704236182645680415899700136977093646184318296660690731354763645120346808804485446394798810546809849670131795494286681388276584446505851797712368142674585475571382290726634603164381375014654291945359934360820628279072531886575179734245746612250274430190922429627769366531587168094442427862498407494665563526804502768435782113627340694356164251537923224666458237107932145904446111092210703976045651010128697605935565797997230393938683961799189869179915938694708632424160101610309887378543956773148297234896594767142721341284004376210660200550566202333951957558645128302417714282371577693287679784522757104237281724684454616224472703666186405467249250144671204565478357269214470538798054274436506654919003697852300703720061418372571130911168102172286721258959485746004353295033114695796564900624462269539125112528680378263752083462010919392052994067353165017583782632764100489944648907195082147841020836669964415554899731185444595312327881988435193646690756191744363874898626362765030272068609251880972360884793801625295039225521028318591195209521603087970922306318243049051361251785266783098964840762618711211938564523325880156358456632568153659741400635166538527833027993327618187316429284343663516156632552808774054766967000188148312992649497560617449945560484056516920660629441907471164740575561955183453874064046634465233320351037784476758856666090172875209822447115645045567243110919573466626779195035111191248491406455543525772549656639267852219176238547538197108319832284839469222949537374931725775542580964794985891026863594535761355146603751391496845122967455478424717793060749999248715039173711331059841824004577425759178667121950517428961196738988890457931260778363161297825694113684507888434449786640449589578177977537633175038686569214328456590763877035085454922881774592134644790364096719378848001565990852627617509773940164374064232153788341300502620171319759124864587923006850025338135489151319984710933979884365446048272891549710760373174451260971717062154915230935580784562839217995093664990440960915699421493871124291466312054691322386063352687404641869764975134600318428982574751202584459831039820140609484687076916183830862387891061468241972832005261503851106819905835176956323656969414236261015215538172503693919882029271368544401062943672645120333442448082952850778650223588668479872923363345267582654613647644880992776331655021300744945298954339617526611749016912130058109554244912267385285122343836959839780483975246451012058826742830639996089076557344363448014904882983571898164096451011928985398760882296922643542082445684123687523325897783852438454227592056760926079584672793866397640133375298174103800839713669875017925188890050961627253029063851224481562947348667180792167574395023354187971463703413595070688796031015016464474223923286729237172565384290147065906886729406948425427159052053409339014321081313854582597043485809314855063394187054375482019170226882317510901329413617187700580911334579616422030125980220431073829668649070375044538684428440879094580713838962095449207599615536655713068440469527129948783097134507882757248448998973497039622465110498777004179371263198665504248264504271322916123093246164135330391676894514835856927021667507452390254517053109406626366824745757346070406527575357774320102391334113813577503322561143900976099469521398177028414610884136085659183932993931358081952707069270927607721717779098763854634460240516904994977475774828730093397894064147369457198504829905134842867070991509330445796639195355891478010944345098270173677240979049480592883846575269082140684093675224888424661256052695097801012605772822873599379849977045731761175869419846747429048640123199674462226863633260076411070293488972360126214984596841603187424531584352054891859045356941964460633379884931531169547583611157497667740703154480578978170504573192258815494311439025793450499895500372704236182645680415899700136977093646184318296660690731354763645120346808804485446394798810546809849670131795494286681388276584446505851797712368142674585475571382290726634603164381375014654291945359934360820628279072531886575179734245746612250274430190922429627769366531587168094442427862498407494665563526804502768435782113627340694356164251537923224666458237107932145904446111092210703976045651010128697605935565797997230393938683961799189869179915938694708632424160101610309887378543956773148297234896594767142721341284004376210660200550566202333951957558645128302417714282371577693287679784522757104237281724684454616224472703666186405467249250144671204565478357269214470538798054274436506654919003697852300703720061418372571130911168102172286721258959485746004353295033114695796564900624462269539125112528680378263752083462010919392052994067353165017583782632764100489944648907195082147841020836669964415554899731185444595312327881988435193646690756191744363874898626362765030272068609251880972360884793801625295039225521028318591195209521603087970922306318243049051361251785266783098964840762618711211938564523325880156358456632568153659741400635166538527833027993327618187316429284343663516156632552808774054766967000188148312992649497560617449945560484056516920660629441907471164740575561955183453874064046634465233320351037784476758856666090172875209822447115645045567243110919573466626779195035111191248491406455543525772549656639267852219176238547538197108319832284839469222949537374931725775542580964794985891026863594535761355146603751391496845122967455478424717793060749999248715039173711331059841824004577425759178667121950517428961196738988890457931260778363161297825694113684507888434449786640449589578177977537633175038686569214328456590763877035085454922881774592134644790364096719378848001565990852627617509773940164374064232153788341300502620171319759124864587923006850025338135489151319984710933979884365446048272891549710760373174451260971717062154915230935580784562839217995093664990440960915699421493871124291466312054691322386063352687404641869764975134600318428982574751202584459831039820140609484687076916183830862387891061468241972832005261503851106819905835176956323656969414236261015215538172503693919882029271368544401062943672645120333442448082952850778650223588668479872923363345267582654613647644880992776331655021300744945298954339617526611749016912130058109554244912267385285122343836959839780483975246451012058826742830639996089076557344363448014904882983571898164096451011928985398760882296922643542082445684123687523325897783852438454227592056760926079584672793866397640133375298174103800839713669875017925188890050961627253029063851224481562947348667180792167574395023354187971463703413595070688796031015016464474223923286729237172565384290147065906886729406948425427159052053409339014321081313854582597043485809314855063394187054375482019170226882317510901329413617187700580911334579616422030125980220431073829668649070375044538684428440879094580713838962095449207599615536655713068440469527129948783097134507882757248448998973497039622465110498777004179371263198665504248264504271322916123093246164135330391676894514835856927021667507452390254517053109406626366824745757346070406527575357774320102391334113813577503322561143900976099469521398177028414610884136085659183932993931358081952707069270927607721717779098763854634460240516904994977475774828730093397894064147369457198504829905134842867070991509330445796639195355891478010944345098270173677240979049480592883846575269082140684093675224888424661256052695097801012605772822873599379849977045731761175869419846747429048640123199674462226863633260076411070293488972360126214984596841603187424531584352054891859045356941964460633379884931531169547583611157497667740703154480578978170504573192258815494311439025793450499895500372704236182645680415899700136977093646184318296660690731354763645120346808804485446394798810546809849670131795494286681388276584446505851797712368142674585475571382290726634603164381375014654291945359934360820628279072531886575179734245746612250274430190922429627769366531587168094442427862498407494665563526804502768435782113627340694356164251537923224666458237107932145904446111092210703976045651010128697605935565797997230393938683961799189869179915938694708632424160101610309887378543956773148297234896594767142721341284004376210660200550566202333951957558645128302417714282371577693287679784522757104237281724684454616224472703666186405467249250144671204565478357269214470538798054274436506654919003697852300703720061418372571130911168102172286721258959485746004353295033114695796564900624462269539125112528680378263752083462010919392052994067353165017583782632764100489944648907195082147841020836669964415554899731185444595312327881988435193646690756191744363874898626362765030272068609251880972360884793801625295039225521028318591195209521603087970922306318243049051361251785266783098964840762618711211938564523325880156358456632568153659741400635166538527833027993327618187316429284343663516156632552808774054766967000188148312992649497560617449945560484056516920660629441907471164740575561955183453874064046634465233320351037784476758856666090172875209822447115645045567243110919573466626779195035111191248491406455543525772549656639267852219176238547538197108319832284839469222949537374931725775542580964794985891026863594535761355146603751391496845122967455478424717793060749999248715039173711331059841824004577425759178667121950517428961196738988890457931260778363161297825694113684507888434449786640449589578177977537633175038686569214328456590763877035085454922881774592134644790364096719378848001565990852627617509773940164374064232153788341300502620171319759124864587923006850025338135489151319984710933979884365446048272891549710760373174451260971717062154915230935580784562839217995093664990440960915699421493871124291466312054691322386063352687404641869764975134600318428982574751202584459831039820140609484687076916183830862387891061468241972832005261503851106819905835176956323656969414236261015215538172503693919882029271368544401062943672645120333442448082952850778650223588668479872923363345267582654613647644880992776331655021300744945298954339617526611749016912130058109554244912267385285122343836959839780483975246451012058826742830639996089076557344363448014904882983571898164096451011928985398760882296922643542082445684123687523325897783852438454227592056760926079584672793866397640133375298174103800839713669875017925188890050961627253029063851224481562947348667180792167574395023354187971463703413595070688796031015016464474223923286729237172565384290147065906886729406948425427159052053409339014321081313854582597043485809314855063394187054375482019170226882317510901329413617187700580911334579616422030125980220431073829668649070375044538684428440879094580713838962095449207599615536655713068440469527129948783097134507882757248448998973497039622465110498777004179371263198665504248264504271322916123093246164135330391676894514835856927021667507452390254517053109406626366824745757346070406527575357774320102391334113813577503322561143900976099469521398177028414610884136085659183932993931358081952707069270927607721717779098763854634460240516904994977475774828730093397894064147369457198504829905134842867070991509330445796639195355891478010944345098270173677240979049480592883846575269082140684093675224888424661256052695097801012605772822873599379849977045731761175869419846747429048640123199674462226863633260076411070293488972360126214984596841603187424531584352054891859045356941964460633379884931531169547583611157497667740703154480578978170504573192258815494311439025793450499895500372704236182645680415899700136977093646184318296660690731354763645120346808804485446394798810546809849670131795494286681388276584446505851797712368142674585475571382290726634603164381375014654291945359934360820628279072531886575179734245746612250274430190922429627769366531587168094442427862498407494665563526804502768435782113627340694356164251537923224666458237107932145904446111092210703976045651010128697605935565797997230393938683961799189869179915938694708632424160101610309887378543956773148297234896594767142721341284004376210660200550566202333951957558645128302417714282371577693287679784522757104237281724684454616224472703666186405467249250144671204565478357269214470538798054274436506654919003697852300703720061418372571130911168102172286721258959485746004353295033114695796564900624462269539125112528680378263752083462010919392052994067353165017583782632764100489944648907195082147841020836669964415554899731185444595312327881988435193646690756191744363874898626362765030272068609251880972360884793801625295039225521028318591195209521603087970922306318243049051361251785266783098964840762618711211938564523325880156358456632568153659741400635166538527833027993327618187316429284343663516156632552808774054766967000188148312992649497560617449945560484056516920660629441907471164740575561955183453874064046634465233320351037784476758856666090172875209822447115645045567243110919573466626779195035111191248491406455543525772549656639267852219176238547538197108319832284839469222949537374931725775542580964794985891026863594535761355146603751391496845122967455478424717793060749999248715039173711331059841824004577425759178667121950517428961196738988890457931260778363161297825694113684507888434449786640449589578177977537633175038686569214328456590763877035085454922881774592134644790364096719378848001565990852627617509773940164374064232153788341300502620171319759124864587923006850025338135489151319984710933979884365446048272891549710760373174451260971717062154915230935580784562839217995093664990440960915699421493871124291466312054691322386063352687404641869764975134600318428982574751202584459831039820140609484687076916183830862387891061468241972832005261503851106819905835176956323656969414236261015215538172503693919882029271368544401062943672645120333442448082952850778650223588668479872923363345267582654613647644880992776331655021300744945298954339617526611749016912130058109554244912267385285122343836959839780483975246451012058826742830639996089076557344363448014904882983571898164096451011928985398760882296922643542082445684123687523325897783852438454227592056760926079584672793866397640133375298174103800839713669875017925188890050961627253029063851224481562947348667180792167574395023354187971463703413595070688796031015016464474223923286729237172565384290147065906886729406948425427159052053409339014321081313854582597043485809314855063394187054375482019170226882317510901329413617187700580911334579616422030125980220431073829668649070375044538684428440879094580713838962095449207599615536655713068440469527129948783097134507882757248448998973497039622465110498777004179371263198665504248264504271322916123093246164135330391676894514835856927021667507452390254517053109406626366824745757346070406527575357774320102391334113813577503322561143900976099469521398177028414610884136085659183932993931358081952707069270927607721717779098763854634460240516904994977475774828730093397894064147369457198504829905134842867070991509330445796639195355891478010944345098270173677240979049480592883846575269082140684093675224888424661256052695097801012605772822873599379849977045731761175869419846747429048640123199674462226863633260076411070293488972360126214984596841603187424531584352054891859045356941964460633379884931531169547583611157497667740703154480578978170504573192258815494311439025793450499895500372704236182645680415899700136977093646184318296660690731354763645120346808804485446394798810546809849670131795494286681388276584446505851797712368142674585475571382290726634603164381375014654291945359934360820628279072531886575179734245746612250274430190922429627769366531587168094442427862498407494665563526804502768435782113627340694356164251537923224666458237107932145904446111092210703976045651010128697605935565797997230393938683961799189869179915938694708632424160101610309887378543956773148297234896594767142721341284004376210660200550566202333951957558645128302417714282371577693287679784522757104237281724684454616224472703666186405467249250144671204565478357269214470538798054274436506654919003697852300703720061418372571130911168102172286721258959485746004353295033114695796564900624462269539125112528680378263752083462010919392052994067353165017583782632764100489944648907195082147841020836669964415554899731185444595312327881988435193646690756191744363874898626362765030272068609251880972360884793801625295039225521028318591195209521603087970922306318243049051361251785266783098964840762618711211938564523325880156358456632568153659741400635166538527833027993327618187316429284343663516156632552808774054766967000188148312992649497560617449945560484056516920660629441907471164740575561955183453874064046634465233320351037784476758856666090172875209822447115645045567243110919573466626779195035111191248491406455543525772549656639267852219176238547538197108319832284839469222949537374931725775542580964794985891026863594535761355146603751391496845122967455478424717793060749999248715039173711331059841824004577425759178667121950517428961196738988890457931260778363161297825694113684507888434449786640449589578177977537633175038686569214328456590763877035085454922881774592134644790364096719378848001565990852627617509773940164374064232153788341300502620171319759124864587923006850025338135489151319984710933979884365446048272891549710760373174451260971717062154915230935580784562839217995093664990440960915699421493871124291466312054691322386063352687404641869764975134600318428982574751202584459831039820140609484687076916183830862387891061468241972832005261503851106819905835176956323656969414236261015215538172503693919882029271368544401062943672645120333442448082952850778650223588668479872923363345267582654613647644880992776331655021300744945298954339617526611749016912130058109554244912267385285122343836959839780483975246451012058826742830639996089076557344363448014904882983571898164096451011928985398760882296922643542082445684123687523325897783852438454227592056760926079584672793866397640133375298174103800839713669875017925188890050961627253029063851224481562947348667180792167574395023354187971463703413595070688796031015016464474223923286729237172565384290147065906886729406948425427159052053409339014321081313854582597043485809314855063394187054375482019170226882317510901329413617187700580911334579616422030125980220431073829668649070375044538684428440879094580713838962095449207599615536655713068440469527129948783097134507882757248448998973497039622465110498777004179371263198665504248264504271322916123093246164135330391676894514835856927021667507452390254517053109406626366824745757346070406527575357774320102391334113813577503322561143900976099469521398177028414610884136085659183932993931358081952707069270927607721717779098763854634460240516904994977475774828730093397894064147369457198504829905134842867070991509330445796639195355891478010944345098270173677240979049480592883846575269082140684093675224888424661256052695097801012605772822873599379849977045731761175869419846747429048640123199674462226863633260076411070293488972360126214984596841603187424531584352054891859045356941964460633379884931531169547583611157497667740703154480578978170504573192258815494311439025793450499895500372704236182645680415899700136977093646184318296660690731354763645120346808804485446394798810546809849670131795494286681388276584446505851797712368142674585475571382290726634603164381375014654291945359934360820628279072531886575179734245746612250274430190922429627769366531587168094442427862498407494665563526804502768435782113627340694356164251537923224666458237107932145904446111092210703976045651010128697605935565797997230393938683961799189869179915938694708632424160101610309887378543956773148297234896594767142721341284004376210660200550566202333951957558645128302417714282371577693287679784522757104237281724684454616224472703666186405467249250144671204565478357269214470538798054274436506654919003697852300703720061418372571130911168102172286721258959485746004353295033114695796564900624462269539125112528680378263752083462010919392052994067353165017583782632764100489944648907195082147841020836669964415554899731185444595312327881988435193646690756191744363874898626362765030272068609251880972360884793801625295039225521028318591195209521603087970922306318243049051361251785266783098964840762618711211938564523325880156358456632568153659741400635166538527833027993327618187316429284343663516156632552808774054766967000188148312992649497560617449945560484056516920660629441907471164740575561955183453874064046634465233320351037784476758856666090172875209822447115645045567243110919573466626779195035111191248491406455543525772549656639267852219176238547538197108319832284839469222949537374931725775542580964794985891026863594535761355146603751391496845122967455478424717793060749999248715039173711331059841824004577425759178667121950517428961196738988890457931260778363161297825694113684507888434449786640449589578177977537633175038686569214328456590763877035085454922881774592134644790364096719378848001565990852627617509773940164374064232153788341300502620171319759124864587923006850025338135489151319984710933979884365446048272891549710760373174451260971717062154915230935580784562839217995093664990440960915699421493871124291466312054691322386063352687404641869764975134600318428982574751202584459831039820140609484687076916183830862387891061468241972832005261503851106819905835176956323656969414236261015215538172503693919882029271368544401062943672645120333442448082952850778650223588668479872923363345267582654613647644880992776331655021300744945298954339617526611749016912130058109554244912267385285122343836959839780483975246451012058826742830639996089076557344363448014904882983571898164096451011928985398760882296922643542082445684123687523325897783852438454227592056760926079584672793866397640133375298174103800839713669875017925188890050961627253029063851224481562947348667180792167574395023354187971463703413595070688796031015016464474223923286729237172565384290147065906886729406948425427159052053409339014321081313854582597043485809314855063394187054375482019170226882317510901329413617187700580911334579616422030125980220431073829668649070375044538684428440879094580713838962095449207599615536655713068440469527129948783097134507882757248448998973497039622465110498777004179371263198665504248264504271322916123093246164135330391676894514835856927021667507452390254517053109406626366824745757346070406527575357774320102391334113813577503322561143900976099469521398177028414610884136085659183932993931358081952707069270927607721717779098763854634460240516904994977475774828730093397894064147369457198504829905134842867070991509330445796639195355891478010944345098270173677240979049480592883846575269082140684093675224888424661256052695097801012605772822873599379849977045731761175869419846747429048640123199674462226863633260076411070293488972360126214984596841603187424531584352054891859045356941964460633379884931531169547583611157497667740703154480578978170504573192258815494311439025793450499895500372704236182645680415899700136977093646184318296660690731354763645120346808804485446394798810546809849670131795494286681388276584446505851797712368142674585475571382290726634603164381375014654291945359934360820628279072531886575179734245746612250274430190922429627769366531587168094442427862498407494665563526804502768435782113627340694356164251537923224666458237107932145904446111092210703976045651010128697605935565797997230393938683961799189869179915938694708632424160101610309887378543956773148297234896594767142721341284004376210660200550566202333951957558645128302417714282371577693287679784522757104237281724684454616224472703666186405467249250144671204565478357269214470538798054274436506654919003697852300703720061418372571130911168102172286721258959485746004353295033114695796564900624462269539125112528680378263752083462010919392052994067353165017583782632764100489944648907195082147841020836669964415554899731185444595312327881988435193646690756191744363874898626362765030272068609251880972360884793801625295039225521028318591195209521603087970922306318243049051361251785266783098964840762618711211938564523325880156358456632568153659741400635166538527833027993327618187316429284343663516156632552808774054766967000188148312992649497560617449945560484056516920660629441907471164740575561955183453874064046634465233320351037784476758856666090172875209822447115645045567243110919573466626779195035111191248491406455543525772549656639267852219176238547538197108319832284839469222949537374931725775542580964794985891026863594535761355146603751391496845122967455478424717793060749999248715039173711331059841824004577425759178667121950517428961196738988890457931260778363161297825694113684507888434449786640449589578177977537633175038686569214328456590763877035085454922881774592134644790364096719378848001565990852627617509773940164374064232153788341300502620171319759124864587923006850025338135489151319984710933979884365446048272891549710760373174451260971717062154915230935580784562839217995093664990440960915699421493871124291466312054691322386063352687404641869764975134600318428982574751202584459831039820140609484687076916183830862387891061468241972832005261503851106819905835176956323656969414236261015215538172503693919882029271368544401062943672645120333442448082952850778650223588668479872923363345267582654613647644880992776331655021300744945298954339617526611749016912130058109554244912267385285122343836959839780483975246451012058826742830639996089076557344363448014904882983571898164096451011928985398760882296922643542082445684123687523325897783852438454227592056760926079584672793866397640133375298174103800839713669875017925188890050961627253029063851224481562947348667180792167574395023354187971463703413595070688796031015016464474223923286729237172565384290147065906886729406948425427159052053409339014321081313854582597043485809314855063394187054375482019170226882317510901329413617187700580911334579616422030125980220431073829668649070375044538684428440879094580713838962095449207599615536655713068440469527129948783097134507882757248448998973497039622465110498777004179371263198665504248264504271322916123093246164135330391676894514835856927021667

031906568999303527296694254093298585047142854083867108892734431087
363457002691119243249637729822940094402075202466206644738391720442
575483401453940803546273409990964069072075622907374684131198586578
798855914252147080351784967358936933535007544672324613125258268704
637563439646390594790703928961259764866705690718158460299360428566
416523782711376077456210903179808657173199934312817969129429620455
695201243315446083595765650783244672112585007760996890714299062146
372250183703199851692200508139102053789839156229925006468735679795
406627829502247459266561768688629321162565605008454294559032917372
009820858817353746878318206447022686127649760906343394156649026426
219449599972627879887420686537748400124027120252747334436166813022
720475458702704640986434675736414944556040180469025649085325167271
670951279005239342170274303288614132330961696475560201838621599787
236096212275704210996873700071225799429518696300412954188779628724
093278461798048647907699890577733886055838249114228316374869287853
814814314622428398496233785577350860511010509261293230994924741892
675131618818620753257403868480696490469827999122161138665663820326
019135968402515649247906443983330031807895236961651840561410338434
799376178182100474351909070654806403205691171664322099215471240461
694247104315222205104885366382704720835222072281792307724378621462
986066830039443184031897119593875880719811504839775086249284520537
660993570351138469597842959544064857515050620849712281233606395414
804523183037590392304193129358047025322396391151279375936015140870
535146062989951940745755354886861982269324695538210219472021248849
044263655395305394353300405061335996832016669879477207952586691433
189909211865226058496088149626836546723498404814381094319857403683
774663709365806373392945879466445168261148333459808948132301423449
736300081770030290154167401795744361620184220760535776542316284564
276440937556897155378728695987224574072558628891627542274001393924
890937205554561607842415395618839990558247974029707414788143572707
914922104355774546093490057376815968281968019771590645796605754942
541814453299247981996657119645518856556568121513349098046580376847
618266013737175683727413061704644380829243916537132963826165307231
060682333889999510255307327441209426994137920110104663208749715410
587445826115995906877240942869787594626942988054778607764580052294
075703084063851604360593125556585533123116554638655410300726993447
731181690786628011955322480995617127881556393519012560771128846947
322636357783025965642326973847446967479381719183498441201028423519
219594794118886452077960231343063217174354069248862375881350795013
019542125685389606693125427932944450460414399751105930683075898666
312089709178673814431062673285729135247335737334139964981622560564
197417879844156021330123029600421489114784857526438625181429619136
898382227494155863957874082378090383414087929764451188880233720098
864012236174176430506370381076927834118901729401329462385639308531
532244222768818571728893885168650075904844157097366365393941138141
433007084179831172343974321289826930882817908454608352276168839236
706701819377906387518688982678985926122487707255370774680909173812
195568154246781758249863590845775282210873370947100534470431626453
265098754187048260408504943289449671151155640450375992218315315062
189417680797400402678059163735949230580218910561624590130191307342
309220118774890541763475234787205095750830124580097608944902014228
321517081786382971517875410232779377854072683639518037227273990724
613281166726202385362958044768744894558935561294166787711514445952
150728011245289783898892439805792611339511716337272476322035613413
766554936091076077542886394153750911283550953891017569927348105802
052447498860656941406880598157145534961021164041299207119157382239
649687900615771795957511675590848192969043559344490289369648619008
661184836408445470922988082018696435010241658928067289346189212883

998058867547233822662874674898276007213969179081994180723645143829
894301735530322511478911846799356943769256545351140882462644178781
754058520462752091194551715071532232290031783786392997264579504136
224307388747429588420386537963008878284974910153671404487270605223
983274654434529356528076960477221503024060329960820144287406646309
024369558220175148190212769591948548065002515966276672115626582703
870721458114744697943698506765768823503506327908552602752457845213
931189712195956022277801052170563123963444611270021258672357370900
902467406949257666965304797371427626570535957448790876199299797549
154709567456866266693317816026949924563786133450401286482682564546
781442573431384860066826797672708278086896403276259416788405714594
114804629164881285200102610561984527854316767343019418002873928094
615194818546866310990795150448813365012546122141354279122839130369
472625897800273316800303122198079222723393023339699024932381363017
812136011313683103886773390012252524103762883920011468747397830196
815162344845844170847115405630671221502524006009566845020996544235
778554788006584453063354586519748
517271104686923708210374238344516214463591642683264697674851191611
783934195142162634345728069393696716440345231117981331367472558103
023300335647091931311841549327154187214539550089621519955776480953
224728170568462835256462754305762000017214990289645138438331171165
858417764510618582508210472454698908552605715433473640776605980218
446223023035764928698353547722866911672550929264234953591129778086
706764009141760472511018422954289469724271931395782511177137364661
288610993859615446366855566406324240965511527124487408138633958403
290683016637505159378894470032446794442050062323805678075861405161
794915675878485475379803613320458867201313746842227773933830101592
787382664035207938573727331878722116006978868491056887673735468207
179344205169186912073390602304035953144024848275524988496686390561
518129135491333068919805095831954745243476885587738310414344539933
978264379056007932577903268772389335970408929063798623591326882769
120482353226853311130949276614545789111470718167829879548739779664
939334049995804451114426378780550061812019428565801151160496597514
709360848244784071510165688191515476308537089917774950031546779199
405543023845556843295871228070339303655080520359026902225743269417
733783302077850881645997749984355147030418993109278863628555954986
664521514608322432702939865354711714611209786446464432561800235200
048641692809012519555879677770712014158540324292727965139982142625
427852383618248974400837646571991833956603365094970785232889745881
493347730994863973831820193244687919192275981928705919005907540673
571852348118683464347785054157828224490798584087005861205284471633
587772330438082709137396889505684555352326182211102373127136881420
583983854747049000536093865761777730487830785597217538818617272005
256836873008672176915441262877549503239222116487221786152262091124
259237746705501612558235461544208744584214964776025595727402918471
645315888525882788428685588949650842703942989935442747557848840925
342436759646126543810920255629982051541480193747004803579667745874
100884131811872582385785448488816682771330320509148049902706338669
469099351248792271005092774614149529146261051764859906929923817599
024416187949702356571809514425854995765179296210567446002214133673
754807986035969594685698337977267285287592506044418464357782403623
776670298728323706045274425550681518095558886835891475330457011692
173081904136821961979244460426499343848452732374619113711082446433
501694801550253284184628733492266134568944139002470663355470788141
605485679668656432669813031783621646465882083005036445481292288496
446164101250237576828218945511845805957320909394620648677509380243
791289601800827940474817422063327914712842095137010561943271762928
685790909493218055568979066286289154737548673198693738466419585618

899770827990016924649425190347377703988630149201118352837232791996
507771558163406176196984072223186782701235403499438429168556505151
743837735392032256115992975687965574856371399849841889818723863447
747735515013535079191081808269895360125247825470432140977846932329
438014725512584376086099814602199698631548473576292080996702231230
779999766892614937094813131176126725029020251125017695385831757933
248237475871601959893096899542945715238027789223568504876418241391
023269149165574444845124608305145787417507387176744173551101307645
537897887522221852058613301075912720128415816099012918291185666715
739299809291799149161000466033329225776086756566635965341818542256
059884318442721810942318109063106059677332784039505960669772102777
361411827206434298538442465877284718517883633752819325425830054996
146148373504141918616198629106706387911951762050556410548148530782
364372016539996520950423890411674976436029024341956762244685142089
456568032327777818280402353171727161613847452643425617030040388071
688888421475795728132416306939771904869321017748493439522088977977
4606413216065912354287306634985649513842991511193754156826880464034
407400319164875242281289908496049108875248189766295443946375362198
306085302662976776229728237088410640306798395704550798642556241329
569706903126069372722704973858490635481194665353366026431535545612
762002245436501409287070103845024942299682469894819311063805848963
834966842315468958573941781004780274310243436170639373226671414047
645055320685254527772713379959897191293351095335218376237814482821
238983318392269540362944194447793933862947820856539160276865474221
602197454162503479486573308748696362146050796657211558568264712564
019832368722180162737027408546123315314874653193936261343872784984
098267899861969122026697545901203347487171527203423784853201997739
410893991832399353436083831944835040717016050019719407955692916944
124826846029055460248055663749256769023585654963056305499094738338
682002253272751014765382015718968917643861760384669147127050209010
166793143538898179926138949582056843152542717333984584967784219554
228496937753821419875398362804151385327095150706778317618492686749
245187887825652388078755933987418992394106466084284159517680029189
967446045623476841746284361490524099614609132990782507767172994873
799300497060952190291591878815655029486924994826388639682384956086
850057999208125916980957625006325227429775722353244646312991916029
527828387808637437948487816007986691844944952033890911819968400143
601937093305472219047178616199521109291127917001976672020216396691
842224134953526405415034347849373009481075000992303709153882015508
200330120076040367400475028238121723533105469837101009657554881166
142817419714224111146539649408810687135316799674897427937226351358
103899882545160553672510213641692900321073122343007239955691422524
643012627356668178228590169033995255886470851754284397934232749253
399892584039232079835993252157435842523037436661389938632000800645
747508807093799846274709665708094369299361070027345381568480143981
800226497961654983924247214953855166490613388699479448764070625660
160551717891131051578981248406744154043863432180804960357763693369
650750249675465965351715008599750764000455954263701196268335042396
940932473254073217465365771218978633545568241703910378182426567244
157818438494538256203497811749471046589508232140820478205399922170
830963792471914357052689273788296301720459841639676597939924684512
021673155759406108501108401501493958481324314326483170638352293389
835732862962500645396532323409016634553497614539777543545510180022
729878166610572423124306235039912669272559398387044682244056902175
272089059731403171949939375760651704430817843584689023226409067025
582563156527103991987874499600566965311694201789033319307912876404
500245292607775735544830851499121604626040796635700429294141521078
517939512489293113108723403687549333211997169415582242253234526991

651484270807496498243209108709130271922073605282398890333776482440
248216436744892838932717872463012952137775840656766503422548447952
734389296263521706924829572233723726052121486755901243751068863616
862068481075325255190808700823937566799300052564004105686873213457
742011004302127479640462677207960288680754533284461163963670296167
636106120956409159039226759772561277082336910179793240276009477905
049390594990355097623285524569201492338038955511453694537989642439
077531543866107961725493579716448034461266623538041455573676426251
445905719258022229306403304943177399110774599480518484341690301247
105284001145301170159264176031004668798434006763661357541593810739
490233845959978566490063310019258076179659274890217308818654512491
561008456492191733984184936400789242534005288512740978260728184499
362334439677783416430328617074655574470958871066122859798438328988
827786089949825934445706255208466693362073645613299517534649909660
709934312563597490296567184680351887876443719273432284967575348743
806058683938732087107123411960330893060235219350237964753015141593
728621182295259067018575855904869810336131061953704410772086023300
066943559822989972003894205071241309633012473989889865016134460416
369764129918551398564133480244010903820420980509818816370765602535
422885206425047489586808991794668611719325033248230241059805584766
380455213789323057235020097155747602593772076776087468148213452253
163020878882398535566840846201887763338893823940059693823475558119
660440536601708515543140928633570921594481160175329658341333471773
027110597090517811589017086602990160512479450702430123230670262179
701141510682002268139997507258321303561667949126100542012864532298
006726890009482097075854102198848852954596019747306361328429698553
852265238161308896659145091248868131259535362960576603197504295041
188439397247053605789479862831714003968480764211909414275681202732
454233195931119523950562906226110099643989483816464487458668307548
577853287408199375727485219741371809429677774117222393641356033211
909334407556787838113045199845148628980006084838694206218527192801
877804248668080299512897034732944631709460038595125453386835579689
058465172300670448889684061086304061213515520387421392844962022257
546585820866986406049865542588590814553099484349384273384217864505
139854273974290958570085614625618349527002281417325367653979469127
529747013170063835415965446342449683526350594853447447210780561078
108296494264788100259793187756392390432917853276342037522975657527
434082950845479470152452608993138857831239117512692255667572885133
404397696254039311749337139944952935680106037969445956859752498772
673480790732676182452335521216214968023449292542886551457337565576
594555709239533428142462903172781540399834155641983771801898211247
608559555189995062073007140345208155033298149750702442677264360338
737539731484313740709266544922952042319900734594639311996535056807
332981486584110919944394627232845367711284847362246063313602859105
963523719387163459869636443906854053223193152413546932487576730463
381703029447983522602051814944458504961203269092337527162355133526
234320721943009358815033935989744933526957874572783140396703969100
170734149325310220632630169252370180120244226884929098195551171956
120838155014485836571665102690866487173238190148609924699131546082
001992705047305688761418932981083102352648281081024854502208757221
283441343794849997279202583543417204425984693274091417821439492801
799745659873698287428268267484421213715468232751128534165237031653
070432582833721123713760969593993754953622322221974659619332529074
042487602513819524269739101756371975343004479617825043115331506758
256273534347625253914251527570478784376788524187196634624199270080
257610839274976226365490018653206454995155802908398513272627321967
284783025385221904792786893895387803686993188466031043633524373271
546998988111864436701140284262026150473882358997472814933432570654

```
7052047163592144830709958430647172750763371802071459744565251418850
0503716377381890296968528440925819317441005557608985029232710160089
8257223817394543698529654769490187384046465643730713195901140761317
4282203883313602970852614512349073071476245340502454237636666855757
4012060405988695563011441544314169698607403220788600313279055391786
9674163555240250767653085632242737847149746037784064609346812991871
9028927597630701598408877817451926901476910300302349794585110001808
8660621628680111091511610409832730880899543375511871831737865587764
8735885449036687907416918253831330636233820582015488544982788382172
4243758054923815972964061963105821516709190326318730933813850113091
3327709303425112210972155056104913870504262198052602476116075059710
3794366477152935349171861250661163673833049956487877936569791129319
5878277774060107533372432790009732407256070388800871137309659634457
0960411750894447911916217455169709648447620461741281545361284226015
5130158952586280695706392444354780239051686133854986192665676227603
5823498556919224890690116459866096156795541549215758128354092025393
1708073781465078832035266513457070655368569928112426567084480177864
6149740454921184312276218452244108847150757019088349502243756988504
9454241057266246094583209596257287203099477654010739855806996619801
9534500506514501054924018861089134173060759238439461246569861605609
0945417729073640093913219567683643332996199979654243480265694068369
8670617587413167650506027135882574433707216458819522613397054520523
4412809231730650519899545028585383522873787286982917275807824209826
1060609015092521109089899643892978629430141114006762877499207817237
9474620916899838961894863077306027491026703883943352474243421236712
1770246101160240611185716870508244049162389434350470250813649155155
1043272507479410247374781922652059335513625530181219642499248821299
2601774080103719884212547447216029695002774268507752175866917993801
3625842241739856076699165432757061230452867280707631894705895440618
8421315713800339848740868094144618458542303449514485210539912489324
5548665930533558457827714423774338533920824127763216596753832665064
3063880695569156202622259469414290079987698344209146999879795683266
7090414139806863332515652569687878992257432796139643702654314463933
3799000819857530797881761337436094839283022338032797320386536462249
8055114022469226478931592944918040382983649034888638697564414855608
5605371772287371482282368642781136572039474969069372399156100368280
7514117847378256221277599591677581403853298688851365523231393843174
2258535628070191544007763016347647781261324380248421865669855274007
6410511901095452898382563484070455808090522147697220428738220206445
1116558020215813722046634766333517561799050089778088560093208145417
6443903235241939731547218920920202474270405857913543792600976807636
2987566147826443389261212134565944568010422276165241837640186734533
9148807900136360352843217478040168314672618806793649304686587364631
1483282698049202278762554540178509177497728294675692935451666098641
9481458364447890960383045828536989560806656661903603100453047200242
2639388157206867446900619298730822484900168163498212554375544758038
8736158658859248597558742783859992954170991722511534025422229692234
3661777781929653313496747757220978430193818445059850750805649483189
6792205834341478905102576174909975820012347244102850794123184376014
7142137829511021873589939170816868055988133546991379453783137711413
5491971243465810931127833266217592227589369934770498435251921073517
5737020054573451305873132214384620876027751840534893713237662907112
0552694905300388822804204808210851928761076772814548927944032272362
8412479431926441924307968137038294820058697420150132353198720519920
3538773142158822114859743182387909891993253934150378768995969580432
7251777689866125889395442013917130641886295106652689606241452084480
```

8800431386473171584468157548365316705126017466407607280959672132729422453610426679280946977358363834982281147669169596955732770282791209979260863817017654210151115811092683287485950646262362992592496943781822291692343254853616041525142063692521477459809419889268147764439053761119174630260741704144922419498001706238668169466079892475169718500094287444466241005863803558630650636095727097973493070964889076063019079233461701974566562487128849867382678546268945120946229054827542330251732132827853165175869340541118457151048346328320081011525254131193679545612612161031974278321038795482539174314098770652570060376719683830478692910032674834420668406673515088586091662976572277996926003868273349641875833211808091686185171345911568508931494044819961072502337896779899188725826867053775743512465081379862813483506643132466270924523654698351740704265136988241433088355263194547542532484588409399378463186733813366103100581146455151877305014130815666018061132268352963971146240104983148464604395200616373542584698260521254538591505364620141659306152413774330372464410397598500156892102790937867860318517672404695991572340529003816760724247701697815214861519474627054614221125691272253815905020395830514685850036024200549094550268261850958754262109739913220950295489582891742329813431540692575705880157365453517892741527128941314318269476902588854786599196898835999043885422748469473118751974887712150191591908885813263769981215908089691823350590797148746133368039706350036955539422590734536933263896961623842541047946294837937463813246124082530690269146502151719655242567884712722975266792028038412966157502184681108869393992526879439503267287008758188610493008597501969280952048237571389454388442416217566263735180336432774017341259427455820267544027881214092868804120300022706318082593186766481490447257994467045942382811148786117712471299402343531934763897789484462560417454780205295530815025429801570903923821472145748055026968444380960380164437261953004403990818426587487953263601747734396347413330292755818200084648860981832917078621243703459404281501604147705818564413222318877939216398896127369924002251605351282937236557373193193898341753843970830903585333085718369311365037008190565450432984220993814900454454004486182214055060528716831341426278964416953339802968795175366678475509672567390773471816933997590011898911139624652061607188564911926642694016069590616104437801498666998284332465257608825710108991959111180835030365079742812307093951984025480169446262059236363510719901481567744000123130725410256056015931684053281699733907071537208809013265415408408486946485133762274828499161274747053222550579183371978299802173759142434471486634380845991013925549446596674237301191215691048473306980508171297273138066292229678493165707003108305567889791232981303751031746783401120813530914660733675118762187223247896837038279501239544495396758853738446641347202470015885201761312154814372133479265672244691795625863300306994583628910381264895233888076384381880921180739792768314123086880946917176552671971341129027158146752944276252132950154301153369038922138410133788327539359202090519420939389794128938076595308702735507730422191143004325668462407228903402171151273168905339467851873822784851582247048412773926828056517260376445844006828466737354482592496916214239033889318117011389131501922285410738181452292592185148575252476183848077583829866667556762888310086015263589358635712742177728951544583129347098008471782896727341290370760715970897838993427926255738188956319405072752087325999456545608586855358263133708493406413934913749179077218228531160094982703667927892358788431608491838953362361576576833422289592702393915386811958152996066452081886184389697200189356785489494221651236101718473760460603829616442733117304025869985215372164775363206607206123292932396240780047428067808124542992795093505143976021192911303780

217039508295099069944393911955003407215741480177592299719328271630
397133938003492679133416508911394642479661320947250898195723750491
133129166984273711489586628649336133970303624716301283210070841520
262768996306818969526666944051675113293786905003412004284173942954
144584447541246771850871030194976373064645661863494065661595559022
203881357136176823477846310988547480314160751831305440342811270350
719599014203058814923214884129356944529643170313190357793773919016
569125771780698512180115203637847182328949534446551462678468202630
396663019906922610623240862574936674402658139719248013316060412804
309403011781839533845697347778755496505226719892039806465088098765
959904026835458290840488743590121024857071905695585182299648163245
701570374833507162561778784026024807954364314277409752046296007705
952321593783926560584763183164539460869129139029707529324575283112
206316530779455997093588019101795478468222029813794967339008709934
835236592521531747809149296359576803641645315741984082949704240390
235340265922135259853080332431785857740605690039517496624783114615
470747101524109270990190694605559239683056148077265373998499618880
204526136957223089073682258369725346732930434907962315838626126344
424086314103499740365508692160500462809793942595737077522421277559
627928584177313693090364594363844939823392918548955367647356087898
997197780703866209274631851985085192685262453825870249573805010993
816323870820228564474021890350370205158682457112552023153087809016
194370538717681644829394491186589768122858228286305889861029605283
254438604524159887904370477377371368502135393822653949023421710889
783710051749817597020686825702725971239990477731332089067421121200
821839867778705656294246938200423781349130666302875875209256477327
316746369664746257061200346715640903478963148186821345980612729682
565455767498591972867379753810674217385801948693932923297854556100
790005472533051670818549550829573316453038966728057387828058258833
867439567213201612236427190239990318506978094847298037705172000913
056269672052843166496490463110313400903781521749243406714466245262
478844479703267080920479660215253622977798413035291824916677547333
861097577175208925700627095398419276724681021271464465163618290386
785573635507001183672415428954047748343284348374756430452744506808
442943288003263534813266047601721422694003496285639702572562837188
280089547602386663065497470453498437664038598946458144805319509199
259451498233613653904412316740712966326187042419884078656034688944
098073565111719088416213133703832847527158014023337408746055090197
337083604720090201650957663760378637218832003863900863187002521159
432783932847926470562069061394036488309455202394372760011556448767
844754083561603998488513772923034323530097939678336983009127779979
497170462853141004353493338226748496581775235312719601590617282846
213714010305334392027034513019491070344617456577179219132414373649
693969048965732825756691327645310834456871594499005092022404757994
214851413924545725326078271647130569958137234896227241015135814633
935985739176240573446271578414318958068767448800803490104731595819
072414641729115982896970259377776750067320383367599920898141119898
575770545690088066735105347037213293489055455497205353010557324696
610538066099500702754752262676383165272527915483314536193916719551
030251519417178123912448276111453221886677717673417406452378993015
037104211023223884776937314255347983888883741324601694849452299051
071089558793216205632208515653007407950559249445974351034919044113
379156636661772461772342505771351604426630343126850879654103973473
927452905140551241030298362589336235834267865532047780753909090946
014043806764382911478302964651821538618920140071149455186269299132
473757299220611525247829166545756956638325114786725256097578274064
438046070715424617738733768071930679719130310725974717095148873747
469503491520242571874386619420798287288310571028157936402714634532

798473494139024208229738660143735485609080295713469226533118264067
965254434989996978086293470385536157001037969986461806802893285729
725290142829019704050111709396040179940077279073005957406453592391
666388114459764187426920093782970522564055698299485145116253966586
742236308393508148778211469714515626421839723245197256393926186481
618297743853241064349252541686264352270431372380466259855686036939
297507794006992109272687025326458806166692255729635224420485867181
204509342716663189051926245014908460068805223509444346582740225829
051295030497996140166077255644874614627097868897325672101929230098
245476543800864374001097572366609241372318680077447626862945263917
248970267674567927348695426329629330779938306069517425895653753303
319825137466974625871617196767115785959129389464190051959423116255
426558903604042611724998909306405095319996461071997051693828158842
491614923639600111384532591716540276495476617295659031279164951736
029421362818726459267767827689602966497224763374234588532554323573
191763216539724350146857623916565043387680021015694435823608634845
735010253174638070912058156818796838583945104236795876705668608026
594829523255942029210268875291154369222894717898363306284089773039
313699219114714816289438323424291014432554651237764299212082763929
737715940641397405209230530964078416316554355424023706082178182299
215828215654017676697661208991448060059529906543762363612026305988
951007723575565165921971451678367435512908451113629555953087586708
377800191996181655683100455702282789191613024036183516412231627090
454293979674128235084330945220216062668426891971808149654830769221
468092724377218327397092955067524792981311566988231593310969899391
225434479357745606611269046429656407405419288226478547979797445592
299821353002931553127176172218631970898223531161069406848815755289
481620820918166724813128908104623769907013229353284450081408941518
931109879654072146182757480558580243538161513918772004445850657984
792025456944114779663992297905320277130234997864403442182496163201
891811804326478311151594681114816472645477617634978713086688756952
976260303166684121463430262440794686978285987418395031713942900135
590877225577053738582054889531696018068632325879151070588932716859
422071560763890785976077655084303731907717669314552391351862236311
427116085445804084750002479987582300587088265491462207872672680636
374537598067416479448481497389238754728459343875279286766443272564
795561569831372462510454288850073348951302504367500924192903435436
010856692082942618503883972921380359209785263348133849959272530132
582108608715627994119122426533013518546830147588361727164882181498
350287788557539659788906106832228618619529827640257722566057247446
559340325286581601438651853736161141630755729703799944665114114054
079427147537781111425513397283496115287533038864432816081165190097
196055850403199345652798222234280729918171305432277487592302825294
802655057174086199252370629681498124204263776859881953959968475068
012031703913507600700194920848368879638514240583538606896803775442
070642716519419088448177490953805258104482473913499622829643783561
309374571147615093589594579265351844458244073484489100939557560229
105606244845612560043821227900342790703931871075238085638921136403
935835401058207371156598072563030772823585173136193707794908570998
474013366850020841298191122359596968222596454588321433935711948347
789268345889740076030693945402135727476634284866657596679093779764
676816301584881677064603897065832759236308140929955039458378138514
377201135323628926420377834841213595082140727120895337331687881000
034208405761803890377556602679235794264582504634285826405824664704
136419947470546611833543879107633145242000537808999550347417398247
970896323065778806615087882093271595756544842616883205894028796721
171980537283624065611890799913802883209434331124691268141432182067
023539066525496563617615132401567956891035488079558258253280000374

0724316505915305906941655244026442226553465708270971438428418562650322627493196295890247541653457612824093535406270144007091148209548333649386576628566250273021544259798453829481913014999381683903264080298234887741471682495878481818005552066910156137025327368239846402589188012033765009206477291158684977059128142539364633256149381380988193472958921912237302958434157505102177387600274257006713072132715585027183263165673229741655699387896932382888666047534639873633285056162548773854356883005726249740754685155054477920664951596322925802293266172296221390772547495226462179754608682099390454206571222320193765629382978088618030619528493132084967387233260394748697307938567850075940939478674254988202420415470667198240680737708036607512546417371045935695061956103385552732209810939122754665443071528565396608248218609920130748257076988534894691543197165697224766161617667065198873444896624586286251522253842901560874089954342176951754103353363403734750381964569616208605048641414584922244531855693921046091709499351230970681243367449929819627438739313209704016695375731733639240662067336606643546095104780339053198170587577123827378025247349903133500461177887772747607203425731481597071163454090918423175198274427763779402582949213724616142441229924935894614340088093891446700320346521898372728974239743797153198307296124285026965615884663944840556777668605819505314833709768685171863066764959789067888706831505251088021562700170042329256525775535714748807170330384064659421313222560288183751881800277819930167409986441062719982527455695968061642301490583712942032835260868134658206866112088199942560871923959865755990688479447989572847189711576011266352332118719974665840035802181340436535578951022409632306446703623964969102740141538402618604002521040701078265991504497189394637653897423395944478094125964467938810595401770617103493179060102858481794286927768025160634698186487615862337015251602363718178799903437959551150865145402587972642320057250444419433171201374682541907085310721520851269398466280177111026424456380791588249566743492875534221533673980070184145968906406936381641534693733580413486216823628124855196282073725171164471004843392677129047095449821771245997353794989592501819266700991923198151396872039240781317332882274061834899368915962588277943418125978223746823356556966161707037739424555386021304334936400753195339849108699633356133084982004622052379421261127386469773435298406141202933574769854283798507314230296086006458390203266832971066474859280280706222894119846852371184628817825233780181757362738656338576525114589769142517386579768133455331259610892167967599216200477777366017396981570051893055632093476805752082392470930750564865354014709692586332447535240023579514208088609935236904979271649529733073121762702277582805843309927781993804392172681117659365167746348586811523913603817239938043619137087041783245836870732687927591512421327930692493680672567164093795811543741008447431807984798184562295337831592094370585987930674314958421280917714693715998839338365967633328550845214487173498729828022872245972111003370678851957086364693267471590612320181128619220703781668982540615318307653843983956690719804851395299403331186528088123217077735132700970934377286450690552483201886405445912131117904412799490178046065119346213835182007919417118694779020864787750881187224477634397287600533491197235765406868201936808677188641543680800990685123604632696099408509903969483621715707186981568085994319275507836176310635228624435829757772397161712443900078527257874425455388265166580154504944151640169244989042683457345626173402057140117938253155396691786508532861602277977618284482128672946293714911091163510449320107800723707449634925691218904931357361767273707579489881987081480329456080270142457402799289156950749771877632590121855832116833245638324635426413663047442069533906874618048742695730825

892854977557044655307253726535449091655058544905009080736081330077
728805917621134078127022071577217991846919996476476550010097766017
279646966552244526699337699312900316835211790528327287495447512981
340250629138109549668572838779529050442889661515542282376017449137
351386273975389837428783645708577787917772282941837862509262813945
141402838181071249937085703496006797685633696864767208107940134332
689922920321079715285432230936878389824978469977558977758416212889
666118309718939561805589888734788223858629836061482819706388537663
870236997190568879142001500589327815766795541265615324077751183701
666268023336345078784136697950895864028584925040484933796171180328
045570796932170830887117625919959435644139714952826220987117939305
641225493818743814308084727553083891726934629489967863754936115462
433363099267699468701922308814839034814606660954452470388376120831
651420100763183375893493999620082480099860462830624324904927573357
191805826053249355127411073586030225305769820499460021459863975785
058321391016158208938102536640432838562215605048187456596545074877
137045502581458868787094158414236369582539050044102606727943451061
502477228062235096633157361251426679434140746652063534546583922261
146323557839399370970262019162213473885624539830201655471462084762
972652574642687233863623153135535568166360519035718952446642375866
198041144802521990130259589349990701834644312706805990043256138711
210450689055750679076762563050384262622513729065225654942654827321
349984841605395004718249856560184731449074269196457452658113570265
107162387733468833725899778643672591741155701655292484206315218737
915952464483195251598063394603742186735024198039034291728820585693
526786658158201195876069825239649292348914679443084786574974508366
962329607901893079624458748437251749229859571678125574842331673035
806137323872079009336336925742457891953867586676113412582471603070
894316874964912571672763444303055435879439672798391605502135148333
240544829591240121878313403045929441621292750499845016709464849534
409152471837782557600293979425014893421463386758834589638702611272
977948184530344423147131244078498214846735862732653628625501039736
030179463212553543423298346504287829926979525076211766041627496983
635949960324349579121030004621920146211722288584413301742909151997
456174122142548151012748826300946974835123007217022380251636265424
573166829154744421810027776127607297892368258611620666702210658425
698166382179841203878410264487735566199629191877992434897623591813
120793421262858085245471431809776763721364584043359927672555158766
872926818986466457737181567137271273270691744607683702302787924243
821403830439878125470062137965560074412906471083531304202892912152
146345225675356683968136278189995449676285459172951332462312686271
207307031686740866015957099166315864652238874562344964515990749792
099459034313921494067984830272254572430514353045490785504685756559
850314186308483486501057512449287567182899127645724607259255296433
846454729452076411844381790283425846409298456812175616938935965975
740429295571529150768297078020239789169692521995691674357951072811
013584487803476120458651798666327827480150636490227523192057452153
557288024819586565693666305469970629511442206125550456869148032744
860281726828102742488640261601893418136540881702751903632798220382
703820898351245577903536165344698442851334548386435254810831284998
328941817291612083907139068486878921381010367965044352512042221465
614364722014989420021438344201240970146623634385827269296108625255
251189212907148360456679559516865330531054510197832165741762682855
934691871492472425479123271529093087674353283875056316259536391996
859258022570941304541646785723973762668711162282093474137383953097
453856228395401248478081748766221827151570135695466014514527686148
821945134110598019590863768978510290474547665743821518113451025024
651282130788257311518251357320827342206955041620563142444929775785

418104799619444293639466973981359366911184515777799264350643544477
890786688566763655567510133020392064109944674869723800189018577123
739333851891151761529179033570357835979595544489785949895424664715
432770670827265189437837948064033422143377240478678694063066207205
650277370025852374393455794943717575948222394848213090905392311698
664096483735068307724944504687929063370538796371011704409037520385
264029625703004417508984780077854740069098315941092887358718953394
735828847351202477751441528641433301523760202571654663931351516958
028866473153834042344370579637648669804291749897294330983715121568
619296541976790875435900658851911949101426075596946536884456188099
014507149283559365332865878073446563488103298867967164678135738438
860393567548126731638346326081522267743361034256279904458071024372
908605563941984409751355348193689894438585629536549986258860485713
492088747953260910977030346501294040228602836133422350547868311753
104804378870682882815000408270466690072446076608132834861647163816
967844605155176165371188763515235927858975553158127353916830280196
887220255186609054347841455916346645871317346752404232051378466514
548485210827433728001348672450682567941253261976165988765239668887
392524079984935281412709665420506976947910909196918431017573132250
706433908790786086732900306398353516515980007405308268960241703702
326706976447029760260469815761395292209643118412199081244329745349
697140355250428613898420598735576265543534721231099641807557394304
984758358977867534082777525594534690720294929013861369117193813173
601646476625497435511923723190337566403995542447783944298351828774
246412327916622487261166153119556515318379303744830710564919086376
924643039253332818232102651629843824798894081615315260981783582054
204757574239190790661089061726333635275358836728525323566999568627
018308121674052801800025536899293369386788116744077729916542788044
678413562009362975545691518076713312963755314816579909072931041379
628475994177907248909988399465812988705098836726884172624560065468
114480637174295084587982931253382289570642635558951957479210047487
800990674713490817711659702787400484855846184436218804559755361397
818771100160012097380658520602274673984321980195069095331623045898
291761625860113602182893530594673628712855570420487403580738017522
416010536449207270031358736274465470777935266484464080671832023727
942014344723474168049821435321906618542992546902783902394693675658
550471520114175000886374437528098633553566630531447427801085599461
612399029562865316383085174797943117702616621107506672591136795657
312610790687425271321020893420430686442656219088980102687881677586
333919879360681779680015207386458111864332223002760660979237284918
816522192627682090188547803894435603204243307256873460617370232452
680436617589619974446116911030486059055319970563600863393574676825
493273026102929726522199970137847456683369278196032684120538725039
677338741009430233106594985975756289440887494502324471045141157592
838543190436560997944177869456505908673727940501810358913300780225
183670471294785839325894520337371154096525172614324582136510949286
496372790675667757029078810824521987372794038199950249285273634074
361722349146013133741882615777539449981758244493738170620495051672
294242853747166776178806443475674224096592080380341627847256942682
902292521252384858533734791367159246949973608350100908415999161378
048415807766179309191594718856909961023256006367965097758829543573
818629851803285928477041372436648951461450501920694469100554517531
628065330522640076777089331089867475923631142492119647379838595425
577854479230575068063861269020901325063079395192221338609185950651
259496621115675247703172613236327036342371720466923617527296437171
794845104623856042582227382057466941639979217811594135596464692984
830670920142577974238009352953076421311879630325003384984642860363
407249831285559398096274868244319578185903888755009025913675504375

891472026058362137766642009090107059305338719095934833029991281
661449330234883242862780945037445941996227719259121297396187159202
084731553469808229179455741106929562277074646829225064076988440475
901917347741466498926352361347170216506656059047328036496824398368
514816737296969861572310728805030865542054653459983217996813769385
218793864713741529934848299188089577576066975496074257348997204999
445924747765506558813098857791532212534825426618429008335815330042
553324053214212483604361281922172957838444800154600298950501182389
646309785607091078801026493793765650947187932258338844608895546152
946266400196083891130912856890749524410992955918508708629877492462
935098430541631892065189010221083687385076860449558367008844971994
147118076441320231950290439872981974058817957555924943246941547994
050095786346491078935958277278660074156011277589271569746691856720
880461585956897392923464608651825973480987627803273578312282423971
491807934995218964891498748911954466018464859793933432798169521734
810473007464831253068075519706380791066795589876645300045068524334
584450141943807605732159724289332940953577690289281327309503733441
374448424049652672649123074576028003390297467260071906965178930568
460857048624192192189552950120649938297904522189071154499699379144
340897775117026140825840144415357391257526878211204779567643328875
396972624473187903295250147526864259374561435487996402355502556241
286564188871149563905477942768587250411279911978525278635555326054
907754163728187879421667567485785614162304336382407650378801560198
809577238048798808776868188531550715521356754582486210745365289042
299172215641243296548596040476034010896079300019701881603401870668
767349301168067625375834101944939599597517992945290138196724876429
435303494038404548902069081145170358019381915900539925385499042811
644320959810429725600989398297814280181586790336160128834089182426
560769657275476319967224533077566356514988924033280178529096715912
155922079660592005166471511308401574719857331559508149313274438609
677728968456286694798136503135660944529386858762164723841947266752
648279780364616646165460093843200963866510587938358408749636141970
634373244074406175963950405430230842045311689261869054774551230814
292867131454787249367226271401948058072089560729540266035730001659
184622869442970735359279779135141545723636680561351033735941605798
693509445930764592536435012949183407465682210680243295447211988604
208897044877259785222935514414359586661710840576191296480249689572
963361908106473429291580124923341595072790860874734729777928310221
170360078554576950931364786905913019289750813682069374766244097973
906085819570369479562173390922423568787195448887076554041753864894
947001250532659549065911585793127048165934568764771061383451657286
356567308144810722413625202977151230115216366436175554153777313531
260736738451161060841513583749649991446718312384781726612783229101
190230569342678897476884530592788790534903105077614255429962938754
841751963997912067148771056699673222538913932234015644585015038358
971663177969650337608726334162566515216857077375304022479440398754
569309976118475950363021128888392344959267003790422257102806512980
545469553211495375827664971867066319707916922605079089218874076177
663472692934300002854445172960164718901564120699430998616936685189
833401054866295948994656903536055415702937086950368322581323917400
113357891228386438186621427639112016762850917015801323940797165867
875199335922606774097310151174226890239632200130165011170089640856
609159641996020206039985951675993361646087251610537383060830138547
011469178633538023410741772497712969498238933474609781466633088906
239419319707976848862203944896652185530855371967667509635988051779
747261793032691283182674862057809265256529034280190882705901405466
374700138330258893616017150154888689359105694404497217202113924564
847498885836600119303093340504248190948084678987600382766921817190

8676966018297820355992071897780429236183663066794572605188960538234
7287593282301195158655518017828313356583098613731652593503080582396
2559082865729213865472195779312243359147323752403786908253602164
6294813204884573987051367246382465059987311460376225349778221030101
4542987511382244821367352673721070466677797424225083858464887294909
6005218614774771519966024655226069621270620854168787239515997826
8612079223288649559947129439200099676084626144529668818191118094488
9919558814713799314815293295825160760107116624951556593720289303451
3965776120215825001792716582114749398982521641313983316292159116
7878631149035037080467291134675664573677603167044711872286977774937
2049841155253422836893646994865992816389029333329596705023331063186
45471594619892117469604407479301121141664974922769663879341360554
3628013565218859169191601265017034734921985779121559965545078845902
8002419792836146886166968729406732657944855141210293316476780504
3203023892916209496994094626811268847761941929888728302045325483555
9564561049750752610524425580938627342063589277682844228129581464734
7367073622559728294372637748776839928637632345393663445058464032
2202749734150248757659910215195448391414894528223832681483980519970
7655926889831544715769511831555106187300676856804057123078324335686
27813031698684340855208154148525567720882895194035878583720245447
8168634608411970360599469204205022389505099905474622524732812866755
0756059266570238196798544369738430363475513530101422313185966598
3550172909122663560683962724430840465236286528423238142349908676609
1580811529985673188270001310938215761913296593145799781199823426305
914119833894125626021521474556185978394081691455551300017513102083
8255631736162154590351291061800153799212994811409282642152270031
8799959583363724143264712437805197524452586298660913976536312050348
5519687014261642737394425538704104093401178690284831732444669643839
2741321758809800049498861502455683932136997226039513159990952783
7014548012759527925657066876033943012129119785059365743941306541784
6820219003381050360714952969682719340824995408242161639655390093100
7144638876203134913407756275562058777480798439458119495188200713
2052594033442134001243688746110151026668901067251610584232452648551
3982392400435721956683073657863463628471394771876051423987077753537
1198854467657293463584698847810339146678478547087315150429087424
5923946132610892919503710118254923184207210437942006267224493626435
0959550337583817195105443116873279605675329829161312575435682524564
6500812280522636814611597532119917066036435974480828856219322673
9368656062542938224900619956077533306524648443072734872088797672926
9681479748532637460821151712272174344612137262433632401184329188986
3711234758307389741059168181881668955953718239958310915895586515
5107939110515371592823919727238518934595024177790551515157264975708
7242876794409731453020409590769173875009613264837074559534153513313
4490003875280310133785644191147423502245236200886553420535485956
2097776348591737878062122554402259252738483077651799963738230090961
9759380174752578307961563233626646373083895738467111670927506441574
7632824210968186670214207763268375526077611108988944964673277534
3901491089616361841443514027645812228960506038155465315732310359157
3592275891134925690093560714797768870073195410228342717487574907091
8716304762388723409696302534078247465097250027224145260335082791
7050952440893757333312319820302154163507867782655093217137817123137
1611421231812440806330981536074376395026542557472387777479516037076
0254863489482815303352021941346466963751432715280802910028162934
4182641082759195249518173693711365151453769757463035503968815577039
3898348715498840132387691533308860839615203879265983426272431774927
6862696354131068465624484349311650896687484469347103405348229548
4404249445480290150087120911679176586065278724812579775347478803166
89003510874585681745497070497359957113398345543688948216100537152

614530056399124244539962580313280228578155567533705179614335935483
126071990025851110866754290799917035370060643683886576034328421346
749363284798134599509524459413666885886365358016564396724558897521
380576361590214842158335086871769121384576004437560818745484730537
879160685430938408101949720610599381353773087430936080254374618360
436914874112722209890114767287794789539517337789411265620467480077
129380534584076556161636714032813645067389621785209078922246439167
986043804019848760595357572978480805364654179647434163208018815226
299727917536184190105725391023526100666128579386128482872115114433
121748164404005618360186728632793253969282324260140007259899509705
126468119851380702881169297470365138827816180181143146872793323314
543151025048027376973590692969718888934545315353695151898759688924
675788779434750697095948389187091999856782139078432637031657987840
807613470059542315330631356291972347631112370126374371698836456993
918020402360170654847410448271684991511973761915708006875431889672
294915351919561931247877010488364900343071220216932716448737890748
697168890556697893495574826859945388159342379901066924553808660356
930934748516271997000837266625946091636911918668340876039816053457
187304488378784459945641066194649282790298714485138296622770697710
186797262748362345055249868433462175825351994893583892640008216042
578581501014868869194413913566869688314459862119923349722584933741
055324623722317841154042257919783776260572497686118088856865797980
045007436559961684910508497230735345599811561675513166841957334475
132922964429136515799663255383479450766103289684869876281724653418
103830016943452175880854960747849007567303970474979941243512618421
371502775616682736902616881508892134935073218269322806605182315763
054160723036713898748377933405015841980710583153109134517143956718
801250305988695063116544061961806470807361237675368395022824398752
860510511351814191973795694693638681274531652810635038346828357384
405514229348007993668210804493606862164600676634483031922495893155
954045816949724136805749636567933079196367152569807186187473119745
962143444787453875567679230213614859728630338941823745049891769740
071428842397668062831729011908574032503385452223540337598944008979
329603741205460354384182007632588093628697893595580850949756506799
535033458370680057239672681329462140754013128286368823074537737549
664487670662370586745285605262876843710312085978307077429889423331
966599337067247096895252498544582098143949212079393436556802569455
922608937043796808830592173549902808187239423500831667258629898918
865217070485809184394517416456009022597384114577622103057688690926
001961997563649835723155371164800780419034073499583084414551229428
177602152104935813182396627678715737149982841408036881971422738830
208745015739858523972510668956996234962835492727345082689120271475
257042705061118459464718285320545427233451594086986241199063270746
182667956011701275258350864738849204602785712850234784675339031318
786296649552070645507689584383918676790667856947756213003776080245
180827345252143205516063315040369941548606130675953365971680023592
160894781404681853314710609546833691621848065587071341055193617451
037101599216136199509708650137780533897593988366937873954822789683
336138162870025109398023076921640166695991773931224839082644791280
998578064053243409137186218853230491028901137161083086808748672381
158597421815653709296744655184929287500406515807192028280529922444
162735432238256495251742467157748268961216185256399893656944505583
510328185664131997626697439111677809487179369657369037614319223 83
476655463730888517770884496832380581866104908902147878769073115712
614123453649639008426994470802551462468087502320398977365460713304
328417053734413903893234842503181688302691387388458847942035579891
651177287920689338631682700432147008424675785294166046964037538411
912376432743868418340633769086563342469199412844660048651217780432

768985518402277518098790110696764581909532313211287890901497686946
981263359885419545567175712676678837155763126702761525977769925616
437210301084993130354371898992470093957528242987296922410711118009
672641573072618623927137157828061141430020171114264825848892720722
572122032771300998294999206516489224551883661421595958912823752483
606217204449590248575680996915586234438165167146316331004873922932
262219431953654657677812098448625420396083011277674528313406762294
368841354220939496071825983260013557530996909691637349665597450373
805541787929382030075698430766954390311872706744220840025986793092
414366290522730548733203749931817038992564161674496493569266521534
815077105717773031882845221753638550909028760052076444558521723466
926597049347060455327562960679060663231026724420595594961738355220
588146676672261525390990922373470014065613975006335694487509687715
208483235189794212322592920100695072008942754098085726809450590641
548218430923520598808493156199837134863086316081775309934313568261
193911313197553117878918315599227750248461596706646640528488092119
285325699549573270133662891020185295971510209873181846394428005269
519090433002235567601977870530906642648447231165417685422211916748
623508214055524163362856193729156026563037485196337983124937125125
846685449441793989395036130563585381465349742945202428280473039319
639196656867950978121194503715184948395534618570760753296976206042
126445050572723736239756574400005188506438255472818396964755953980
263000973297590967668112501920350862913964308563802687805424226912
067084666956677938084288797590663333851918316580733312886410779699
802773881216321646181972297993823279225034392320715570655636931529
218829355461902365071690769166585341141620278688145206333435182444
586927795980466279701274348819999836714161593004038549093477123216
230853924966881367722657236650480074286645066723197281356544204314
442205442795569289248165106790636054314227622982051060394683907359
874274419025274298058005737842467721405075868932757823896493302953
931675436558263682933422786679969086201880895596168898194833665683
085048883617646031216687630865949198367526097189379850864296154278
013157388422626909720754930123834769332599468506448850358448438659
834930060179435497954684711880581531834838854670090003623901656750
909851974382608917629705175162644632104682240045354860639954363499
876179329543924509327532211671851255361710720256583335435269680085
114896326771841394897101579682237123237779981704531349779534409755
378224041468783103118732043028771398412667428430830099637565151994
256408235324995905280936608017708154802977468046987301844281722489
133366907642573498314953015491797186038684886416413668203152784150
730160382876090478550704679705102972376240049732679654044375743461
492630923299395310213857334039465464313803932847546016682894570398
520289897056199050058566331681025424796122497994161923808081718695
617683355593082034329171118611940661424593023416858928325402281122
375144878383804375333860052459837715219808574847409115965605101261
021357390877587448522307072786751026934059525840044387691660888998
376929711576754644468668682258216315938415334021845430200445567878
616535387790068479648161154614774256766434666786165714321395605805
106852492930203486311187697906757373061308658049805399624983088516
297448323420918709324865566316859988821807634263361803803309370806
856552116308357042659878493962774257427221268652608755377034324279
721316247186562181502127210015338143527640446743840817921398524717
656064532716663053443606587885020021620083222433975474388465690808
482297610959890936151291814324694190177542972016826029260693628677
593944130399807004856247227074616381708027260932449104071616943112
721693245961346699250726493583327369223637207752705679531859869893
932974519234065803202897787347160231747227460805571629448861637690
049880287681783446200622032710191351233577036490080845554425576826

7109148155967365538582594382219513580156377119716793670206222456863
3485469059877803465897456584478524088192688309227770546049463556155
8414097453554239477673427530500839360542653385834292408867914093688
1442328587900922305291527999159050090651924744141709100477084998899
0303385633941944104864392190202256628773984626697129628687375722166
1881712782123563349047154994707588680405946324980613245053389545233
8627255245644008130368634687268413792193795785077498902782431583822
7323503310967089933015706641952401640480259861958294666114657087899
6993126661948681911951688997857138260889720165022115150000459167599
1831413795284292401626307590692810259150476906872272115545836998477
2444222110127830084214948985427598928757962681408284246202734481156
3776158858566660628707597984538985888539735915067846667744073558899
6211764902638877304631228815008241486403525126070075059042685515455
8730819944836142517500969198812147177694290558463647557815938880666
2395420189374908403227510206348923283093410282378940889895317257077
2781466983925467615050901133856273474822826535359010089528178689233
8513054833708012163757061868763785954802148810067207114027051724466
8180760760129942225757837851526237298044375489744603759100293991488
7767534772424112309983814691187957254447449604200948758632104967566
2261710980657037545797784528755506706813330119403702836846318300500
6680753062470647522436185743826977839833542129935793422513520349233
5503103226324224496188486719775355803165214710619958410406221190833
5966907897963380812136398928437310743158230935296944233518933301344
3708243264385327899082614976500246622336013479717464642080798369544
6513329453775576622725244620507851699672019899993401013335072075800
7710403826558364817000188252048738626947279776868198412449713014366
7444617738360378450588677310415521380295511482941381968111653016233
5598910147759480759411771570116508692185411471004985307661544503266
3624343420505278561286837117886986242284537112278047537319214048955
6991086183575809549838444566013084746527651201522516404600863882544
0892101024615835361898855615679545559434153937745029100661215587066
4016652981681454915048709104536360233696267900742045245107874767111
1085816038833504907301984545965754311516272250132882641327327200455
8647504003597538841150858712366767330536716517667235923474108220555
6620738971458967361244286013184480990099661941855126591403124482688
2150505096821172238621231152652300713165865427360924785213735015266
3087364504209708686975277506519538734683752316717031793864707833388
1254703056742160675957181244817241549006779553950339497434406511011
4026688323398138407299525779426860509803857100388472248482378758700
2492339217271606723237837596682806558779053366211090973343419979788
4433485265324350722533603987194609870643016788954090843383183273800
0955490568085092791321896161996636262009622696371104592305859857933
3213945718121849704692397468471194090316286548267278160233466412455
8675531537220698657075888456159200392763767615955391438114106410711
2803664182738485702131664766755240750094911376831891968759459457677
7975505054359139588476862792044160738994386605970487557360320761899
3992907847325701218938635124345040256111531106159816041062934721255
9033810339622107691013592064184427275725636244992188419296928450488
8655716808147667896609363464271937555630095802657166740606967045077
0055976452397530935669267213475656322310521709797267882011756733844
4202585858563099582772237670124261950140214124151701706257482391999
3757315398056518447478149788507156871414235362443327302122681085477
0827797568225700932570140588369636640436028018657355839449184790333
6263501554324008794445528858414945115640942709224065884058157027466
9906282629123540380245799757493270182391447626286719728006741510644
6157842608708921008843375777396856008810813692857565080184359309966
3392248748227346979480784064776775774350810648131138541202668180144
6216960634863869351822833302134661336932791685950307992469114783322

513082076691015178579716958660185792729967194005523669049880576522
883405403829572329495666610618163109789226399407301157002457628374
773576308183798161638119079736031587212198313819620344976981620642
398507522832773373258332437288216059788610987355191377855881403729
865083817511632667454759408345296967092628169980844889010443639929
417133550491721853044413612755372740438019661670623338297380240490
629825860951393502749589875052807206731837864730248913057905481562
736693420315336593480354522123759323198928884374803732410786208605
622462175566914066496560325627015246939470098514711600994089128351
947568274832967227217668299349444534067907365525417174886583901039
319287413336870559291380027727314775486977554119840138896045288554
164615529596730625328830411855501188829841586911577081853736377506
071335054326871353825913285097958561423027436039576496067859801089
379556543087594553873213235007603007730107917896408379918870861963
208679051158891374126204838541157848822039592359433270452459141147
350780545504033581903384706481014248689197256381168631933216497629
814849539646083874849543969458515257951809116266545236733517450163
608333529739655892210921156911978316729184425792575755004030749046
664227090664322428513245440807605030997258302179025792182169082377
943883480865654516580322492664139624733960511524759583935636607443
829936677311897124381434507144263490668336746267965144875760421649
004657804566810897623735246630448651394648231966040726712639274053
714806002028814119479141742109645063130559424100563987780307683154
549550715730058201479251595904602466115178393678560161253240627524
204371192506137964850891981909587770580992490842085600103886023943
155706409262296585461940599098696476507408909037102041073773228330
001943441313189898291146076879347947563792607660871483258647930931
942606503765031220789605973721194264580743339322946456217152992869
877572374421745243510593701522448697613048297775612559901751454759
543095729786727400721774104428762419527057513081889666990390251341
804618791396391751999610688039579457814086994355792939886140071141
724989460512023152862732692324098027034403935721351903458402887798
823381422660989331849667478892669887427451477269584954658077358949
663204319684227293905402342069936593465517594644019356968088757365
135655273782755845600317812578547257973901689044629670212624861217
876626477548370653350285918705939390177804647175942432362408204797
958379349852045596799025405159404945477441760839577973686737690585
564791827948871367856297647036197192772736875817422061934212158392
214716939705250724783165836869092120660602977814422914515406119505
865265141288187110649187969794316047996512884740365408399967083236
314679272132449362698570740704713058434826329318845440510586619403
755873250257315665275392902152362207415301446575930722750888326706
370582148843196156531041890179125548457874634053818750561404423146
119784513097056545369126825579444878042588295007696823218986262381
111072368744992142469150404040031332186682305915435954967189013559
031469691373622407761443103122648603764791939267717493540753416257
102558871122830098045539403337048732823588397650909521672656474725
802054809063989279095949756542290618460679476258129147809703755742
924808981654174485509672708892033514269360692215766563468441714251
418841300734634211770913388497281568269755648308409133263367172477
250081903833257398187374366036429341681541608528746458725565350762
115073088506880345696823291535406069272424136126383496243025222734
822076674328244625195343816866329833335123189264153753902689802092
306902469188086616534259713703199646563655178418522520774003775501
336093586141637044909219989129808172359620295384773165945159325457
512762182658826965023083274877952126416176015483903845869845831937
117209368199083288932591601683152378984175095542145657354244770307
986890738854236551194035529997686151218503691185811806196552576296

4582474395494638849523052293904319208347004085959062477586926739003838567519596357188381477340819136711013050785355589760197241318688735176938889609209708119960863932376114367610703567567519545098566772609283078303333849478963026102904982778516773496566301792404581418889016861103714921870755368544212842300294361846299784470990238920908702825867759979329002406874041289587818001551262385066983520761015258102922206918505898797807610317527440586023149271114853812525022120659086590178202587801042247426440305081493400889355369942155412427940242883301635560084548061142519736000794683676743928913680375452258525035643374923040611881347898649501791203952139272589351885032274797283045553778306485639221525481817233208066725276659643162054180594480709373376273851917793186994730245908116531708464778489793524433572038061469870092327006538735123781189355588073827614313871047732133065852429217695200986522155832270762404786747201853835783437702235042508142457128437565714796893103634180893732997220191856775283873360133944177263815622028490159604474228410691468157779660193578485360756676916433461402954821501762780794026532754016988343543419531698259725727032376723271043885021457551160781078490367273413756850547512833318281681806879234839423579028304653257090996989339293195685942215104275316137785891330262383995724924625652092380982163100220099385178387862912847546474997380846098849157178387404461981374483906873405912842383139922683550617309537760077334415519711344778011363671293049539990019837057605614721077458334412357522587981973201543708493212917688204045074494910930517869705671177011419656169548746956995545794664902459880581820522678280554400630551716201091706454966577280912312827303711793448888192757525099425680572045715600308835125483545318216700312020988817566560248156433398561210335218031222038724546176492771976075616398999255588448247125397544419957528755338160752382890749579434084784098190504995472234917678089989735554281691198660291050548190046634137155573699308168787927857366696462826628267844983767811270499161680058808699835076429727402925188906394922183187369232560399158730451177448334267912516853853293533405574069546616393992231256623733289130824466051596075681864137729049551578853288990615412343781361699480463908311094735416191894425279561023915503630620246485820967519520566863096088982742416026648684310185357920109341489959132940129700809792093387285245068367706352271937561210705529957700568521654739563093401459950709068494399952610995037011491148583565840356944392440181375872950668205071295542099185087650711181272383052550942709701808367332460479643916371196248167128781688086672286327543572199760279585502124996042934207237855661033521342883540324720884075362274770767226570322161459853084807014327113323279558962703353311330191389309874263357207481014426060412219499873952098204314094001173960150011637464937393337425009359679835495980816817494264116175890029744537902464510011573256811536605504921822907182084906673068373474135408768646880362810534636606363963705084006032259444236806840701090414791408074372432537542045294549232054885409447277733086772895728448351993062406233360148946570102953727693192398007913742461289880620940425107182168700354182717824693588528969368387123395081807112652032316069421394350441642211804586597284701987470587704917452361406216647182684227650430987221302532920806244101356519019242976831941842247953230488722276199981305934343540963408734149400983611379929010404902165198001696534510509320118652889500134985548157976486919975355400199623921830371557049566111407490698906146880316274565043092929685649204939142663256004909546724624506038670277906597782558864298724579515518278895492937913618643354949560747960629939433289895607101850074129740279027598329853445366547373825443050649218755849054725366313745370776589299887545318170758033901446976

2612080121923648496013191453758738420887032773631045444103175204 21
6817020249242468543393848183239720117336727143759140357324974219 07
6628395187262094877364228903555507099568064481385391213949440769 98
1765712217675877697741374769300803853092414702592673097243425147 38
9794333082207091984442963493682734555999275654308492145005895949 85
8388972194908048002110310107746945750302798672045602966829024330 04
9019586565615233850660108608524705125925107236890391442604480419 10
6885691711560554676557754131337846734357281913557921521252441634 26
7287935145798726236068476879424492454329593759322568038241243080 40
4415170323303546805930255459808401814193883913099131378039516586 68
8564053442504597686306846470385293042281253712888124063837191968 25
1315506404586582122027033445251500245556971852044942712661755709 77
9076522016314099243456249658234632741999669659190963031507721622 95
9737941330449069123546628999767870639644275439705307201015678667 65
1462562096649852069770710283319925355222101790715425549106689098 90
1571435225232088249354832640582544332289933881814332460776202119 27
6353556055401812466401180644907486567924930457530305406358998581 03
6430506691788360446337600725210593072101922922418953223829158909 34
1282287505109801463162395736714345765411805422274015650441113110 50
5863376808693783303841959694538613373289247160253222936973132249 37
9972928005992675567247975790953496341417002038663419614534418285 90
5495280534099977630879343399759789105686041339012606507861256082 32
0438045454486313325106637239101332430231854665265092939118054325 74
1690275954071908417628867735299986425460585144807032455850880023 74
9219853889982576356904694020369388274466818401399841773978038022 54
2914925919574927731783795745010729161896999328831375419807672649 18
0013064389905496374484069145828464261242049616787491900287079969 79
1885341262481087950554207435264417947194040344796510910272448179 19
0320555977587738427273813105814360328119623482496877066808719029 88
7234172895279364187064069484639801089564315352334229003147981016 68
4313694791961472060973085801361507055166513339144332448586946901 20
4924773255718230581885747169914636031877933013847597598611209617 07
1626757731572976133468570481409796915486001251242060298788553533 76
4784651123129289985200406105400658342503451634685630309405097898 36
2527739341235446910129362617019899713956710393977453100963054901 05
4483657021671991872203463575636382441116170937888938549393556125 00
9047936371715422408061034381188603305755612486733268456069417527 93
4409497025997743501467019950271079144783210978457155621888327410 71
0976530634168129793345934674275670724413469431451621670473376735 82
6853171960542817128588708301593160488414921903243630580503307219 66
0182809009404327179179905769932354438810503240674918606915784062 89
8734473767092342257822449454348280812656771732125804023869289361 12
5565308204506530663405614901036968658562063810884162112507243746 87
2189242092630265334764851006488240187531107752387990519147513017 01
1555125936509502776638656047599326777269553478063270493039489348 97
9598091177892565374432654393742278506271653261710049130127647658 58
8781300480840711844142690290625420067897649616996620372407499901 83
8392496230801679713342360699757087490626659373346349331174723891 56
7344412867677628500732867428934742984354169140694898144641854134 45
2472851022266000796138096075270104077275968926615167444366066579 17
1195189270912069311570060784613104860095902797295146546772338319 75
3003929982023067750861479378310340923251667930588580944497178020 12
4060753205842690453320997327935806562077891445512444048255269303 53
3035133590148144451647017174096780541342209529918090602912607156 83
3927661689267445611553209280009335523419347624811687507833375052 14
4864123004693591272455168409493435643592020840086639728874526447 64
1681222431979005740376752041131459356908294863650288665138667187 09
8401912652087193821461049629177160561124132622984729181973501923 26

4693476067359179204734601950214685020422272549906050039052717398308823934696132954605823559616885961443855177325568257200408640667157261428742158656293653466676503053994372643377711755248633346546866174710294742570747114524084069357459330581066479105872577087039412159583497208014347320216673200295921778311465479415676323580901593449839343603113947019602368064364710155265483303322903249484887401748716255541784323459835831317368567411948216488283219830392937820890066686416563563299900258912425367465987427578424453135056811633366503798136161033428769139977909395653758746991782052951266065435874802495310546207908286923825317348709348850922029899749216770759304665115811333830478324658453779995964224511055205282259855135799543314078046873828830918171688650474647342006016159445817592764878317510061540571533408027770017339348691597254483584995703116908905023479004182696112774389101113682498365112435221732941277060381302634181652357514835007798682610689357810835786811158166380254286394548824474315217144653188212265603371686438855279490837229672715058599839020073704352045196213006682618638971124575397983167483580286609024375336597037955301742860170982285254346628226050297822819229679749507068469470141114718271237717945439542475455823176370720020939524803055024861529194258380746445612475663032936211194064385351261733638375785199093033778956350709984872647563288157718587782973537054845621840616651638052116335535956157136554883040230839048499053464022753635053221312628579847488712087974239197129806515715622514535737622964969957851618947258601930480188788414077064277898211155038934041729907842887791396031909094756427746282346450496185589571867270893050509917508259060162303560854100261506599582940941882231711766856103431094090151955603594197629527151914465465701146273564760341664073355301078406612170688048776765829603445926238647585574773412285590994551261972615033569804674294654465241074799894679978464892745481446292704610242772494517285427202517075137258797350890288356512373075140216213117026404825133197504282209895138455281362688417732670088252543172435798898927526987160398846333076402741020725428607539146320563341900517801695481641794931734887812244472518611941008835131375554904941816728642441721880263881656275398573336004110595994336004451093825257880276648144257909554842576325666768618659127051483898044159755020202230844231648171457820102308761368940062171591863696647566801150589395069179486179388110691357391119291194764571638424398506767609270156013876254738775551308216149178633110756769969326398363601998430563988679303503631101462125926182324329202305048739735551038806183963033839202244502187780634180139029250800165476559906039088069177185244075096351519581930854853494363752693142834726012863215569558934759037521733395698605355818133404046834587120394074492363546970801353967029689205641532705761785074369410421620028138597409944580394843717122378085916106254729128941850143373320113941923776992728498792167957048572084826211789537450995901605731914510330159219504779986163997351363430181270766361962564218296135571474141968251977845129239951809480276157790515052996245646768594108003196552477778443101848753683776930481208594934751495753635190836645103834811783830402008724954329435980183990961161442520808102461548343772159007474654697078956825591781021535644406013968931598445159332120198273787780746052798480631885339359255326405680475871392736171274390494442064001760465960097267419950111801944219947030763868082018915210696080337311479275450469180708663306830471771276993888434975203615083864030923677079665962407466530320588557954353589859477754208465854466213117417321363819376111745267198453771073659565949816596887595352725655305225867778743619699554527008888504312759094143302364295198309281704636367105770040823556263457878747852344823998469437807398003882103551971467793 67

439484397950422615555306393729577597612139082829477616090698661599
524434740720825487274741637905412032043763264976757883944911594852
617550814823643520144900684490375525495045287112775902440268289660
239913812266687287169439142423356907167219748529854906502866938531
390027800622281054659491949657677340871738526222585319584765741013
650016883548356236992354401223596936006051229704848086706559828336
254616249888405808056838747680592416721489925466769707042797594707
443992401921713587689294557714824440483700282760938444366726557959
205333286363821183436202774646087176456018236920499752142614111949
509141365939939598884968732539054561689863096296277455693159627111
291389068753371428581683265582291531673412889027734335549344788683
553410612823002184662365260252030829905573599629412128403615848769
828447672166506050843093323577916341259867252410741162855560887417
648349820714209069639040582853918262162289982686959759493805904885
753681523517451496466142696587956201997664381005061504180068707658
470453477147005963307233577907943767064211961192058242544441864130
889629668960333915001324327960992277835339589184662575993194526690
242146365986846158650593407148400860403033855263822463815891581183
633596643738185621040582013281656985403167355638163019680564587348
039675160571644904016838278201603100326068032668396046855898129134
031175368012912557689000970360499259145265139775772983468530585536
936351824757233378044007504755143509075612721952284629606722106216
074612377151537118688504003714786281788426461390580536475028946907
239289094722636256621257205691977369329031393413587569782287912428
335072502728595632347802504078961201978921641323874369299169139774
347271497800996496729789539148727048958122750145899044623890586964
294927230354129335323876189211564588764429713638978164132213843945
580346265579131440291412501168851998922870799882033327458850878739
620195842849169998809625666397846140216095059729972870961243304576
253129268156432918037383948191514649529198853619766896498777534700
409893333797271594905193918030312440938121636064272059749937430095
796162204706746117408573410974428749024072224071920084911858181518
124276338523114088091933869905247375517969791533483698607788473417
923759000206964547789804654420961655824545657572601098292794621201
603586459098001214611081297486526766493775485550163800936391440387
470440680741730711149120395595564763786368725212586641996518155272
682610249104716189727921996372881405772954371894830012920612558250
088095864823435031158427250447144179924088583160443635426313119988
381503447473273977326572582918374248682532213362019148473697626755
507600478474750713026331527914424648458310542617927325595978995021
636498056801672170239863642215138491367894696651895996369818952892
920910915814558041583029638779178693541218300409986888887076505606
757845234883714489299580313972269250026344239337293778361219989460
046080519291815736507140605213243665711748651865109586655317669933
181738303448325237239280960676905236851464558272384358920906669573
835462780112429104142056474580713944479048166588098158783472998391
031027522874694740469677382116151097247127560918182160321327115448
287990220915809954467179102398577577600759370662369931528510617800
162228001306895034828243805988974280780978633732375367387515639962
500202688917156087205681980381321592713346498607978324698826325052
172467732321585052767727690807395180206332392022289351307434265978
605937025106926387895048939556321921166611355155598132690575754094
401663689426009267552040653336553951459594430336472986972522461302
873983497304830196186945556575297910677872775472113472308106665122
026618370236590083531181275297824104741768120054732854088244838854
668374142336505912599422868792294835077262714575470462006165094003
489129260399554319578326832004035426871828068254965253831583532577
307988741429846387393058843241116758545328754899971955023003383521

326423565271107017507937488068307856033254146019433209677063749357
415395330037478839909900702531462965980415264558977993948764754107
248509319276032948979171741362137841981035068496164039387135610981
878533506494822506753456264515252977403298927537561691817485375550
733716370480511310820927684935994530695581210082285314541817055339
623762767685236462658936773733428035587857812808211157430619791553
712435643547688811631808683393775827893152246419954930016978447909
000797664761987833614645661921975754528302389984112801986210384988
301577437087384108280801447372876668190323709674289419709340243364
458161318074772282133775375992468949488568872590487141814602376469
599508013866043470594351749860090523183122013945918488907530401736
869961254394667213996723140303493622862701101830211066751111569744
130936944850884308639209469638005567006340478765610370824098048678
842658505599647762752933451721794819545507384938113304238594644463
901683723440199071880860774745846502332455205724897116515037354612
483953355037071663354695583359220890033148110931050356252415751554
607393244446202438951629450718397676169870974697327731850083632859
062863381325773471767970860082863657847710142436557087371372940575
360685199619901423261535191218781832403826601040993227680387025182
826899050139287494337547628268055926443806446358529156983797510240
859940571555962016906118060638530479462781011636883711150185564208
324098816256980545241961108050107591342574231162743886126499208689
264393552121508479061673596495341792033572993192298700945731199911
697842268853665105393723073414833627765946108202750720135484799053
771977521102080214881391072844348389583374523960791312644616573885
318211704659936653431264959034724197008910572073105140310031420016
078368342775492638478125557268114790797901786907065870634749514416
252532134659135416115937735427112748784426401032091386953545141751
045683594010162267754683709086779176383299513414680468895693528680
453620097557985880107544175928524296410275443941749831975845436916
715453758318798583064671534276462601661707365201502412509413289171
747243577279364230528420491538431367186886237867006886699026954982
422348265355688667764379757582173536817241785261396212923528146510
190330402029598086319943328122202989858917413312941254825530968687
233116292184678213100262026568569686333898603114906825151840653582
620284920369110801300451065828997688939862230200298730202666823959
834337214834359411418680094410242394805971295162152859580318258362
458840738919247171307562713626974288333595200543374022971689775651
438500239796312208322968868544151807687575048509919864160038519290
649018781843282607380365794153750889224733309128902329783915701654
709899025909633775625832771152197699012720654276736343144359633866
983789906914273142987712102809813540389905181965902575287171101772
553598109188971805969066534622525599610871060290385068261037365951
903659809459038756802348958120983781845663284751012262558117615391
139727878696566364760387833095845869521297413602123039262307275831
620171532709809176029470213889754474404764535418138440232395192710
500836541126144987477629576646131529273040826246467017087921767316
215590235210339715859547058024228382702797149401860222887247744951
501920484063908977847063936837638424702769184371401132639953490553
916092843649937862708149230848515856910453657203421411183827241925
996098440307151328839084613953670714121052722050610253405101940294
074975957452717492953907938586063863227169758830913157754808342730
845003458209437567851176238291813322850072395652673288180902382192
834149414495655428426022137905886102004188339197317863254722606967
863498146897954811292456491956275748589910851167660235201086703572
062410419111398965080563101776254467899402821164892062993099395041
626919363285250565907122368264291345975000114381266244639619402922
612493139664600821783860242226340290988260707141310134022518229251

811450745324961179827809809090405986688873946543453374152928352732
068452037422867061801875774419308457568459008304866895218185054620
583640072765206482316024479229457650350271610240236048276091892925
914186544310797306158572168975813014599779416671685835670145627974
813776287791201997077337600915488505485437349191072444887826850797
672742474988775037169509964568506621052359813315597357709655906404
999570137621979292143842319021934015133733714638856997560257526096
919920416796982308783513389340972127413617967133318021610655335147
840122718050005605899625441087429177105963861488871216534202742021
940010898234916321433410966455236456415744254761628061499486226281
979471209953326569288357570768742314825654762139665761587018860883
087352063421381805508095387106264331097921834012391015587323449789
928640434008566433244035520634294570835086745978222019072043491820
981652741547556192053287163770669883912653893258830090785933097325
279803007139032546111667906126220914849586424631374604742928512122
584090588471531943843113310747680446329529101441178853360841472418
307882287955388926542866644843467401260175278300532377950471739461
989498412658617883899732766773092597723637251124093693571530993445
334363159572110004778061319562566494190266100292052756670249815648
374796640972093861428742821806717729444668642296989806010450055271
820474193533036594764842861974188173599121091811051783171735572336
204876797734979795164258297228610893435015799839631133567144207775
122452215944588812353931831789842776790776195747512520272576345924
105999269154185950946053770947153664423368160345377494478203803147
994524854190241582254730780105109221383043888730097415958976243928
516827241735402495335256497883617447651981462148737973733502013899
631749840480314174733112535768108772820544027530157949921224822818
831585990321764218085761179589830507631045793941516754013599164596
088966112035636072409926071387687035353083602313716182758794943707
880262354513499947100575161658408314018141609641484855569557304840
323932205248542084091772149915796670505394094970913094260358442410
735665967515059412976505726814953177565470672315031304636084548358
457214462467208837762651946049230729108575517180870401192629859967
437399667039842997856292449157836794560501938232289199784202291438
461928771033981179532791964008706484999273641610192982828364441987
022831823536960133729526964003143205504271571656300347801719246420
651854607568110387948042645886919236548593033626064402769482209740
683542342439780194853171920260633603021898499877395705143192427941
574283714669171775653622153838039556125883362532556198988813839413
151905940783614415697879733902202666436676056612603417723852733817
170074654328762267357799173442064014595759856058119852043609907487
862010633095050398949713531747581834943611833358525756392124646558
514617733143009987470829349366305014653167457492149127422582208884
946092094232114334628251716078318242748223680631197587626810722779
638741191448120760796135398449987832458778085584707914035804032279
332157013895936581773539678475775385919860590770257149851997929188
620717554066504414367406195975690246107524513634966072493582493815
286236865926413923632758445954235165302660337023066455584086230656
244569711087919783006102976488461105742426529547417648662520787040
049090179046710359849647006034864761711029493672651497009872703284
799059993478928185130602369007493095737937181386951682139546812959
146498623414918326207550263876824895095674867632026469345517551029
281824983911964679091823935241871555252286326831894208769977596787
361174983485889930089824631185447842241011310191145821330652805811
241230053589649036369265243691936406940486516075632836894857192461
337719895892533652652570482026720647698022098371415108748082727121
455265654004946322613711755652255785578543862048439727451281124698
930395385132755720873858613633284515498099912162217608194229832953

752884308497481526598950959603170767549866453741376304678326072883
851651589828190598366244240984123976754338199564138877339025561910
404340709254058733122719515004390733257007402291089271063985702642
339450723016625621780326505250808879203903983023905630409308301813
017261457073083950018428619529012573812442180643661159699702227693
367937704896765160022948925518417169030129907212012965013335062700
714227663549741111999219819664698709566640066532421003947145178129
100001780324540645368945014739497490056690622425714606805692549462
264794670488663628935046253209784701286810903059627837913196010909
078160372575988890915668049431931958905969736237831810429437253396
100728725746329776748022624482511578553027500586014154190875372211
315288767244349548893937126811823576507975737559186226095475879390
068550537922635207130175199884858141373912082390955291049480886320
773452653449560697377315653885478357543068230985809033063451846343
524211935900991772519327329122989298239984803314307134208898676864
918317664827645516485097831831275719666859409654673991686667380311
428772560547672156667644589756821784995803697938800350918275358548
373510238035096603225525565991415554441736919449621569243311265081
247949877523397160098964043204516324156612432501455034316605675360
644354019814710729774780115502323050776586429235572979795505513976
023219507014587792644147392121187155759311788108567349467436775790
869700486860076104855396740093966826692529948537691346709983406583
106232213642074997103676648809063665818289088678365447656052399611
687466435038854549657939336678299942212390575467578961132002146388
875142770428485161410379085362672854329928261900912400426930018423
089741947233718827707653645996344376750730597244890946843733502536
860175083172039515236001787907322728885243637033300444409278129059
345368663141470104659341883474689282629988236301306013766926988217
798851721245414573378488230382467191665951105174632431279031560874
148860708155483110213254013356868540558834310188708893876139373250
234088079659382014804830316448112317620154024345025897217767005259
876857529110799488761703346812320199323113219287434184661259987018
174656117914611868926837025201652991198988874948829242061696496543
089442346341753064626206632041270524790465222259474852629882180166
510377391520956925717676051391512907908330630891313846707678071360
829899189944905399843274940243889710601762751648654324350417468217
404772053579072978819030064762179565605159378531746997543678504299
622806859383605835065216371814375812035946389801353857890087538637
799944252751397164285764558538815095998654259961111212635252183537
375408938382994071476719479556565333810335609209165135879604317564
521490021087374521940701660790742117146389209287184760160902492319
110422671510290601789567464238340951983591142408642645711070748530
076249802206736383779844598841477515071622932192031026050055150907
697894319437834821122313171976968730832874683832939868019319165370
266382003482464988882800995308021917638041975946273043423705049816
862663146331381992449951350409336852132648622166261430456380155416
702997567018107991459837143013400320349765295216438577834202480497
460481356556278767001411676453276570915946987857471095170775617589
719547014691405289876238634466607521691840515292037064341671434458
101488124590410883667693696301612214043030796233418792780707414554
309612195098803307323271225143074674379294908470001118157872176047
256284368744402999034907235233647795614826072754304750733835794l6
952085411858141421163366331884361393046086404438120305008737474074
303519812587955651210154379618540181768351639553142978892109793350
644218922063827926017080859661513409231014455095980500497093334182
603462822266136524578624368933828748180808311663214088601896279333
796917967023892600395108849232222624879146995246944822132220716228
187633754117440717644082563597774910049844113158664565521693479469

9385345895276480298615840226409999421000433420644939416446515860822749727905680465910580231998140418166646897107038158991782599052443794164767665313637038164955688078417197066690888187111892963554097089449350838086720874087385891678280578464638730133563290056081755657051868983518288538558189418761846431885541883532205586551491960840135051091304296438673726701769209462568404821695594224381628363176054907299398382901877071378648219596279582737284384930210765170111412097127189513677811336345225119432564060929092039892030311428693110299616289715741651353122650976566387254150218818945769606338265402520174627484331378659366835358927288944137222712232373031899762712187563590305524059334406804067165885499108922339510318522804003163077793139388788124263739945766173505780454864709713365612269105452680323351709465782235132634711977541566480012164761891538394454382707413571109880273825243581929452706387243028983887862373972699481019099564763398726779743818824078646963720613457502040438651404814540848637229480891874953336845383329185692611600136090526980748507788080971992207905493846449129981160445105124820157681342303697584597931352507649917267189783592046244135583539680204390112988082084879270599398452081514555277160455545091663910861465981010943648552995834158940501132217591891822740788585450573707541719807935765713476422566400785520271235964984181478091852475405178298598358719450900926456203221456793600320098036589140365924802970597042389340141784940898340588942082813754108453271947659401578491808798841278671288346973044453463300113407842446974676100521632523146960746179723522751888911366108472825044733387698808997882496174571432653895931989193809453735620069779566007329207375987877395334012242628246381176046655295493277601516544433987977965759642130364849538029733629734054095365660271566620956242040189725410026903088730688596758323634848608003136749337880469881081792434870555858612604433511134155068347210280388630798842486479599344269107098070530828950651392898724560947408991150499159326607612639813504186421268398792438281063901902442716735076462402457681752412977343770472115340861680417829499676506858062512747529950655953249881866111872216992147209556065475521976145504910999069756842367521533929735597122527571508766565966450271917178205293885109369447210327912997299789495953721796541482204684847107971331529242256581065964907688575121283151575700891568839077515923394970555715543961203428758061751886703983086783340810134816839437033921934197424316334687716755401028790595185546970244107483690998853159223576836050185875678557363745857714841063401334897590877749058334553977057953521359016826647738272508556557813548876359883200278577063164224046839516165716965633117711645412249718086216525530845080263561891326043596292009640323843546372129515947537029341355782056091034650314827936141065660345442082853716002311366813190919641028730492085004174370383374462810464620942776963785580934275787598784183334039966019354202671488261281948862553950438151533608881983528174947354252961305205889894745297898192765362302146492716408632029235659299419175245477614408430602237931856760648303943416218753627041474976139638429863915287083114593851766848536922452479133970185679661898100702047221233180454192307994392215089918339722129466485426691824780579878782653881338779174799929862716454333930424609112847414100761105420089712585366728363148608986343464569341024174867567648864999320167607691395117456163032737449804460907809064030467634944431558869897372150230602240876896280899677720082954009728621969369799908562863781892068734312434251912571665860848853313223426184326055836735173575594624744942409189135202074248917184402231826670207364676861018624736484927580147358881296157116453077773099187906102988862871493020467922752516710370807163943912374316792866821934446224765726040705459985968287895945

818122960996644984189543550512697462222284055821601781563848932415
629429410235472447440652982759565085230803988104176753109539450829
566866700980596803972387830710888730991670839909866670302161465717
224784085226233338425720816810073396534603214984320697266393091865
149254801370103838705478495805692390809071470146803194411882916774
100108676071463670346097016587747938619865572514916032126199719973
803490164842267544912596739312397990074831055385068661830482906443
355681392530449017556754977224586553701311488545214557527650034001
289474274223755834032167742658602941502854059595734178734907098015
908582653022046578069213686344182383358550580440690789048769469523
016824226895303019503849045740947723785841308094244812638676254526
179071856678494915944757525890432985971556253916870664050033869114
702527528774632307639477366220502124317111976697554070733311267595
581143076643508377661383937418821198728140243019592577233992497745
653599173737048234552569017468386181605906850252368717229255820454
717814319918580749491682119101061410175466753076202891546321342918
722601569145323392446783536092923925956317992477364265588541429930
289457142976436732322262923602401555030564320283705186440270320700
941330893074078971459341135466306263658728571889770055691796392094
089540494967577669166831282615198053868579516388745693396126973669
872220449857426520785733934500552182495973648387278103946120544515
637979612030291659476574699341543271014074745772892654422996600802
191430751632012114712233628868911003141982697620811610237200462099
132116432607069198868028640972266780902380740359354214499157461979
683557148136771420102843682700410344318799421436138119770538705702
515776750087453539287747201965450490621594472377056510619675999908
569487775939149115942015050991367741964053191223539274975510275226
212593290315929202063227431563163988355989476949127802825984508358
367998620353352020685460559216786552835764981566953231585885723872
988882219155944803787090891648567299072137386053604371214396169103
856951761602847570707412208855744548038615549299960111090089529305
615092834665028803983155291889086590281766493385503602113010042614
046121856202729086358517057052077500603308295180906193350336573369
268872311459864004662237348473629802877988102147101924585493748777
453115962897925405501780747491964778406746552790393195565813896692
543928611681270286078016492471757947690040713838418710229217335189
894076408089714318830892216393659687537987014204003784913012750100
361893552864648042380140726687789499470242525139568329366720126727
746887603228486942873013499735546344984108290399024614311248528848
255524681487627399427149890898964065884653827774882015498940055948
650851084658197861933024860833800725503537057526726162627120895748
385707810716790396321406114798575892731652066513874141839901415240
806942716415312484146575073671621014372856615067280484820945901214
115397057048462215390455053205451408649083481693367506628520708504
476168704764247062925198421823405671193175977385071213843566161200
541291487091099968133185503456755250273948056094553333242616500497
427369923689595571203234581644450618398094463681201084189262133146
656721599470819817686659148832682318546016554172883453416704493091
663748465689767634231201898326438391034187584136241967457994649202
221979834593056563692756849359777671093103041411307312539564248638
501455500757943604266544947470225968985102663374383018153260704636
104120350698291007740247523365758424349259806781961067661254989366
947457932038348011891804623993440204860547400539729198870648908353
273846254259781523770165393409066396161418136993626227242206373381
984306775264803874177190613456070869512882942134188943261411559837
419843096506180799248248599557473975865979178350016251247911768205
661124568789795467228944116120724622182150361118719603867594046340
815340520931954899452801363923920455820705023281591771107908638599

4326625268337083516221862790696351346100018927287897223967334211224885525379496233480501745645714169688636010053871749288214974692896253474032490659110794774699550166290271429846508839179574390119154423166333872790505489315733714008430333877117939845502881051522538785585885276786724654682252601394142126380025151105253620202850883368116711791314535182745826907936214338287363671478550254061831507426381713513107673935765006518722579662135584845251998140046504966442936944626432535342270481087358438651531657478369349438175618439389101920993392079359173023513361343336174093788943324363676621020575206404986003394762611773065979007173384350861190466728309191914054876182490354096036111717587384282953107129788741300678157290071872028525347373683052683882088519006528889920671141417561482180485903016126993630220042457303654506308344452127181404811064626550218334918087281343170005938945464777807178007554115944795663687523130280968563849766467416423979403809780240068223930439751487761855101468074924443130493684240279796638069701072185944469466756952631588382852626134002780565139541647267978472018739287343174319563427146861286870318680268051307783311333649705142434586194339937603831348919536165221985717340060262681642333152627532561526998660446742821000163078713356756417605706103653972440343499640755239144597000424882780700901824785204769730606818272868950111230402025965464639168826534406245138943800868582630992637073830478363038980860109948994125751256140153446384423708749095624413019599875638910465209667545877660086590395215269307249475934637655249995739813687046823835782221350227515627717439223995541345490143078065888871451328133707614850257685232363829331474280596688096462099842247620743942690027942917237589747893279856242472965908532159472053323694904340279662663074027313164322304712428965781608109046022568044881972470679934948937439150755051735578827367466301133651280628067638738944351073404778542844945810324021530268892670928927343216222886653080791725525366482531922248604671904011881497966918972383904899214499063783422472582974487571387163937660383531958221258389950053175670095529364850788840429000362324607985108094470411877669656985270022423654214840823074249659128990965088853630872543273215141598918162875678113070516256851055815126713593448321780267835089604725800542617103328951883638910324473716748320591787336509628297455969434624092556528166566428133690259307587404400234673137376777924867261026258403688081693860941830435421605123289943113775339106511731742579190387744275557746660304066200990406304260514920298704318460132738950909981527030643369446904100445712022354511710113287564039593702423317102983934900820727390364959796732460701174416574343254996117806917646759647468797915155727815162473060583345263648512898167784698088189911321003939555111869683602326765781946083927775887735609407559829177542808611454330139500455246551242910049113728859660686718953557118903733300649089756833516500494824375020133685157284996369674642591495360373941154960982344314351093202218097093597803295497595988950811043501360621642003040542535251820091558762332175442175880859419299401661600036343910153400940398613816141852965918958274686221760040075402240523491448741154144506035042563623296960365972082364925594214765207713745747951220023253307577273544066672546063855660020024685704460037275403923296087432532813924489275962636999746081980307612158694436812543464760058234517098658868757896434602270548007083790041330514172192659415761568791150191340297485850517148608173156097398981871178896399754385938514812712285659202786935286076096100145004686282143308100288003423799080316038850406082976294182308278380860352272498102367705906046463477309524024902511871798642433919025304589573203908585078719522550177703765216266421852819817405073400266637251528093405208116710112696986779372259856933495190

432693201259024230765182777135271884472532778020551144835864447823
011547118441835229325114932572698861749122603284020727778843300201
824351288952626434850401801176692189400301384623039255957312898153
724381695307731589478556460254890123359844526030584211078366417704
984380422727756181463614970822052978940468419642105195952976344279
449380087623752745873654043686032392568120396815397806203184411751
734063549646449468864312900565992397103980260552719134441217649315
767012502328215868291339917094347218601990614994727041937223244811
365777364784334202259996962798552988234835813451982181425675924349
886313315764355498522016187470094484862457290141545591894888707730
437495867207924838385743401098250062896166079971094418369987478443
956767929238886241602443690271546527600224939349036905471674482965
770830739242100152832723379609356923990338824465601298007919176430
314202194237399396437444250881398720311047330446839944062988196979
371975773532541936499970332980309505730194490517681341165244535932
990515291198614709570353745265578742451856888960135130446546702707
588099460903301835695366013279171879444954101560343692286480222247
044767586960903220968422563613405634836829717434394913450350154562
712113070691281968263867332213184044441497703738450944461754830545
368993606820580388987724741195238929242163746784562498542798503144
932995331585543002766715402629626516695809146078810174714306991744
199865847329040165535665857626308050241495588477533489852364672238
934163656532479436451005902522586321364641258499846796161843552340
352324721110521226636091573602713021329448208976614103780709193655
802622181784957122075851190422878000874592867736276332300969043780
313708952520766671757271829986143936555118371669223725419466798082
166668111039566043933750372807554514848068166043674678943264045371
156658637505315120812713275492053068222000525692985014308858791838
338588827226166775683455460042038732166503756308540835999973834420
318792535151098838338539003290965874054873988529729683799722936601
292312307160205509733930936050345903955144350530779986167924716144
327074762450851301978973869927099332578952464554750676366826464527
152552254333880535483627391626239252966766458754894673447577273356
013838273729005393896656592230598571048482774398049720583821115538
200989209661369468931771991114747170373374826981059627061291313996
060882187721485255788982496057151197409955071399286692015456583834
310142603080858688493271922984158950926435718314092471047051845128
758699884109287359028743120393437627985164110324412262926311001109
691495544503094533576921409803315676548064212577277675625253662101
808506368182957928716083982340214720353625982063645520085231280580
032671686683448151104637370484997348399072102721190358008843242221
164334445080022597795281797172269973237438645179469844576480639489
491833438525180428786932632752902447890475937940428598452749922277
972100023891121548938382391382872989931731194761739061150447827928
7691102376475502522257173219481814737063013088417889819598162999541
083390244410692706737595956997119535930938496110286574076506367694
490893018558649870372897272343345722492789153260922324770228772629
642491769808030278236217239379885400503625715548875361008901145686
498282437678150512482820550492067614725271465218966300496885795997
677522593974060305110289858039626218819712821705192632230895174681
586477249400663476252399854173196026161036924195715977601971694902
399328727439746588043656593649688016852863977515522475999764941859
502680405006409698435113073797110441197918005746465493078021521252
981008731406046947356590646892418148391263600007362471055648198258
938088745764536277429937681358765419179735722961270008929684713696
493683678963525182303891310399263375859652579616496449908909552435
508658902553027859907755325901273060023553112413722883395464048657
778331615768298615178650924137474237208870130880543952259278853023

943092165956490984077060959426129628247967788111335332629528747975
409878835566787900429195451576744148678404482363922335095660072754
793914016971072318582441279892338820023779406397575365725162501335
163672644359159774750611925713016230090937345100474527618016380709
677370094376805966714229413589600824755383245974803932060796044905
017692070585123672619845895683093796806254340250957462165951887975
505779654919550494928671233251337556738716057356380028942990248851
218801240568679236189247556048248749553282638731464641642059885385
147743343317259129731711974000426498722243810614211032749924136371
337547432406296672518156579138643702562024303964790489005044298526
244657566236218820854094942368505732727377622836552938642131946178
526062604999062547968847458530441305937394727793077535078193557627
344106921558940727573628596944966388909215851327061017106149797620
538570852812095752763294985766777194759352152421676877868173437055
674237402436509635179971533020571431114640135864028290245151732610
767169202252500633762431074161787476243110180290133180972231123824
004466527025579134333864823384782408364150914263032146655473661759
625616966594331206659851276704614504355650567632319272380345140253
542128096185364065860659568650090054298405004609354853062062677047
658456432303555796213970401284145071551329589154551669286582783894
033915299223882332902538885726058492433074205048077496581896610609
181058545419793248020379655682803999261459692054638058771364903148
774404809112742816548241917457211102449742316156169247547908473075
166326609819523795676463876788253431508822081795667477147068016351
059647568318898324971204609208556997133757414465046934783022432710
038421476879322142485713565646283401032424128263276542081608944807
016915495419078899085838997387070670154166653384195835071697145193
419203744574382051040777297327360839324163745628589224133765386363
749550495430566377084345083651770046466463815328674448226290496018
468605036883440776084482397002567762132357213772691392393095925237
942205667698370439260789034826737347545283328576599177610125695553
501926205517993980215710312414311453023069858987010303589421288523
151506444142065284953493622024421560328509445445462874140740184508
573337343507763059426122501925255325129918634214765821403830797952
738737610527302639241822426415421509064600988318441525643072600146
861460116194913024036693824750171418942259202080670774549157595384
542378138860870217866424786028682455382570607007852827332226510563
344566490874361582295226450690960831695617260526525349150207041380
219034005701788311831237419981786872388251010597475120234906541684
015733501431783733524819386198287179971086117048195607925864281956
197702496700421100095380047388039200472454678730906292796860054268
202283888668402908313352076886505277918656290128921312403151147840
465000757126177971158769600362591788995845532035287764184783978631
665070737509669088361316739147668310680483001761136059412558390261
849754766696217285340358592190345237671511643133726006710559414359
332135805934319651546323178338090818578233195716802322563645435465
739653891585126961726835665295452993366536165073980298734018388646
124416365174666669893892482737826454263142720386501175530970761558
733454310267608916815162421264870580775063592788200735717780569088
816079843345659706100924240360984178262541720215278830719157976674
288514505877381337614484000839126439568917135693227613352816047973
256116480024364781339419493919981444634503389773048307901722189787
611415267584913782767136404814522241700976380245927541667269859014
203411158804515183793647076944899216519582332638281683336325513023
426351694440084457734264891932074127715509564322610386891038570095
852192162848418489882732686554704236675275075298431229087305419839
504408942021416678082196809827976707749289849712423880933541449510
829425629732782669230041011618064786854216339301274558922324247867

4916076157695411688302134542171596584609084801971964948722854229 24
9133226957718991065219268235209734285293627988609211679170762948 58
4749614997835983430872007004771021856897441261723109103558622624 99
4683902497802482310610773890804303172905984770452243032100330495 75
6965955759098089718773551327482963398864571877846910640355644896 12
5273514486823105300277818843106768143634883686881519793591948058 64
5183785865973102712078058781768283476422045804174854652725579259 32
1275422093550670915217460741863450104795444847280432287590427853 27
9892586453224298523386332572078549434410071304918160075095719817 83
8095600028747582755714595912142379824103442011990429800083484667 98
4779173667633916755981233073604499817833000271462079471539626074 24
0190517782696828793073342737263554559682051321475779688516552157 85
6382150610037574421068786981759087972310547187859794509341635317 30
9713427557368480465493684608589327951938780548353518384579551278 88
9710753852648125918197952271446731488978306681441294809043876475 41
7203288367931539487319278428206140837821111238551859257372026423 44
6466169852063384534060085526687169998825468531836845011643354224 24
6766319745613460084963085605745373590303320585846047421171983158 00
7289300135611575620717423148933047964474680649638129316429235235 81
1390296699494680014506883857295049880031742947556236767437649942 43
6129590188781636342231949340725849731738971847387427493550985064 72
6969684412652067805021942042876107362888938588503873245685564388 16
5788462809886618203203578230333800993059130072333413234509602597 37
4605200435709986002981455095784628320015135735925460273515967564 41
6365301122647127864032448240077379969176466506023873396656356793 40
3983565680722196540488511932248820542798091297110077004500227754 61
2066171669155913980976565982271696317371323802330818946438128134 86
6452495995445735996027340374953198103413735458599614954983609176 12
6285395307873845707594632930371488225193038171751154383500267089 95
8265452638110372525488774392601354060252214549198169957987371645 35
1325509905208796779944078225308077581699560271112775854486844027 60
5293945142888002909538028485411012261577841491581407749998414962 92
2401988913083178596669153882290099469474502478449025713673569726 39
7928304032860634546819859014808677414089210890401057657503110419 22
1614941874314587847613671477391830535314383226695453832992239404 56
1336060178214118865092922079294966409121600359051153880564921627 05
4464191236518908206532775891997389222939012002682322223697736723 30
0393821723674653052650743914068309474732126032088400989901480267 80
1994826858553514806570539140057693454713673203875772451306975960 60
5679539003726584611384511323064583372505805316793447259943055217 50
0853177863633981947217743849839416646214485505188770661689027887 47
4197775072785946167848196488792383924297012302195264384876917116 92
9419136764539897530221318944274689864451195233611358086995256573 84
9951322723448589323113867978311951784387713506482307870482998034 47
1550701418820531041426682294816008160950246823597889332394676950 15
5947575022359260204247226384941003113670440974536586103080120593 08
9275276107285263942575292843621863776425354278189930648006656963 67
2751616971819907226019375711689259479744761248762888217986501367 47
5075006383234798839649774004884123575666865716142158311084736091 39
3450002732005130798128157022256169065526833030836656381413470070 81
9422166484822104159343491908204056408595224038800378073492616503 00
2317179993148259291180037747446595015659399813862386928690626523 82
0612336236745964072098350770110829907902806903410917509635731456 18
2319044477049548661871606922803035013735952241233169641834879908 07
4808040868998221727551316195878096775216539898309620348940936838 56
5394211961230810211034710517424164346551719207792771385295060267 51
8642439692655367233447841006814559511490367828388175705353800389 46
0027690705631270232301414130663168017467973350972541462609599578 81

5941072780659665422853016083098094827980877799541513306341851977872303012663922253999559413949621100419546078252067442508032881805033938921875652444516995541376477841671637307558479723386593926352240182260803169276708468269071288406191974911765628699969084970730823375647797687484667530526919298507928036681821437679607305087380808301446429759825417007864397304961083418619696619596322018403591635634118435818598205141363191530912517440662404939092451358851907627068893662709905594646893766800692046828363046250164021027437917854480248512861821612512114570003573467406925367903689095025923989154817162254182524520806003906004061554058929077320687309580061749971920364712097884192446667092044497498748240886590526693588948775257516401354367423792014530722023535768345446812068659513932726355927699659957377744110379091071568358658465622085862107139549354539122567328062952751900754940904896394388806425455707262211593631243949164597256491018425754082200472288846634512803044831901784007401167647739561543647139523558199976935901084127752197336203083262576165968411936614585331142075321195292766971704206705185984249976283460412316390812279089005602391472762547230446561373891934829119845493060944629450961591153675552528626105912712122041446841774978186305011129740039411193508189083573332905511440743044446758533039081986774585870646675310587332044486643819547370848098401901457110801511114446629507460652330517345945257725758930786370071957679284954220239137265682599318384963573717455405038735780540832235428668250983407424619172124106592840528111662009232829603017213638492851047735852983920870989263169843588574220637445799561054144370524882233580267567460992544022776840093593181777507857673345320731185308379736957382024604745009604524055600641568355404686418106415591598692574489030347146086368684207141529519539968886399441629850126219826547899506312921479605647184999313392444952972883378335522530665608113911115575999790713828924183735740905193241180832753210575834434078628766429488113359530078115142595782796409283781276316746885253232980285679247320453209385421015807147401809479461160486277678673437757511437592333049254994572062768423364394693270173361084440187565356931607880312701567743292110954603746698646305896432996195799083916388510735836553973586858039475629404022863520963472170394504703852571085313336244754542001052596712178357874633359416592323562570393312801881979934876980850885387379015678885924959933804104507095668197806890979130475312701446911990817138057938235367271579787439956478915490640769381923678366723218190582136399034973143981196742174048660691965065868831513483401876813467904264395573859006548375807152818128951074144096045017043965485359053827804348135830772445160037860973737431472179402649530772942955247321642858586419313390462255573142876790225334478786885637977093220704754382447137072108172807262161922034516766385698542146002937106613178446743494946034274590970779402571198873753139912381326000956320636823628583078987415322744621275917935463121492158563100688909580778060593728283740660451733814756940668968790744643728903240457174689316227991526076700874957946365529810800605632053593234614913221150818691711550065566655477454978755929060742276192490131286777580134210840162883087292122615776509522101508044663744032978226504795483949290880913783491105270189158666597815395224336302038194307797221007492952919014175775251699297714793500137189964488911520647362966761218218393084899260246004189915466997385231967567293096434216988980634192953311166015202689067552639251081017259294741159705724670203623791445765730763105504694796613415062949874761664184570782385557437047474057201870953392723123250003365765441821501626602351718472672153312107509640158510189813774914265452998666920870889036949102304930463034175089835146799025697287651150445102675835608349627733345414395387 96

196861122862718377702626499995437897566186384522424473949249215054
851012216707555240510210387330028459361318458442786733823142617697
356364270842120283188436738192834713195087173122191012031672114110
939589992288467480174165676099378781968770763447597018787011536350
704268062034322219624818967910905627992687206315734435950789785292
306967195110433305566783849538409612527795838910548796848486208679
717493008452144359425346200112410842665667586897808277627684013469
829419295802033057400474913978971059122642210407325578913140477467
095216337310954671071478824347469732253620897184341401689515209329
372895796179900974531322807631818289943318969568953047237039953890
583965748355015081947010033649460754156809390948275449981181003115
143112437162060285082116771605290150303839981778749861963500489080
522089690682794915503815722397466511442040712132800560653624460196
857385732581380945079434740660360543591168103854745541390100521085
682696417436459269757301112314276156916406438293041441912580970015
014762604508430299473977704434060255848315518370986210437182444909
324499909412396968072735574990974754392902557984797093482190328085
059102331850565958856934103697521787966167710423049423523510863007
281287132147932780402066461426230078561408403259834892557120851189
853822385136209728791951877465064186101050110001523921401988115501
033319067153914966127363813534906201898801186062648881416943529275
130201207444850693949715656963700528104436457965400855804416248425
718544837208664333866575252285810948289217257839158191476913646032
684476202255833788430706626820136562567060429166096739937396333725
598175402369018835353007990159396724928774572310017813388850629426
776845236106426208547207080605367376268476687684621043656625525457
715582096848955125604270948386990045370602363886713679104249114919
963014756467260027940693936292085268041593916556942831013701721500
246134125553880321201748024661994057160259811420538497330990958586
477131121900577851682135465676925436958683955395922697911981510567
862427873863559696351596525780100887775161394859476530289336591762
402297065783698536071100495347567572284079338746963969820527548854
138638091280465567957867380247796245580749357238874918172010300891
988993237953392756249295143063917541756523620565537533747840354751
434991801696242127730575175317271408992841799710543799766469304839
985765697038891618026889482866498396473822403052368385787917654987
361628471601522751105553564227093034129063412140503747065387611044
057631277677687955828396936068797492992473055757014507128648776037
216713666399647951681218150895635932214508085348626452441380423193
763653355274835333215583412888866477801396224946024358430223059175
741552754477847166515158060159683143469938602241167033961033143444
142152378121270529004152968328358142745720548076341739976854032114
278702709946582145669614204935860051783203074959984999453677596390
154433298372959877021587984045304241723688539565431132491280016688
614321335901814598815345115649693087226879981544016379036258474494
027676223140583830246323278355589704912287637551609935228638759482
647092345489660404395528296934963273296194539263412540443583064912
727969941442577153786602121596283848008077648600684421195128428111
186065633816275879668504679093930302438194147134504446109962381417
080458893859796343824476120094314750139145110290353458464233986653
377503403288751178445621701907008268753712348942548452679529059672
899114162168717207252789541303662531312161687184002908491401088247
419290331003958533280903056898161959584146403500881838354477661617
640834335657628291603652785505334292017344423999129821560656392333
096831232606113498474590475348175724793522899893500943495075396373
482891154711017298440790711638488229884179218542831749857560164435
622264612259464028308647766387359459884245047099086787716750091393
003821175198111842564994499619250193938047253373994593337731252524

630640434299251006362772644042522129335363983888712558650282148393
519537829121923251329550479427077479817573069098139817583642674915
675638034024163503008997458846644559510526377303887533487344021772
585481657032600335620490417735579097347598439475995845429765463674
121075351507013851212617101709438816386818003253445607801138931545
723258763168759141418393365682229624660091462055145978337911564647
926663543627823302548582197820709974731091603511064700974874000731
522287664739629127786218446835550020204307191420072846279018318639
787025702772268782391036972445486411058889166911105922029444932943
627035413309880526880087934617095630484602882766011907089407300282
006643598669430991288386695237929866628178898427269704888604473767
609420261537717791790967751278719747104408091559906791907723417208
080859904286004545456751422771384737823411005311824430632388715284
408626875660506972347847736219620237658441103372159043811894698293
130692861159856453139893139948999833404400924228379125175552174621
891287605139468988472677187466504785270664362574831916908491537125
880541454036326747869539649103724005746130220293199503101987750602
880237975002552157499644642453349885915909369543958084528045004936
639830563782254105626241683217301132374663650818321551390498019391
996251482034852336035202979892243773111150091658570103260035364444
751774246989158935734705658751497625632680396958169694903975994610
639763432305422721308762466857346704606223493784199198380130993928
023652274191986054264249711792282050370537587427136667271485530946
080779629080935838546546834984036355521684570343035006341023502853
487766353047125068844087232667590565579334784591133212607890192869
809935963677578312895726970428837993551303926951240589199844906046
319277629905646039476875652776188987807508202115485364253791970754
729071126344281360599281191735709921525555198027605603718090518902
071857735055523271391396250159427253930237186445017661783595005367
424528353346296600400846807272853318083527248634331602064968738739
216160955927770747209186383619128575571939484457227933909841306594
059996512638479997332898272447135236300117319458797298546955749 66
414860678319364121215726453407658070668960253185401478248747972806
312019166738072237763920872324754210132174021931705246883119566130
365670703052191237861779392569076272238477050523927062222837494238
143061034403779823821088774143963139015107082031275456607954643371
353459928062871969469725559248762340560859976025423380535602919869
909560761373682770704428667046412247405699674920985983836128293650
677449802452216780957009937928110107393230867893546477556514877507
479466500875692695049132556126428060059883949951551625764082779816
057275574439612018147497800451321782129786374375109974767337631313
444066932216989790648141522459605796036374929385390580458098260355
681952893952216695741564220430366437229914696760438644194213013675
900169324226934913049246702707782481845523113461103453489273315060
001230285383342303638247155025551368745632166936656044414642455569
818231927411945082796887046941767029660195074502498516529806186805
463813475251438433278991996092710858122008935766222569799359869922
149925464717682101276519959597824700501422174754586194296039239459
092882841814687748419136141898738281264835534324101696493465526295
345563461708335109501680694022867505677634457147413217775167306207
782187706492244447520828000983425570457784919191688417677178688630
332126895491976457394075370098837140048702542926032963877928754377
069560437339990100294852638150032628997285511301036985819326744885
052228421918054554082277475276074653898905064374798169834717774907
610376006282890619457639678129928802002759672944986154770652047321
954189029670858378060569502688599152022861683177931813811133700846
648282316429401940350166748929939240870575647942513199958612783309
735265384335938416752475309684246977819186243437711155992528282731

329513697878214074254516312686452327578082174391542146889435909773
181565802205796404419421225370147991892788537903377743287340955117
413783520191979152796500139386884855693747482161292716729572785638
473863246938584052924674904022413438951883238010793146018342916216
331965737001597942739004500063841651314517655908597002647003221302
219852249741391577952987929096349728985117601811374469220942531031
313834496135599318178835441647145038785547166597698246724797440311
660606198912250415690904476466245712836382061667427564702775968974
627841810514767013580425903857530620365729378401649166948271359285
692735430676917886700049220273231626404070255027956209349621622733
861948681106084493589560178708588313384417288763890931537407240007
280253256276428402648656501968697974430425922584958047417922792534
005525247449502340839265617239093094230060936630323480202108678868
089659181684792736833014327146956844570493654212738523641976274894
176037602975201615359389448762202357391354683427259462829509057651
431942095955160726127413535983319184123578419642134288725668738970
843831141046585600376882203246308656515410799296469046577706523795
053459602461494020615605448430643787299452582262636091970063423456
958121081018804294851368286739852132534519851986806527920161753896
561841182522425296893463498323878626573832248821467182212392161452
163325275670017042893990524654825877851241251856125788689455531665
497546430475350319035590232143812858179275339840124608238907170546
058359605867719902183465283057186827751076250665370945298883021196
273029318588927084757014856899852966505733847103862059963894320941
335957796447699221415378655112464853794392540736219275246848238284
997312571864576555150869582415134979815701743782733663799343065090
606498092983863033353942502182256638127320974354662244588764349940
735538863587706720633681111322942298365405268821561270245962885723
548264218314546143319125458331181255979147364840129146862219867375
818977195188232785208093327828052850343881380195284554650513932469
026911560267684358544393507628566726126650839453589830932080370010
789324365829155080132238129887146480913564402924712524410024525254
450802324615782206358687167110556932854380162468346196749237552277
513105100126776057643879915719948597065176021389146406317350223384
643458394835435027981902728797303202850846884019875949870378146179
668646287546670399896304248334225490492446701329392472583236231531
199712398944621765884271933825466621610382140069902302774264438571
417574587943978979594815804972905977726218748279191542139015671O4
042498796038383987308071550425303932011381726233669143418847662675
503258634492672941635544061641605812600689785048902465409536738448
508044199481231522643777835892801077052872357981319176422544079026
297752232994316244056822824048979585862042095903016753077009841255
041439517370577205755550817551260179018120073513341772372476220812
000860440795123951426598964340376424506082959966615608890385710684
640291412717376571513488794464268910769410895310119099929995630930
905035222777233262914701401786445146353118738378495543882500856927
308783947452874920176886447311783104110199160063149881892999061015
277816870842162138183955707918405119806759776959987531377577268878
910886459165446898313347423547929805191092151468307163238553510387
271875446767082952974905345953765259319251659451479336850638167973
478663688327078954395966772983270966800627905395999829457773168238
326073880180654102514617216288678835870661909367729796642225593369
082458671032121453015761406563848832046204655115731003310627177636
632725355105114011372947974242341799659537348942142140023658440811
338831976175255058900924545313775605884224762865238760627246990302
126704707809451241471629495570270401899866663201798423005507008440
753327962569991771876542652570331254951397086494471914527294488305
094601841529556251474040952579800990146338379776902129394085310248

8561567350606338634923684489507528233401007520258283062071137190594267815521410921860570542096103071329372555368225794735587462567776516453310929822876028379225930251318516581337706052092108657561743012334289084699223497351511631421745254267139789248050025172232090821245741107761163535916860465237641185208310455600513909589498730970708723111425470231216733203810854809201787391487934488837268545689214877830390016547741762281260580728355415331369007931396300063769702007625350507261233441510111428070936819402236989991308247424654012701940113222999932048332874671355383494579635836899288623290439722584493817107725905803949716259506636916042428812825483869715966530554742543545597343320165017471694261408641380380466595322388060995968930493981398914417781080440177680412631187307038032840781365152378659505510087403583849737817232100166230527219947879907436057423140992833458661530302659108802848943882627192860592688546252611811506554314391860473863832014952014199240165101739767409226043254842945659258581776899771652026749864198907493364258824303008229914088423037033492000321094764235749370825153883596128554028571511999684121309513297601060622384467853304303605283324594771517521109132184692968901359920399067517466637717540893162635269159223166758528381513309573351829442340194857599928875715896113735250073352994468645177277810729355506620011166278640684583474212201535461842745627781395631003503800901852220399726275905468272699143753600658655126345316534223994033256987619903270018293229045380216469805315530988295337618967309534457130377128599254581802272613746556905822595786920989804611674009391732335754451424181559427904164840501217527511162224841376487939528948768911062083467875763236881995065081723493681850049201395396931150450840631833169795650011516330083782711074977286046415193311497771862005817211835717658891646355701844887330656741216711045991852850612219680110732254829518774076669979602303847200725332760059467869526790514319525735477141111573062837948717238799010110737197033795111387902442285766119513470938240551686729869870945885528098965550905005839479776816362135995896454669367741167952365593301962543171459828163763773483041585352887106282009286734513178670579055862422877697703803358671896644007604521050778010902637401436327800462862893243121698489569692681269965570096116297810488083332264011584449865788691989155116498775950082011654710794954761627253597443140698950143479155214870180524406888053182445054861510557508245833483060153051527141034013461587176204932737682281179363822377263676950899606005764576074349083808672495330340119364736422164031877350174262838309181603371305308194700548145666334229294394379129613611797429979597898222018382043393751513900818795675780849881967116995779814800468611110202998559769628419388687612327451524627733080446573369546365493840400819777609706639132376542539186868203566854276619326843902885919967881472483502319505887747564159106418991240691253094163125619541095435308814642343408331609704950449309811673539831293735539341187320088670867106762928026623131366609838364307561568243371003247612866087421391893567521305950626336204982646550082066501877463331840481096537269399354992508460932223638918187900587249238610783215779790260035562226643917254444606289329459454295831001567300550754372474262118465163712077024599682774758902127077460823281087774656437622050892211762862594923337322306799176150246435991356381620607408584397425133159389863383102724114385075320805389733801159125087956234072913945303862707068178014681947724028939617221644175848630204516488795837610929850676053716776401041227878179550018233197260574617611883779468454732039891838117019778662208080181016483471431403292545031424952200821114330744664013624225319259875091575121739132432965349401209539286534708463158821504955168014428706494848315638437272630481694795792035556

684457786382972288953534411852061006954504177044547449259708669886
360993447006199388647273449927912722231658528362329253648259342107
355249952854844231273220467471078064243669958423852863743227324420
182834397340003224185901923803059005872229289610551499388306141350
064936910473902129154397749453605108064872080131190490223110707230
770624283391952893722091148778390879044959631522298968270822053048
965601639594558607553522222159573834959609286492041366112049876816
520941632691258940484528222903607027727550910423476071510260847037
204995330735661652016080315883563879622431208900709419217345047787
877409407146870679225942590522751818094928229533182148904042084393
337728589025366508426327725814394860195937648754924471152085966166
588308595533616071705852042479775057905952120494899134627373339351
797353749095540180502086242529471556100879915414706965354572999224
070932580384255389746776351480895187698836463589425494284220731203
645100502716078039833613170002276335732205805047209990128777689353
375985741664585200763921687804857367539294950338409822939797065831
425555392829591922969806877722796639729390777908217851732476108735
564189670849418232302926913249449134037576977880000852120689948851
950118704243081970477677656477051666007382064988485717104832272345
711925918065271167048969229098580751536275170955052842903922436500
482488074481318574364656698445218053666467548379873567916422013296
190351708641479733715775410615161742404449579958031556111579103087
313472099035301894999465821929921964771056882282986141014201946390
542328584438171240836332265624325123838594776356730120676441085014
753981446342859310494968693638264004641625964695149031961110485447
759191706584392706760240031145212717647083332009417568873875947770
632409980920684630534774332419452200217630004662282023808277804197
794933893918898522440855068666909872615099934327559421951361489460
327548540028247463818557430386722484814571204128940220014152688477
096246122211499922887643919191089940007645004267363603359564464427
081897852741770774513395840957604462114327655989257121246407049760
606889475217788884675365773130888483170413084708302811779259466701
208771841286594199018778750963200281102375514363561230486554615329
882829990461745177485814776012313343417313877109055770936706573655
020317579004306722930314450241991977428096762212425199286327402583
704007529728174354804106393450637326750684346881838874833235411216
634188042412330340349096757794165377908412586879828861032352788573
321553381519883058804585315346904313089809636940664181370415485931
496667115981308994408254571535523006525082284961728723967465082519
004532455823574868772006674794971216360282085235430278386536117112
453214864879424132133170085231543372774607680663766996188951228804
910891117659551573649847388606952471668475237514464152133654892467
276122585393614841651438581869173841675434827813176631429116937855
646181716609663402912720536530254447638306533550451146415224708651
212131290099900196815169592152430391022949696439063552199065139432
163036534539747151573501445915609700314795373825072228643267911802
228545445051006686838264972907481325848087102088749505142696429373
925813677184169065452156108761573780205352795800446849136136917468
253717280353678435036189012457777583386467700487187551541811503714
129454911427269687720886195290311000652148060479389430426121125047
463622256753934768192222192006351687668258215068279880160735706110
805557861686704947486404202700061440097794418761478549764582395624
980544495512570910640270832390814460092511777876520638039371357117
644763292216214126564837394710745132290507372055423326208635230121
119230099328216475364369237906325033568253135543303478962915330449
231153809199575553294498705280190335116740752763665598147220612180
438573003072978792173568500012562331806742598872401099689698138523
973061919559283369486119032394925359441583659581613839121854119515

1992655070437222451106336712668962567275866773882387907913364509386
65117220139628594478654429621832667845170020278188419240093649036 6
272357443728744856331087287895845848235250820156274220792392203392
045082819462261529184460706197582285213387796963236786401331304109
955645374774064577757684903513791673253218265004401500646241636404
631178277966053575667603716137442026691617218914299163923048497382
542894221998545478689482567054577120830606966401517547011398438289
931933635228885729811548286913596032428511636219222076679819006388
404842715260865907155370497185933522605711810415045793473539632763
998872325606368960841031544136421487826119285418384995743044276868
385145914918918742400661902831447985922644596334799531028630278018
311500893000765762808870776888556671136106186168996249963879937029
113619500621550959100266394298058423177896496506627565297834415149
733882552363364465203622013216280321350049361959750702691727832370
138371576432302888101329633287393824573874624509689508223833084417
619240847605102724686019144743039151080377748192387105291127959055
174988229390755127440803641693282921255378008849192870285467542546
669735739705365362454007222398956201306760481133915634972776056714
496409064045114809482486851179621640442806897195762975356223618168
885002728569433652880013184412121411238983851952785119481467901665
284068838218695868306612959039774599056148703612289809841138200615
859142471286229860041718906453010082032794088580385760890512269876
008642460648269485048618629651722184875183556528814663127523768706
746675269172441672973545695673316677184928439431385995773850485061
097313805782920944994446323943060687595811900386026190492109839871
969933647463311406629451114715205569480402798735824309185973826397
713404114160166237752693577223614775634790555275216648241460998148
628132876631187524107417432987436275385045777742542057576627393156
984043856772914383783590182351736877088048034374286323665907452288
539522835778681347219537660500846861989626330050330936040997822291
481514752577183795281384891568049192181238360412271358296116497247
108026125459205952486511422783391156497546662744867185218756181629
471196471380668777153608541786941838614660748536539552580901696623
778000065583681884719769824445873229844530886990299337887795265719
728089915979394193436752271866343782907936824440320624639601866990
497231903150243625040813055353383065316109237895273339143697925936
907286974042623404247587033791700221492583435241015718645398347845
451758922412361367352913626017121554410849303321644230075969711058
854695869571172632037985137192940144871195015371587916332125383079
389694412746892273986101183720851428693197150286469098732848172073
873815201591163794512301010196666203644541295629190355481051912534
387131600152412478550452245480417085800974416436084037596380188386
074895352662695035328164809681679448801761593992935630643145711148
351667475654627759421672537822961337952004829042288165999567050760
734870429085908499684909529491046326851763652246317013487998937688
779842092948512962785230159883015336126834299177661463925494770519
302050031055605493677633916328395389557886997769714313544610132496
189019170570120182066702116577665414605136853434511733028437410975
267518355759247185151889091689894865760416453321241702808114846759
077301327254794609415509728678967187961801220443350296879621964404
832788637996040983882536293823039583969837394961017123582184217743
974270364691125810751594526635546467514367886879782290229559947715
764306712652597185561511035747661042096417841247683015840339793601
121187811200823174503714075700409271083743540108919934594983756706
127409717699213954110952101250839813654956204515024366843113398973
858736113065124745242321542571509170983114140086026489053937071277
441244066907683167085424057300361178690524323205442682356865003270
303065015077480473870024026729241448002051530677327019111454898739

634929248920628971290478304492685380028314875358105997056148380692
737300968661098886378902955732473773218392968237265964099999476679
557605307182161869394599277816445253696965812245009356458994432191
727916867639855789253996966098824872270586902492001784902712148635
395532805946828846993357856689685299143803410528336938389808106541
631250749494609447408886469083652165710685290293790114001017104187
582043926198243372615611135685841730762865206310031274971446997819
103881992609016646179754069102972565847469045071925244946583352764
174639951788616905673265931633412458545178258082449007188501785142
887176672083115995559292672621178256119044595069348961757228325071
147520441276776575050868936709803366686786798668095854516356450456
700098282130467124423758025533584949678345502763107561618567611024
200622908622178680112434591564775661362731079617325234681409070011
050097945863457244190266200577367179046123274420853797609726268687
700946422725868500716364595723606381633847494349975206543190488268
758253505156910742713458684179456955827177098027528168620203131954
435974916468654922876308046662143131853417398650352264519025805451
724237931971335479894431301843024011898085682842876356115113592545
051759006609029317534197237037316626763104552677057284108266195395
396768023500467639232238198895390409926916785915668921976462053713
869576979999088841317689151137434627023755761356023595321292951339
330688041066225809559753015227590711430726120998040611204959544702
529022458298103260460366610790784614562477180721220945378224379608
920847836988153399857843835847623311114552944993163589565451707434
941008645637568236524522552890279717199986729963816213783471325388
298076612871462553478352971463013937978843229852958454395196768038
647708119996215817080944398958038250707113582707983347809856610300
809248363340106643785171705008672856057256658249006303816659250276
035940142072432533503290715340643720991049554417219296172851893187
876675409135877109975306839422832961708480658343497111814009227825
302615934813947551736035584089426644749330958461998620681292484299
900690495309560199167359270034227705807777257942989192483507500026
253575382687483236342276724808711441393032576445162636301415773729
913585526476183106154750550543500397887915345327702159604456635703
067706610019201793214017147969716738469733339707056058598922830925
312952649427953618367607928799401770860176084753039347911247886123
969453298233627503274176462432178205058631210032808102535309052281
121335769067348278937719290836686403502827994706248624768867044020
859538534724137046928259372275289596415597499167757872687009614379
339091219386991136301973189710945603730161110976662442440018178065
055572462339859256865538611682612704334070095180088689713989492131
948076545616095446512264314966934969743616983961687411240926925087
916495101225186752483635616057123486384689279645666764984846476716
504656126699086548540370528105028232541582319648245828614979004180
345759695865716578935991202692404754468625626567071441627711432070
572623204570586425486485386428718325923588272100501781925910321862
102525429061064196493219738482292467214508802767736310025106146589
875281845672592050079006099263317935029302633914975478998055915983
874072282012111603184744607313092642236057201406831874074143684735
669302813859684496367816355466904575253185482659911617918864750699
221326838807888021781507987526270959165282807676731436874076260554
027715833284666790562252441503160456864894181125999979503530707283
995418800406218376904052860463782206835537443655485694778361506263
598993634787027909030997462772184241100176482159012705671182087668
782275764676994285411305424284697967712936556371908112434752499188
941044238998876558149098163133828344398678830561422069948656437056
345681695102097143423812653705290231148917416269759846890675493815
136882312553178553493746115050458035667819443184768513482917267953

046515495980756411689982379368626545225447682319382165559881689756
544989847223360804023621521263789857002732047970983507338475588088
526560046011136366410695357973449086862143285777692977138838668754
917368355359148530578245719109983029831395137570525256209580954017
897695413946151720261764966079521063305486458118903323277535560804
209288079548131044308314254117564693796449370088051788439064650598
699529934562288497813679005684246906689823480376722839141414639383
441970505255527456615243070311689399586409521846800689011361913009
089342682782883757056395195330125180182350049293106972725805703196
643197756434141864919570951944115202257960157942174332998712495398
481643215884016823180156768220588903344340576906238372620605419470
830269886808831940015177925067757175174593722384717722050820930704
159311736220200038813060788840091117739664188836733320446529646445
934419768596428126244512562577886323153831909565428679230834512402
761688835965251028829254701745885087785467432335413143581398900523
422703880060771431783425266802996652515958052673968256629757854112
732459999634827193710570217727679088300584907363364013164799368378
780942754761608277906353980263547708948992177734451897291008461649
056912644584004920708308526064556605541887941017689160277283372571
529263391248090560002823379201775264068935118044977919973237802038
054845153464214411215740926117175317752534252312656527895654799495
249996613418668561137172657535761612467563936363465852902198835935
831392192491393418642454135934428166038405794303405858305951612584
120866417970404500557090151031427979014579958567197456136453537244
757325971762416221665609815477651079243328459730350221418019104377
848724061746812837199614628391664253480309667240241178483790511869
883383917902679307649564913279665781977169564574759333133134262607
748971367197005879051641105756086803903926865826348706345405515763
139861676381077414451225941285507544942159529485739898630568471553
514877119332279431038600662876069707269223883921101042205418231418
783870028474888838905667506331222092051480787061361084283744060089
044614667973715826272029111684229324748247891789685877780597760941
818644316340028850264536445513550671213401188690785557499410205012
020584369359438338431421187984966957966712318296941911718158049435
257952406018375850997934371130880264021542881644344306720286303061
244985371567180909678367412752020111345413499839171117253538517021
424306732100031441372887105544078958247023764904752320309705960620
761202742331730176569032003677492694227330322757762751700794150691
302352338229522938042374299195531100137570087357404896049301491001
310514828563869984292941736475578552941533379493920244023171942716
029023127159436936461304778015704697510260615433560235322727132552
378164940552536518894649839034517885744396354358013434976027147384
385511984781089286682294577257535978429545434990952690776186980501
260973242575667396516641860943350384149618387370350938070380101695
366304616092407294362211337355372256317992452086818271670641961004
506900017835172681539217865847481240698829943944692847539212047696
704008917516980044735013401137800552105630498825434932067996417341
838113208226066190871983360217148256562393710727708004426030257864
375469141406467379988133306274980434048844433937285851420929071406
931327851505346968127345232046363566630091702633597632388614244380
188240491008510158252293256556727930099661715167571103702279090057
724322645193485395815336762018043079066346938595228276348528073739
266954153406812865943469956911804724376608938315621924386564103131
340589115080721932867391688323814944776071992081035475388423453896
732654849684408063510671575237120307503288768536916612388601773535
404009108803958927512100254971669370797871866429220140145588245623
424144670313230350128103244201632163057653771387099052725968494087
829811861525888492527218603289522182402628298323270081763455612997

145774658378547242967461824664384902978653007763159379196442547839
488282680655017633162340101463270947259720188233353881335302545653
904634831051730084970426153736764062033013790647873873712146452555
336583558282531298690853960026366890725505682714100041735228984821
717599940268074649141888703063814711532964678955931808651223026636
997204711317478634905277669414273422723279580870002598280525801378
538783200311808721509846232270743162776152941280404273173876699769
554819153808427735709328137376056176693707288061211955806890081599
398264876451200332178564869843209063760256299992888973076124774122
838326961611075164891522825064454626830641722718033834317376581971
246395144787832000935133331865522233895566025164708101900224467477
939875010877461626988940950281310748569751286370900657719141437702
967548562314383251595038525173136644265502859810576418370704824060
732078711770552954530969818351972929411541825195835308336347495599
789876319942510417438087738564230771733254051976361123906352198949
174390525500247755923939446107776141318676550924068922230286443171
562375620553253594758883418841103183260566160857078012412132972749
166000048992747414021583437012481574191298877094711693311104113046
464573198787106957011354876601684723955558088729089717469122036925
118972468059171125745739439114565180610608056378653089578477353911
887809522439697800184582364439544824465655627892290237229846083255
470549406822187880347317899183416054026859967487218008132077885533
824305278895252897097708026085078817526844197474750300894424270027
730324729381969579912766676269025359762295246233268788313894840039
368170446872185101395713247675408223230437822817504981212218492381
106072004410173724029200225719476280314994515478334470303346453163
731315274916269277287196580757976562490296241213427394749494996058
045828871922218243736016280861644682942178448664330819416549098050
350619393537344184449884818559504965026322264508622520870983941795
161372926156316664116379086259966195848326953820640610268251704049
489883857919242168650982917047583952932601194431057299970094882000
646282506414298088578461198438509317434993315875405684618443087248
169038284969445491491211283899174426980355442566720592375940150852
875841205623818670543118901668180971789190221229351887427909219551
551808886890313447708457777145492835784529617248746573332043166606
413022240780940992408614861966164617915731781241135208581516991824
554105441256762164334410701735120323683692247023947522319864680946
657670084414776271127818203706739477302725271299203114384513520199
463470898105814668173287078775610244204807475308420410443016348726
332678834504450244842310917755167367076052841051752145229349324802
842648388484690020994452862646418420347221657095686434779811103662
204260528403203412153418624039613679265397630572391767808239419738
793739699311857904544687873150027991647682588517407844000561270342
803131241594006256008060316896796077526780558515844477375732040986
866150895683261795987193471910824256221882689002334105397189176036
305647134218859359916610040431195689986685470952963076078686347518
088766821399004676927370948474866326257852632299652649029047557265
564262817106736122779326818851162138242414638514198006184825602021
620796477460122499669672286852450602851380061764826341859670837 63
616017717878758478721135242571274834527649231762646257436709226621
130773355520568338605430705415874024492900357561116555571699857931
310006819332853800182877747232509141158198505769545095128707200576
522663427177149957535199423758520108327557255910134198306611780920
840327015963024197359149660681090192953878864962912091547913184092
762346231311444102527801585364435130263119592438461455078334368713
210905114872712309587577971222071836071360236141627933630276200661
513014317558424364472527128448286083347674941120679990018473193190
469613014178604322552671008309502966151612339140072324812740694337

484813194585701844194851954609071396254069592655653623192382949857
2212861264509463919495411072692219061756817721293282395098163294 23
6973124724084346206764151658372429522369301743268413874102094132 23
1590431123090085591780898098639811472423431259772730725874965450 79
8846085036494035563606421360246367250297582588214239709069638947 51
5852196010056708757617434222006881840186782896407971342119879894 24
2005426232263910916080832822172062383218156600956376613115070352 53
9431370438476406712570736598637047057472995577056332924928706676 71
5784263974841648189874644206273262918091863456951408211126211183 07
7842318805504839023018238559619868972586637538368548518808900275 67
2391487967574758714477044963905966647638840555139832085110518086 94
6733442146964388936519874292945007963579336776306583547913440943 74
9848749178110525952934946088696039617923756363527056828663233569 38
7547828496049149559435581123376296279491108962728456306695903129 23
7389498739064645482345265301124597093691663681984294019757039611 05
0180939637076776957461341673659491868407159799409774921295144753 64
3557051040490718226804775346892192192239665598892640138338542564 94
2877508240456558180339798752807500932132516595562648968432647209 50
8457269426761962032465242536113808128554108098613889932317527610 00
5936827481891930572687927052706681650947384411241352252246216405 89
6697737766266947223060479259376589040545108908727196692960261564 60
5692234708307765972494222517344900495881032587117349851293340748 28
8962822868451406587059582208854635662223796926845772708006242862 47
0839100012713227466932910775957841160555232539427550960060760837 05
3344808069613574798661003456942905048742886541585205718247493430 29
2664502012901522850857613745521097273911088254049019546790225373 25
5943701093302223343533679247915548650890561031509201053297700331 14
9099331914415825403937671038561512589708413151515283217981162509 44
0783844635992698248514779982623836715422818506966671626620176160 97
0567094861125093594209257672502737040835833893160975056783035442 84
0000397002028642333353238003083672757694716706320727156328814513 54
0266545280537056163375657565561560456839735821127233493026657137 37
6157808881484169546069518924508977614582773056447115603672340137 37
5123911334222135099520103561776430770908044730482689991779764687 60
0344803644414864134634689995078455510203029888633384832818107269 92
0008898285719368413891539811167635237013599604776794327352194624 94
1833983034177275287197321653523974783666159883000187013548007254 99
6779481564122250731982077737494936934051592615121472524091331243 28
5352266095099178862420621470761814443651682059066937026972848443 03
5060252191404977551261450446572569712315957976429582179683132072 04
9766762865704713117146597168994141419415585527921329165534108035 86
4539423604397634616953352898463390144370977103171885263352978600 98
6693086698435263934184369703188574390628654710685190800238247922 65
9706695796912922827797760081119896191705187657704715480407623420 14
6901474701230072367200637479414652488488002186972544546370568600 47
2326742201969808221422778472139305509963930666588351205734209536 23
2706833519053743235096424169280243492463752913196824007038389209 94
6820797082050855302650608417290646789432924890422666237149391487 18
5204533360509281927220026678354509006721929344835718740048645258 43
6195009455202553385301501936082754550814836776294317346668701839 89
6632736870667173889783817058514355616626575305893492837995688371 56
6190511746934019853587525067274615771754279297665411243066168857 92
1466304826537623629178636344787320604811685644513319633775974521 08
9142064262107733716871629057910904737837639993403176101329586566 01
7868615084132025994391846880266009194120701567367052752104460254 24
6476622255379685622112998222213661949692780512346135893880093787 00
6445888955300930967808022056712229117324496219466633608501509594 91
6809119443188318875572728807260492048422572695023477727362566817 42

643079240727324305549238831405486394592952167346120593473310812267
074435854430639051354861245846923527729555959508163391240344880846
155748316511028056596912604738220096242987861895952287301938329285
962799349844728509583416182154386610869136544064259089115896981298
974427147086063734408895081120792563243439121604867098037269298223
248558249957089431108637654840373752138369775403725350930868402408
318659218947543425465681993319202869736416876380780559427264909405
431860868835870623993038093248937760639588088169278821723284010080
077874468552849300953561681336987821881319879609157078006065875120
413681055150138991407216054540732098424457112407869027529556371764
996922732097345339032749384224697177560429121963794004392913939936
963834312381057271231601654623818608034169377923236080648416685167
008845800707990539901913898692886788685012546791682529572979509077
814007758372919595259230778985290095374969314464885604201830724836
681645348889006164184872131401761979206738425388528296023984285540
130893392381411757012080830155418887191011712203543960412588136888
048929394096629476679405622656481939225363716863010600798228698905
371847071955248636144245298398605497226673813992712323269791352678
194750815424751582595707182151747793330838053854225259358687900112
017697150694846872323977569602190530277291344179895332845717225695
951393984500808185505028461731209346332389767439668678411629825366
441222235054206323638997861301160460642764252472119953879776525470
989913875733370958754446068818103330791592335169400268050996900920
168136950287589393771494933112157241590212362251549726987858042362
441274878998655931459382697569314305549817950653144186683273288195
130275199567217538237057152252291788997389839063017399172994785964
732882451480573157662397274681272069283546859972049158200654932142
637378536537776577631054256491563772800189810994441267947276921150
956072802362828488060198203017705573603550343134769074127602842407
02785624501867604936809217966630506285791925690321609215503829815
738436504956833388199142037563207628943768460247661039435202294481
010140411684096352222244652669840429640253001700640703723317193522
076178313500357425684523102015448618474112946240496328898364055154
538023000657624314766841533398052677981987237595528463455943508759
753081225407831152856459138266985406198907598592026853727923008414
753034616218457484881555487518028075008701148602056630051107952626
218165099970462460938200305624610355311040342287433668786969896589
571460526360133947402955027742886004823589976160326074985704712218
455866713227141006978846958171262714993858982858592146253686919607
865356133568450075766467433310863247115800710250863570422004275510
279692222982886795051603903278422738873811183032977329682416042683
236925334343948056151302774756556422435386312834139392655972966202
638496945337587293946902596287388740148807094633806597996983165111
929088025119256028581730499108412164179968432523640204120266603390
613564414013138932212028730629445326131983133356541252058211952929
321494053648823003371378130699337526267525734272054718259851934397
122847549742322825404076626173085591797748712629887002266741047470
611468698028702735148198879330690780405185216982872231913175583775
553832930641934332767058711857276687145649647500467923706677199070
691387000871163039442974822989518150941074191583828498300089051563
374997092344581268411890557203901380937449267263259914733455221666
514058384741593179377470138831092920729725748072689274863625450102
261803654579949694183630518520334858306138860891537475768145068904
830504743231041756725444567867754056535324624426384501125401588113
376287922791445769993245490207100571938902433231581806774444726672
317664560768234838627921390923872430415110888337404826020756447675
776852313457857843464984582933624074648014159794645214350928554441
465662367260070571074429471480899305462524615053954667294574547752

2309885938349227321181401218199372897274783639130242229542283226 58
1272993978869758174293364400546233397984792366191852240226254562 01
0126296227425623817530051776328637553954276860419576758487868293 56
5801648475164681085067253073869350186544318606782117480706710238 60
6132897325712447823979443239268514685587071759326786933587685471 60
3448961677611716341299667336450589792110754601830123633360691144 96
4188387934121842194935378749966523797515391763059159646103471521 21
4656427472391853965021316325911230816528350294868195657135887476 77
2396290191693199167768645873058755288747597090159188858413402314 94
4535163715820461193929538650063090196878114872520024632500585955 09
7326566975940880924885276626744424667263984945494096333588544944 38
5507082778660432637669558899094233921334532437458692369553584084 41
5785410486788230887659149925858776134937893933817239566126732645 86
0588609983313023881770669293348340480489448144118284346563752808 16
6560294729887698489519112897279950597630327730517769376782997599 25
9573447528148628394441893629920990024135806002423326855203253436 46
4935897752706903435988090388776483528114172898522061576657718974 59
9693126904994274595857748900715017710280050510218508746333748028 18
2672386637611059367212151178910066446038280924572444314173858083 14
4173658768637249757501724180505564815645888269442413625697379292 25
5945902050613210392317227946862868956199674192340073901316879381 80
4182128317285099359198047059989085315255060611129029607455276541 60
5554323462610167088128515203185163523305468450163097876868862516 09
5539218201938430432208578808474078482365115168524579205376362858 30
0472488949906232202771501038935424792067272180390323180854549352 30
0215703224048303575168346141950451837704613129450220640492325433 72
0823309566895789820800186133500506875378551738982296086457926134 60
1229336427894129347237377241183471056719947498813255663024174855 11
2544613530731382508017759590020575525688293535410332940582476334 59
5489119148002630594112285629020991982970892267520245118222001125 57
1579240733650574612758838238123948024110404490718897939017921349 17
7985750287517112124497071036628890526745050625093926019965399546 70
7668561456593004672173104967569904556248663389377247350889687245 53
0867838193979741498390808700712484440614848824896131026973307385 91
2498790374187062600105142158390408876123703653713520292948787650 54
8885182853428248410038398553662313376457670923704477054434480282 49
0886232209658664498591943363119205831620545310144578482233739950 95
5681727341450850356502805930648292718529747212853238817532400077 62
5654889901114846834623218046070717582378163621157379393336563514 36
1265578824300183870597868274534862241033127098248694442254632051 82
4297733063193782880524736277279010236575158387770445736380232431 01
0591330252239746032544228425304763924993687412259412525344274571 91
4357904261985598032355410755517179602225731007940379296322417785 53
6177805088049496315873832128675554297068865951555521682850849181 84
6715119760879072356978603646204109197206493004123150188837770458 31
1397629187009238052681820978751665089521199378270494551399204447 98
9433788484231222487627259892997895482574664348575579264603758622 42
3085401606220231239583793967906186082020738486233385006386037167 95
3946412827981719113119614782167000830808950785445813266954818166 38
6256950925231648219178221346241417591335437979028341423778353888 92
0595668461979810523192825766529383288091196687610481858896454424 40
2205427489691120693335345523517307201395279545216123690563455727 02
5608582660648538391808817916070760661386419533907534631042316314 34
6145514954683921525457176523017627907134670298964119348104686248 73
1091290588286930561548550109865716994317948320400204615028922428 32
5398720403509193836454205680657981934923777973992361643392328441 22
2912416131023681238412309400859555943921238902610148002411843673 25
7218566928685211711743895773556713694403655341882663032196737122 04

778461396724242393438206557571014652108390411058025499057430927093
6597181213253729453276129255716033811599327295120079560078112053331
4636112722208931150147192517953524282097046383536281323733050395795
9214257143099602033906880057835187398188931411813030607371891553693
873116976049855114076377510951304991139839324182677506786129309334
341944600471092978714189848380110311027418843783292048283906220114
592454197652212689145317872121451331267798784556044561595952497545
298575352221546817148984145111384250814105501308548962342436284979
104194400095354198478294098804365041833757195053081693362546548668
506765948602108440320888369791785745623588814599483392712832581756
647814893076301831884503175187702543381512172174738086474183691780
511854271392830592551520369615442655792743340951740879029284689484
767512818289936936963800520966258096309522447841141623347063886751
542015041317535385000813287716280810470018834689655555443257656803
765673502026726529968266884467810665749563509998596178894324430751
583105497717634532380639058204773181162089074001240149200801923869
544920213679302425802464651769831056714850197856958739975556787502
966173286707949082756613216383025793652727117235764937197729881838
765016004696019807626071729733871958649870990996872471749476111410
147825901880118245361021522042204981976181635300339885999787260560
375797598680307331475224756400757151879220164485890554090220069215
710067763592498572399552763201444121684915435778005526595484999851
247978090199909088142554531420111741169391548331482057382804145436
271804646866262913735823496648287522089699583629003659621294068548
508879007970609715361504930307097144793566401490311205530808240950
244384198758241647086058648990739406875441311017445035194648723317
674626024900661157888610971626885032390425642166579301971423307771
445355387862843251433602125018990945261445400526852137717787667494
322391925935590651440122799532365738785963366718867982100511224848
175402924750136695005047596708441801263734588010662253040018506698
181097190490881302291872876366011259816340730258132566262078794293
937730883405816982294048034393246806755200085304321405383703681196
744973642664329903378153525502252468254277644827260490649082697721
388698375592442072377285780670194840754825024590313667177787679458
280971200747091410334539792304070212470478959724163063167939042624
636533153824680278462999849295582970677703673617809127448651096131
947837554214734726097785221422069774417333930237588915451744621841
465888174063321925106639986263693688001561290539774891786922752271
866000006486886069438433416956897619104793155040737906994805414239
314034277212056417610191846519831622755248470937887500888830317863
573072829187673035632713538749271862784783688027045784857087073472
818436550765235117062567559404561475108174628886513889489019616918
735866440178911396503312482862821277323414956625703814366703949649
642201742696525412503138667154925779242582471839304062211931552939
524729565961088349107731184650661085378223722765072330072986433681
676406634827126596611335519405462150242795988260339343777851368966
589207774770380912306198797675448538884934886151790229895133551513
178169447938984480551016044753872934602081056423999956531329438181
181794916820664342298614174888862468891091040157838357890464703385
062422545091526178581720960154959041901980817249863332365323996203
529983358128818444003123166844128209215710365003177932767165548592
678876429119443885218233896509449510829929260775654351813998833350
031659441874901573825151534911730252921091190759492242574830715970
710339463947729017711117954838753245430821919571097116313655851113
361677992674500225456746172702619265822809693016765263536797856218
150440685524250206327743120789245928073083160954493228199057876042
271412546989922037225580941931329073979529846907496688497302828612
768342445094025419733427838637133216298271886802877048859039816542

665131221746650688390343880540662876132415147364980113225683524539
644591778808774654570209538890575598696204337257885828876643040148
602881347856677752231677008528156703135170869936872283959797761427
043044338753674046103409626598140079595373380376608206680710192988
098346461125352172626474614809062450843560918326623328614765717887
367070432644964237558311181086135320630452731355013389325846980396
507589459952787098686773661665652729409662709536189923808343523250
645354065363667008123808431421291608489392606475750424360773370076
978010486226866973272061313087712428838079513682325960493546269875
478506976087090621394446709870504353997170682855777324556183385967
258415295843880954100363797690748326392232269676310863039981067968
923416082594660541196578250833791005176430719600669142267780803108
635894954912666019682168984872798099736518826787773105328505919123
036083903834063911136575293847387327396475869511903909612921446578
009656278230854852576498319237213841985818804914914081220298366647
625638953591530954244210238114009422010543587336021765607515378419
898361830584080950957923288723131448256881159002930870419488419522
251557714253601969600071163474474393500568456344376900335741760282
554910888521532392506087844859788167278966902631720010672064720740
459424890050807628233972035138346041011685589515018945459228325869
99099358318483888333495905712543219713999285566983682627054269275
987881498938895553277521957161988467584092669237122724294525243982
523836417638668895023936487670293181515076725859767848492337458449
430892319353106512504708445945982293233904494552069195194753050359
746278833569846132022212993897490693428402335529626356027699332292
725089430202063972785105589058443007897205048804412923489474672886
783891502622888529128742952750435565020962942247367934640352551269
544793314931432012374843865246657866338104602907702516740675225470
354471781686148529224587986189617232446589528197151406141012192732
756948471047837285769460924518169168799535580404984124110491975743
929251100959314280976592191588701463865514029775960129535847007686
116829226443748883090358961861126606849828180729560094573218650739
506343957012535482591503536895771466916509576445043362651251199168
204088534882183992149037468832063892195839677180721260732788561951
313557657524341345366570848659341598443546453769453590903140796550
810660494682247214518383466501317890674110392803618900144192600451
772699029465191226253027545038060324885188360479364632249905482580
495776019740854482383090288704159731204703163934836548926547230251
529080407727077303618518251961201662424935513390875866910423295025
219622274262425667159298289304043119006795704120100840947826597883
650327603103028484313256421105347720758350314603861753293827376171
032952138127566993973744418533716325063558390500476440504318465550
507304089903892699362886204694026958443150631088978643382529879170
065955281987833755417779176288761977750252490672963806218479450140
143711027157594365404088338258179644194308333085998477598891768522
529027457842168969003684371389240369032563217458039652881882758186
777509035040778567728215280374417828553924639303017935575994696195
281418099816035856227245541275974536963753619557098053523451326657
464468262355483975934275684583454305809159838250894738928769210297
688742998711895415173204358065191585613271735168111490983541922118
283470365728867448815599802621633472873434300014617603878659001368
802558127766026346001449208170959073372666103607362423085988438525
246934735966099379697569178349256036933896156460246531209080086702
727847741561036000878975336766394492456229509234066383358361325146
393239122021141047583736381538827391201590659474884818049489836365
636223485421468944007752508002307778637642480862819102818203471183
174373034933442617797543695377194812821564298546505610964355938031
147766430355133888829485169813786387513074991074046317672875953236

553329126594764048735628441175079713904740055133239389619555128569
785980204776976005252060680548352054715812429473628012190883115120
816936817092279617805040894557835991309347632280801514996988913569
688136111051402341987841705950776564048246311888590148083377611155
237681938306357143194096425742267143549818473954077948135087788954
885441300133174612510004307332540107531424253624326777980106057834
007304622567359488688561912669235601232984355772650000647431975147
022267052828038895005700215205390805702685168817909665359775812636
888591476941979748297001132585765878572669679968003208967718995220
502920349071802140169102005628654772960086926380272911146940147119
550677284396877248732715812748042859781030245051328997441075754153
570752061036910290943137069230002758489532125119784688464230106804
490373289192260583773886130912107117705877528703187465304552555815
403887856524173249573697748876076701195049289418324591821807077006
224950287899840599660972843531403320074981937074803208342455636186
316599741500158189254372358414756584357119349433485975635464919175
251174560830819180438633576468307192414107500409488685010605996185
014204586165368969551147587396066060761951716262525023678143505929
992743236542178715953380317592362788987829491326901896017961588736
995835363153030503048325861920935222145942082074130397798738379090
640621247797034395114163048800821786819413639283924963789204424567
103999952897567048617902880627342384547397844728193551281160733812
632791742199283209721893969611128149769722842643702754075034764079
013453331331324564963028811571355381128154995238826309068009825704
032522248214189545731194710504933927455593513204206562153689294936
646864090563109585365986127655029652850490850254774598215669295127
371133217455907531571802909943258834053869948164099672325312246083
210899317524899234438119482066431676943205728014253352052621878287
214890835065610874192721644374574088327016759943094565014599538832
556482404066112833332293469447382298147582021317975355055545776242
963255122256361396446854997894034439749368161016591941235650173277
012122574966526646311730711573483666997513141825911584614328108578
758134217376132983820767519555975541957172396642226764547465200211
986278228782024036452599026354617705964647041183230009657954902487
137117915631899562108167542835268880911798918327001510089430933502
265509880417268148908812291287082952109766415444761833098113691278
977477122980202133212746589852634671926624555388319277144999140834
101059524872569030654767321855578141002838442058890414682109669054
337555206532913423204111494441007724598556415289012162258426352551
147078595553213609880932895069644837618776325259427108701314679076
750276325987325766791190434282627051475245364942311690421570359226
975853210231171944589747868172321561144409284498744812255972316167
944415034458684482061781214268917234825687477876627836168367743249
408773244948988691063112600380478865818134246210720845544736421518
472876874593713829842803509208275211768993845949109624624375453677
491827465418467347459270497448425473396257541989940808113988809617
887011452014324499095869529262140219339730031606517359189393323587
541379546540821388091274730654354517356644909327584900617263889764
584218459134590451977008403452259990961887919339840288954323562790
608249368941546323611677529403826834623555194286330995651263696800
115648880489294626619364652318434098834586027305854133387446270545
353835020397571542522128525212821830130299332661790412226923870614
682842645748068445855575824866887213502541730937126041796912427648
131997461140265341380050202471813955693977026651364840094792620145
897591381816690481772316593805505270690768397746184553099126023123
008399979841076047829385141598878724798422390914376454307318765451
890953265023109477426368636396795651027995433045501962995102898414
290991155800188855761777033734674454678487614755250966994450659033

739301500313160838320599729844570358016495399357568734065602219921
218156704559208658030544608103653688805197213057760336977912821309
760536858063007951517605647629964106247398349695899949168518709749
894440934598359352326346812880259885425319651316359589443145420068
1572435605398018596809437149428655669444149003491115569797346153042
368660689450470004538919216435945629122204204541554833978183611551
289833126210250769674369298551683203785418677422376337169514555308
472609142939192564006809487739878753639043132845210957787024572168
036765425369755669337691689372159682642757537449497418005416994427
085677468627943345741278193459750892923701449925344551435339662061
684616286946318834116381687326356609668442661816850878358302201726
827279515949196298894654714150300669941787908569445027324080377784
271503413853735560505077429615383765820055942220959407277510193188
046442455891294859364538378440501595920916918099209322085122672924
749205144015172216109792122102691572302365693498551243092306603059
365779850296871735291334499724162964734825673764078783302942775073
652195965175395117328425384809993466669701185413116405189495265549
500814393148266488144464640209407832353934236123495389862481501646
262872514068441540323241030381292120284342758340771587145838104201
688625456195295922502893750379438292604020974002976591306686065952
755327510698794454143399655516194986920940659152382011727678666162
834438312481666252038836576254126699845464566305626956300524145350
816449079591479909647782002448130287670724944687100703971599113759
347014971086244076651521471989330630186021305048086070345183800627
956418712285362423280665412324259297091097700292972121994744426228
543308684127337911626984752085482300415690572101220265546980356715
481577352519046914043093788933407978357663892407945831053809339861
714613321315002679579725818962290576615119916240999193263215059982
468159612035086196162185140955383507737097381655129651352026318952
939243562395145390317544223633130657390191841231458942990560684292
537598977918245736986947330423936923373521115413358476203818142692
458622068160517057737223332996119298888705213003969668000446336718
096923708853050360875843204027557567729670050070934456292198755900
985174161171010498958797084469521957848116947199891507108449626352
546703160579177421169382207722093235260057182342362393439411167924
584091526061686413002559143056811111779708277736151832307889385667
509092809708369331327427004104019029461443725910813483549170741047
323029756221693280169277049193289319141114908869724389042049221875
825979970675643926559694021381142671289816729574280407646858336987
035423399064120619089258734901166405703082400764912261285010187293
510024604618222568315323826198069982118384010719482289989607372010
907570340147326678500305625838998049807198497734231901917472465983
722720742108557695432505903407033405048224226143087628088834201672
994184495616288695892134723418576797210819805461815839733317831479
781392885064825825088189657616971791871415017060915450039966398970
448413092125469123200475844245375100418446477104934773844112579849
619665499434168902643712951268056230282085270878043999881177877526
438418231405761652466193118860690405132423239531733139514493709568
694609307469314164276161660931859705889566352596023775307483043131
652545728120657912367284982185753474555387547682602700226624748154
671053178253911427371720745827165096276400991584750433634318265463
544026404059732843711650722540326513665539498189050993479379819858
069782186550218132703931536004981536976329865055365877436341521795
877561245185803483099483370886025489091525828237798366834873955709
870662763298057251921879145360285571773767137829961577960518685217
746247279774458945612549743742831904815350809536033451762018622246
012604624880226814489030208699540534068141543851849396599038459111
745869040957358146683360303234564462806343052444122209949711134702

4145602396931181520775528928748789363961892837294870079331117133257969760228398317745577545861299311051810723969487657942820375588446207770717323413389041519555404647557632702765338010792659045019913999111486319557132221245993272985911849735013224173775276724983321615737118005690929970080181244718689665787726800545551153681979719882684807894522754243348703672318842103520350485308724115445938475496586200922220756979932033803693719701186206867854194952555633246609115865504217061767650666487720230673352128712295728646285932926867706130769255514137249340019276522485394327005956196410907039162084403283279898661939875924646768939114950318280834793998604665837250695021325692175649023887656152401852170122116212746858948178985028647447787288052235235682285203593722441279936637967428244839643780680849492301720946887774251866079593245371729080261399410969767654663409431271434368806589206992388966339923778147478935028094373660275900364624161300918734985959634280478473695673504965124278435885422229450338001132620039715846114551960415420912023544705338201478038741000722524862318388819378019216582103744167097585735374113408393955228578747068218429437450448918247514388781233455586345077626270217716576308624209174705004017086640127559230703925990257096154954742574332757651801896883382854955713228848591640290722819747439084329162549923360313093772591162447164874142268157619944324616057295606007775572712147755502821957088598761030861652699743456773684934533075310785509520368342153820268676367990480875512997682554833269000521950313002848256936888101150090070833401009054160543970501044880124123142517279266978094246624616089531136154168053097688007906225531274958649556987999188853625530061132458335482857506369488225573730403719252797800932401965754539426019497499777946963916736741873698798782505943711750613684512558358007146559799183227867154283537193419549062224893595622487235001596551595820360278287417452051357454840974327538257555219568073477789127242952528547537741631054371227392203054066316539460792954280119497228598688626209597057713657674576946422985856740408549931614689233548567863314418192212213688629970654031711152797921389343628299637884132778293950989205495568478674730118373845505613147815705416301181460518125831815266099326685656447494864272361110063990201531934128825983210736949495291608627029779390422363625159140824703432470368192463982751895269102279503149337308998377992792454394116047159119733154123209971781337228886015363032800532128349392282194279859554554166792585108532064032098488506278156616214881284667672704162016898436880694859910074525753019823764538420560472267979845632601505549341618925533263566771752943034111901878705682235547120159829502893609563368361185608376927023150693340265942304195620459672440770113647316949691061304972832176408469533539206481500583501825585103085490348803833748183194421881309584774220376952264714441724599395934119070544163630797214176855938286411209471656201941013076569395469755996400829760053541886617882311102017271557302255059529033501374025869285427719746698446360302981411834518321932934662119104972061197368362956703341942779167623834154639216655897206710021103908596866839052817371965278139213450154756878629131833284334547580722012475467487682898189234688032497121578622185846523818179160338762205445026105352773016917736647808824173908844525891385240418917053930136056205025322890909131465997522304654649626487270505042798563552499956894354846562905335283890524074768282346944251242205243214604969115159950274511921156829136198777574429975081994571497877404754366466712183671863998593864775848049737957431031355610692264889332379418862103430300996712957562506033035903221766749415535633723911688903236692739237960555947822231835025757695690484995783553419927080045371029096730804871931921873490509578327909293340979011230509535517787

797425730259126615147330971950212836644394579639594272143484791826
563859652532019504389724159269468392524242186038072871266166423738
352176839344656894225000557581315184030850597256194696235696507258
485093481378292837221706822897942641654257844542009284748645428513
828372778381335135819408279535792283988973991137735931487315643412
685577389382827974403739403858089054758537646920265681816610003762
245923078253690283197605974266134558262895200074717519866139348452
285860670989229139860073767216427450218418990687512821051640714206
606389587802832431977581088443726138355462961246061281033813110128
147483064925001565512390649263760563155724150594446447578971934930
991026680957081484238104884185792550051367684291351141125175793902
226258740088453641633871277575302581962372271590742555475734011936
308352925466276945714811338665484639021230369169058049540678417310
559465918598303503065831399068316976483876101199519366253613888976
226616508466733759478466304203938446589038464883472822845237953170
699411613641081944696996724440746928392364130567933923015291910836
075357808954407226578423150840588341954782356879973662184218680496
876430753139516367036626375443913518542939738639303504732241669528
654384395492271047001209217380103835760953115694497464875956857723
939594612549508301794218645124976069095855328481413039828332195744
156994123083932663795175897709125491779731184593992615345221399614
109834910618405773674148244441335724652915031005080446257502537452
754598551539294860423302145828071311737531913940893517121551804301
704622055447973766966747893173096271481780432052415373985016954090
511688306812531486809043952323243760231622525043883668986857066165
306618190456852747466558716177265851650734155065945175613667842590
715951852649232661198941992769783302883777941111750144741042290808
902366083391940652111393237267613597885389234282889008977305473363
126803325171091298392614849700054268616029942960963848864440600491
029455039665291702342363466065042222960210987858800694421569857705
366984484108686731050696302960906193423160663874899001882894951256
155911024044628205315937709684265648034603378173148140538450449907
592035509064459099090914452597256714953028272645215273966122182604
134614750102864922445726576575452903240547143706012344312177520657
776455271900115395334250541807539121915059837985261573370516229544
119703078887431739989145794825349987158969437877343737675161566395
464762435208186531597510366689707775832838650575235805017140232362
041820853877746798302950844233833110374580653641907944749706284770
654767948296218866925906821760067313916066581607858328425138624683
306260336679546791249223032388929568772894761215153639603094062765
311976690109113804874054937787515860827573236354566858067490627165
972159902992186708613896895137370683173156800919554827792046505577
775066888521186692976521103907786318443369590672352028399996432396
387686732036137916466867950311217912679120494226863196952264941526
031838460165370450925925788004244059443494676228555085452556602069
422687732189463339027671535820222547033122347985668270282524598801
234683953776355174689122536788123818919954637844127033241873737633
256130422332711844450749742422378266197046104011122653275889557238
498505763648049409248040111162770871555279884079036125727883591241
446081033368567697834958256973338399561692398900085029353534299965
740696377950214085190300644954777473548411684055102877611086419856
726622678722878593933583970287883595451263588076265414634051480829
780026991158114186054762951288199430838665320484267405555510253322
746134221993103619477720495243388211785032250462292214245695555003
313775842857102650520168844871279318563741260727741141786989397154
226727632286584861696775818739529467092943452537335065005696013460
731802577679099155134503484158398467526505737423974714748125400935
781953545973458470765007447097325493931035952440878024276669846308

426444638673638016286511939259851334195630392655957116711230449799
232117360684897591755826316223902262248984622865793572796249103764
038231040804819749428592702463212739006905074988773173218854689364
243894209649285664424617526423270727279330815571558866484163195316
141173690908132019027389528425154996722430348902376328054372916139
822725299083690889249738852299592377545012678088732611954616362090
049010456572723812186073402918544779953528995921194551880521382956
261774080412733037942090367807531125676008042604950740619056487808
050234373558916155389927465633076200800996912907610063205373231269
372796200739054334374622102589957213978891909703175501973559378692
404636111602720633833840903223989418916722494065846038688359912725
214200892792743098661635079572837905143855138729527112983667896161
918139486407368579422616671178599051403823472914009009792249130249
978704003805414564064290663790392252304492171609589052028178161599
011704358289134091358894393049509659960511332429448250983651074985
148901512823058472152580251850119996729092301953415910466349751480
976575854195444193866030102393289408190947927848078304524045296572
175086034722585941757724710712442707798690720995580682225197899864
793298473283077934323740885248180241259936076280749955223104632199
106829980811820415770202403786723523630415869096428133036466253394
949402426868625765918335222413541020048878505699747168524067286575
194342570625734736675473650312557208374721707773173242611280307759
065469239502290675828206127673498471063519764466442716720938461560
725568203791375209490395388168627203084219274933011968906502185542
429917425792226204508851936642892877228775513009477364630006424109
929900045932466773531374440548668487587771578864754288251325710249
958355230509904068164634131457314484939386983313477742735215011666
976795032349208702667063334267712152648808795365929937351899242067
225119480731326592334357720796452977775486743842476892616068173391
692812463591354946096450111565885729253824469117197557285107865529
845427871658302341211115530007466352555048245234541048573187791003
476233058840725783349844254988058458762713484532692040225683831642
322237097645471405091236413440636195922050000387149001957833598078
100859888605597284847911653207087459343563080218716211148958962680
089733082142060670637691930990581223716105016338135993918359318713
030449261801897108928204261003161499659858596679590094337472123334
438951788213921316896137345600884721183405537041739088655022155389
774248059789541723435365861490158428369474900604308895256061113875
543858496272691509108623975548059013623804035826260053073574512487
107741426347751250977397730939072528233632100264802882297270641095
201052883807867694620371563957604766114500580804771764272878001313
127741324430272340885129896856038315246574743679823704601279075780
608924227128322264403079900615323141020797123800533820530512191173
870930198385367908173470015597347513283321602546122502724867125834
953157390056206935329540235971715394692464427438802759614864941292
475277337413495071487803353778663631185623061534867577670632549018
682958003825878848517645563087087772192083183889862253677915514040
364912819332723464641041042088110931260983961840185584637023543594
672485749153090804859418312609672994066748441944008425983217548033
459861740020043137850828618329975681776571031196315818659950490672
379791787367219075156406534377644763062170576698567087492270750280
876724620422535389447414241343631101664366396998808785733504534672
454066921810426881156695438451339196056976913074521339412192950876
911121752555502502270789147173800685423650303237374867787025583890
598936305021008716646978571180516617315670294795858844262719841035
509462151000546726532666861744293511621047567390241977309661472422
087045241117074333885178666262283897595052031288720189523544821739
478219424439975899983500607804123879219306567932572487287019768589

2465232506696114739730088004771044338654372268843956825008857164813094684461270956543139557547580507935142675631381936217024229731878481227768318957407048462335383621659586843308579628537616031367647478472565866092667662925760874527312462223550441586597063699892768170434994116812797928212969226927665070866724480088023309789282035593082888597853534119785021182140046842505467402549771417333366895230491683144437629773482720505996184279601382438264090054050050543899562203257694031141072966938554123861845341651357163376862041548822631196037539535157965115615791687520081674536224127084386082271549504444160687673507152786736164990684079102050438446012308262194961869244003083142930447840925631186320777512643059419959491771706413089068677746385439177938197576164468111066389323583618409159532116359788090633295173261028171014866694864396113101717553654033227890444701001976312123410764636245830977326893259327741919571499244985596469761730467291526992474055576981593496630773754704071040347974975517710849312135392429329973675722029345975758802576403917756905121552802314521029313049965333937302320092050540968566150405868034163793425219738612936589718569331253741457848278270086059892165403956081731874297587191910693275712179952087320826981863598460067276433876482068156113679122840973794403556138714059739036867993047423692350310653421466641916361401970239342325752694299027469339644081593015791879515442905334626141049155315688450125585550009133220611127010184435006952174617306595986416881093816304249760877061545783954493791871917977821405022579049672212207781602380149238129400854632371404052972776933521421226846185978211744358524205910970289419846246899952324223981655590466771322876361960183848743194653729812296335274062864396104175916247182063541974356857130071675436810628538358587361141488327923310460061344514130864700024719564721679855282367690019636214043564985333197688806488862002046177902558022709496834234610834925724753544521339268873693931289850485958822328147510381188809460451439780369235292336510900341393458467850412291741350494370619605866619482665643182166566296456063733475093289553432517497273791362242708239356455824432730660077315547130178084016931960752234418399227001189768898484463505526607269164237837900804794161768433438322845853768413811825116775638388925839402012061356744691967075752780470693445818276915779568602324958841160289598378405115255979783881841029014784762345089670289667288889647657977823069979881927088351349330445898170864681797111283818200204419749911772154103420516542702703199984663491469520761598650684885220316328732259038150648355010008451778437545074361450592219424024187262824903563412652643119265503040632025753660531509186269423040093331024829860919247004411577693505738700428590486697364230705586127356402054048267764388090325802788508769099480925732901733117165305467839238783985992084494944282354551111075567141794156706197119864199564089819746965038359856000153776319786594268470653207291301834658782449250781987830112350426129913080412937786711714417436494170413095998427749615238523119521635415716204185542617257732897542748964252455836900805206971989783405327952496283264156934156139985728673098742226214028851812860019084603231750133307807483123616417826138051177933714470414812118995919098022374678777281289045303579959340056720430564404493547802696346463971915660009263686421734741176193906414540231793906317382851890365239862900709983098625224892343712260609858103040957638851092438950976336494774858416747664414121044269967768519849816506753795395453287080549905298268091627341027240261084195694627375916305703069360493049551760291677145256732350336713095707817696108902556901313383745115817342608720809197689623761582472327996967626594917696890350001810871147385743498364099092793369454718156290345287008933177864626001848458350275093946565743519

966246939627896713327530010336216705802806020327174553177354932757
128128972230169431756595295821313773640765032582412095839422766174
898757281848476302875186074619815953091451776753761422878153832469
158203968556471318706824864146943554341864240898653402265297935042
531053074556474458598776518827085892094219965186703306605540243235
853057412498897188396465826949814140327296804009856740944722029444
144102681929134110942187415132828733107982329251450661078279210121
801031119704277690769430354497528305709508933635333331851096902389
069806293571634303881507485149898279017106593644127074761129851868
177731370472344540374560028621916791322136578270702407950996791418
033482196255862669519743611248846594658272876564498977159444421182
349275517498968274483846421554452254335777101571115312646908838680
103921379271888579610984245440176035892736009257986197434102460936
289896955972601919021738070097558023753808355062111992331474090884
681435199033503297762255410262247665024469857171353572652882185686
112393022122353021905162773569741286349835408390327735481050807116
389655729447668457921741120944840199655614355306898308612084948698
645171066737076918157254460466820282052648523358699249990805955192
782881458694136822989769294696665752303826034310488580760551170830
091971011984925364807283615697678187742215071103739953692766152452
449653509100345288959777029742778854153397858810660715689492621308
277949265494362107345696787855068738556113335206511000951672238444
435331586671798353595357645119668095745274000424340154400490626661
379629574706527407578371460940023326556772078370794501025269804793
497599536312147248881206599950447324540529569491611354376512759235
712601428038891439971320628419546819688588417529292266886442621434
063384323544903587183699050534709475844958153940545298839146518631
815093973623321405450969089453531162350231626454242878880321886191
725362579809217969251069766315534866721289025535342766321179954188
944080231117124107891063911791380111690841561471043264120324352019
466878197773271630704885109519004363450847138672677153789281655621
250931993308461735522700918562419626104815284640455268181769832360
332944871799050874485624405258466705707820296738438699369354654364
564447230044536227375266531452992196223098811764568067671397854379
165847320950469443043848837077186501843256927011517134949617369807
824269408040645009225068239337956837176121539057082385854829100457
156285405230944770923367992115848871886152442294590478039610523547
752945184329182177647040126741902556257847710396875503365299622144
881793526405867813332621592091349044189412312552153995977952657068
257462407649548661139248146700657227719602495734508058408702213078
799454350725560360218895109924837033130076618530852539127407692038
969809967691756070148475757791425271380220941593877488494049646831
202572153355580648888199949024491248744387026964819045617856715182
153214214586889657263425577957562652700629297845964506174896425300
021092247981808889462516757327350029677147050327998757989962912792
874660757819647948475239203522264559406312628324841753975481745701
065709260225931723208740932749140097978186970260143742147863179385
313081557116718016184198850077060828811446496753514621719552138523
925110419841081641404822044249417827953801735743096369443649506351
874820396050910708920984431416784064461910936771361505772143004751
642927846276061258793264955586447186590298481822016021004997719836
180377821189595651689225393418263501037136321157781214602370936691
063219118839169527053631999425358449086023980831429872436326258410
691536617915011710034582739884502300190025577477431214334215609102
142563004729295216805128221564798791314836123033461102775203382979
698476487510753086209628464767863531202306860509871808751282265505
281989169994450545825287427459706340229603356853886949799382828014
710998600870595684841215196473343476633762743513560052413635371901

147912029922253153318866025153770683310154889099953699017173694298
447066677048279322448234182065674258791467811151897477490725584891
554664700433292220398950649452770755329322994262674332988281428255
689657332667865756193371959390806642919147503111328446751487396357
728793429897564133604265814305092134813679852888292572515952463686
670526471898396196359849211341569531986380455814847907178090786952
927279184045229430102942836711083950634842506382085558653923276545
286227266488341850748741393854930741763027955503102908136770694302
897012025317701849418209149792382963165602229836198034820405795231
938132665460972135883930448521662872143525751137236845059292982495
611474226829218305794265334692259325226372031701919998447675715410
489691608200556092938804385615393938861080003889314155142519880728
134675709393315430052569037837144515997302344291450569986414675950
303181271527278942502383466726331733374558809943544395003677567031
091909407161141154055734088483428335359931921443710600685936602646
993465720093603162344426382151424428391715961465813174247317051563
253482379314749449200697954801646821898843327591528070953988212683
989364231810654890448048927684892185471067697886135033352449811386
163739573102680251642313394115498106267049750201361285327167945064
131000846851005211213639958390314022619108072899098163070206354131
115422666078695487177382021410547401402403263196456393009621855171
291432652442745944144182756021445515344045400287432495777058449408
098403302725818096317965249173507014948466841572546592899494762363
700598048903965763138240276942366626226901514789471169049802549490
252587822206853726045733035973090775835393170308887085423495312230
318022928795990640501773733442130627431431286152209051808992390909
028791205122311868460241332786097500246750791712605774045752925643
648232591599931664450989019146148028600254823438719690337398705360
037631100238503088551562739426231573907145932473869850297213477822
607116809960060440749909341853400572171993409130820882341230507612
352696212454550842224014986166215731602997904014966994917274623782
673017894094249596068843773894743154058947327147016961985825091282
157140530857846619186132280610365034200390130814037928612843022537
881128448946883053710330312357004346231659076876308980839541591685
205458809476237119278899074678575366992814235247044117166525289100
360537755973558843542542374651769418514470731452547687692899700595
461349680397382329362389980198620565056246066813102252036805483248
433496956896531003852375466758785679940094642537266615054675709129
876236192670083598762678555030286306687993922650993773988460839066
820820662301246129973037801157482964498260524705900201388562441463
341058145336313282560532052005929264808303132120350787766295846454
418916953734242813903516292914225048390362614442557358107486834892
518539656531498196305571241999503513238219511869672381514583393490
828597559919669721704534304327316193571896762686397897886844388611
904180944185825869579602634563037552477583111543114762035938698507
377670742765108248192897681013385969978614342443000676843942075955
495740760932501678239929600457403577583331065860411555508146908508
483995958616766251290531530322439401172444887403452240718053522187
971473853132168896260285312534875858556725137078186863521169633110
572031722370604678119508611426666560265806290848488172315565972646
242491374667624461437195284569352215618440020898130159548600299025
188585073573071529323123181318499178936373414539913411022035565410
510107724359860556317482778472401379834746309724538260700785102526
381660274367109349768677960974264539232419633798976824844220980191
330548596452868142462908735370534116434769696472966752583007869309
307984984271168790440833818632677597338032526824224332968073950641
806715656027142682112303123429485516536532152694576106468247007076
793291110119601654687799208741506218692435255860566081435007054636

5728251289908986529882762656823007437703610580622266454799657003762
7144257157725812686825466294687239956406509728324077614389187704249
7676234221451309179368774606291685183902468031364398974169613588291
8597365538906457751308119345519829181491284576515229964072629833896
4187393802463224707363927685951291328962522675690312898982088924713
4957014294468682265559575277458228284299728097773832127325305141699
2465522506661465591357598655474894260157123808779428472791143722367
1115303827477533414242407179488602028340577593858604743995094332651
1364877543242214726631364859057771943537278581043718337900951798179
6311350895996055705684217532144079895592318480852487808367550068345
9483934247832856563524764189493364922492977088581624231502049139804
3590449810043149323301059449024747184763989316316844434740475056811
7237464392161343183812685470940210927957187993693138600003846851436
8458211668002182820129991748092639890989566539405829588860028274480
2699916980106424049242658107592127246906309924560742166116941838972
0402845531707331523151114205678509496575612039007683156747627210704
5735661718784029626248721159176926327063591387678465536199985983182
3045145540096974268503775012008321332069588601973051218708541843880
6108838143347134325666897926277024434199218076863100164857338404082
4075853991365579081178390854206293306812294729335597163936794620335
3915270906372069467342338151383426947141024398224388308667689896244
1571506447479056693233180852464295629231016005111864095688006805225
9396549137978135636662786584825404606873151675168565871305305211587
0559273837534786890808148447224684033527700001191254671233038886156
2966960158721813302472824787916219784911643328606722244635288310532
0858360458051478498520944223549847131411318817352440769169838170078
0317068336361510531335515330958466236553926008021258336209667955544
4629622348692244319084746391935174960159592862044537991279244395476
1912799230415844328912730521691240813192340886616809185433824240294
5446101708827516088104274721924920945394697026860285263079403107069
0991309220431159645042981910857994426090015245142428334158154882276
7582003641525105027903580048053428092302369982089414594633169359389
9070015559835064469082244498619717044307628693201445951564848777012
5948501151405397979633930100438405319067907436471299584718332138871
1090998235667221263805486838741894010107540013788101839315914880874
7865878795001515539644076162577508186358217931011726144046668884818
2194560440513792488319715457124384193438967247551444236658936471851
8226071883638750704592945198112364117809723074340955219414052446422
0090222166686548356081018514431715550874520307066813582038843521549
8794496039494729606547080639824562739470639744064051378929533572858
6689990085990869067039434348708008176071405043256903200881536947167
0760999678083787526081073299247949817333213953183877832823591673701
2583124289064735474215088782551498414142350130167267508217911100255
7972517427600739373192142103926032970810769819058988038916181382579
9144434322105291569057878847458211103826605498946693764305728200945
7407529253361163657509360063673498502843887265817107259941078493030
8130117586486115832906605732742706030699969482400586086403325051074
5707713620352068864946234411905631680698924069541256556398194247188
8922035224322609723175488853172826677391037572371010673933766380898
1515842676046852036203672407891490187927355618076617558781815383071
4572203881772971875938956629614975056785345081345976592828371823671
9243140097410543729784451016257543958762457903820890293493150148104
8364513538889280304386036901642658923162207218966651935591627497955
0480092555159519046186067918808187226970297962202088800248778717791
2581021178942507276131979543109624664019772097232261002686374553001
8619088598256565626527936608540782281152515511826170482175695615416
2785139637782479952394359312469752436905921867794263833307337392092
318903473

965575739664909820526187772708461695587361455929521981268933164433
823973142384726159453088525408871886577640223851051087973101722522
845299769476173758249952632471411817301602018331119181122281350885
698210048181611461676048518736999561147171048969514528496967581258
345361297803477323132965359652239063357421420204990479779497140368
721532807064357987712128372358671861232848101428938668857037620723
484479675929144219538642201210820779887655110250311529371693291996
515388780425162560257463240850422884499776238946926963152865729374
134301693600760353293697064178277877377930650105697394561162192403
366764030907835414551383164348894499792375578158750144552064869347
130864983300738980880568943460652073487595248873814528563311886966
371772606539290129962451213621618212497585510924608606703292781986
026525232632282648357227323918383492582021275938940571530411273628
188361027150253648154993700996249826124488262917986728249764948743
752377218823270232860787867414464676075600182094890673465740757676
455255636422430212091679637261321178552432544516726617845496108737
903746066329417692646345850239275544630030638667795115643109766170
095008053674662921490319228931546134479522051756730507822406654121
643097747251304495228849925423816543796406991093037267686443698771
596745583878817823699939072913629511356585921262470605123799207098
445306806751021923597625803820917959940362293224160002144948306949
888479840257115758866426014882617247760922755278249318006124546804
432703694942805199693247405676562139868967945410457780679381898678
743896031549308087544850870841396937769738770431774286581696401871
131875555139837434149480436738506910996892834084554865883150782980
444429217813519681652065459628859248994721733329206073700725483852
165986589889077098835465950445851468112519578727298313561434878739
357789016506310758644069444839738465088854412407993571718927238760
433926754142118240462413909054293870612823904674491829046176385817
767365973869331068507357006883922614400496752244109439386794743104
730175735221430687079113899275198337431708032830179968893744273125
215821356317063512194580243208618521125215636094841086274928276882
562501686289940598158930419958816428678871076201109560759741287230
522131712605071600536828028137772342999917730050094327364174874089
319317228940480334599855226410892982215449118147136348851808329524
371156096051075765109787038763128124395958783094617349039834758336
946605791545220589591440569186560964276508524589570587748288120413
261239285894873523560338104314000357289123785627432502504636500743
890095572740636949446188365286660543726152454172205829830850965936
482841958696169115725932419827358324585383144016706816148734750861
264875198978077618791749682700117124818918466138987978076114957261
378239841534622457347391337669061778806054634749248021501000162411
212171394543562120000272927378809068638911983351608116287091044323
730701597837680528428190952153900899546814560480934772723899921077
127088311999408311819022041478882923964542865097988111642859821695
566288129031070650000932023736562567548443741799311021817623100802
014046166684899134498942412244970191755576492854239923972482065041
525370392288105886341191972034439414870165759984553384786876567773
477091766714680382367313401639828407880205349546832959001441780461
553940129223593044699854458941497443909752401223342354965425154758
972141848937914350277894140168852816498143955298295540112830729059
554776122500088200251227256136937374376926594900873746044344756874
791717024790073562716781833766636490104058900490891903746456201786
773831536991028400082287598197485352472502014372421047928487967599
705577863311114044746980788275213095399080803117091873922137847363
226332007664951635051082589788722053445147150516363923455862358747
744662242736212811092579582876119166254153684657318540677262662818
074645612365270570261399929161653055198046526503055454849917940375

541108398913905394966914995624130865765744020383965166355787306178
906142079724617567147885358069361700198445798435091165510200227365
198453484798249644618744625830476746482617361233386321767888312504
912779200926688484238181854808972739446180692274654904673210947494
745911730963142655993105118144649849164657348892990956236056918074
537823430166274316235950034507637186799869307186295069833110875567
687675209574031434859002288112378236242261019467416930842492958188
229097115975301249041505839344052802235065103086218445972740352939
276989646953854696237284285171887006227343371697645499882868235530
232764350376352044233031219185947358945604266835449612689909240130
899833787013512452394879538459690393424706092469339833594451428338
166876839060145419707922473108628216859276547233642121807873842997
293257587252274196921100888994834956422769640231142753912447443241
958951046868251384157117226630817803408346944469160547679329388942
043396620864727227134058503469476983784811165925615771535228404276
482497291625573147864088200401471973450965477703012847556814101632
861813732440985906856266371605889070719967656969369557771869700083
478267984657239831280624856641195835594668646336824101470663496905
455558926516296221007684166189734544232485587531576541157523508244
197744050894099009451558682966610670967090242188567263513022076946
164741961012483500488324399099685855418687667730961363628557499053
167283427905552087494170066782095830556996249477733433635071383166
500916099445860307019912063620947261628162815430505579730019787551
337883522833593019170150243774923829395637935896295495380106173507
005062115379411613692454017793311067313680790629158577115343529063
417402733746525121883693915935520571525480391410141160760734932551
286532242006682870834932947027112382304481910806367017055551681660
516170852577538970942707296686930226006489657414662072615047086137
372372464532543782557590092567538422619308420193014170554457973751
975300470416861185897332422455389379399588897257396558578662120421
137705850100415363247338063484930878170555660615334112886073767434
207140406875176845301024918944788228924670924764545357811472726834
119812515793747938590136536319222853772590745949489165008906353103
732934604820812754766554480334689750994173377227300800883513866121
639608769508033277446327520211238336638907570173658688340714324957
376376131890231795720119331834684075713003556003589101914426717948
486409750703489157155788012167219481317424387469848401936835506425
753888595363540645784595258026348940536848524566120713520282183609
121072824486035709848598621499110720560382459194859829823543664747
724431874436259385244319349885976950847699830301144277916640235362
440694889364336665191206330022706964314289141781565157624136606794
086442773931508764532782000051431144169384502136024538595273582412
808228405148007249950595042158198022032119117038060227262285315277
822507982816018484804463884542381877821710453476165937626072970888
421769980440600360950716305937923533727603256790842966377149031844
698075924229306159288757880937545566441609888103227332432611609552
152055865572880841228200008754923083828398024930370185292062238770
987704278101256226828645825820748937450657346805895735712702446993
304075235454486384826117949774906522988066396915491619456757341717
414665666776155891302285172493749891423154404203600837602319326891
099755462658514424413069169998816280890748596904830687724150091092
482260722119536627835626888806669171455147204065469867244613418392
754666639492422275787260538619426620379390926492695130808143167659
282504876472104863583938056093091373069208483881377562730641491551
091108441252688428224477997658280451670887896100155962325457841694
603139229878054528721555967498040850462270826259860992156066567760
591964786619653501186103387301206818626988983175553856349333496203
625458890583539704996418141434400220806029988614134685842391393401

535642324721425301428591205657638569356277624708771726378207859531
814331466842886144057787948439418544145539011896337566375370590242
787878298698149149790489982181020529068797424909942843227323929273
792293589493756346044711741346181187637755675531713533286944678866
336852529977913388840625205355767274522687590942502698802623029029
591980071872715439050757092048947427880811216175370570817778606572
327394436896759719613240934977338071774850676806611000249709483177
585048961316817104715760132811599797966971179648937493658018282931
468853266706863564738459193816300614124738538845590045829273118304
531949237507394953597139353273415735585162108563929472189809593934
148208358495693346918311894798014433218146335283910776628884496086
869637826510737550363616449216691808790491033399484824422074745087
379193547456961564410207232332009042650394967045947946721723223141
046197781792322796891151603338708596999475128428050100075079366668
644883362672503775902521102625881334848928731366640495934702802062
967005570773406220297557733498918427780848503294924150006562102268
729921694084302378690711214874594089050476760915559737701406543115
136419939101532216931977506053626463245629378601040851025972253251
730926509659234796949922483311768116594806088000826189237040594980
770400173967458218924295489507168771331001039785153353453147106859
570944506053733591953658617196314970060182160151167816445877799372
327454843702167196983240352158951685355139247820559128279615349166
301514580517888411480971867069842722647325396278568856218601357551
480258411912578616981540276091270507419511403965637928982882852274
086014503953521126538953883867793995172958629270947841273911524331
129259464000726134649249972681810945151440336063005287128811763455
747440049685242334339885902991306423665017309231172549322356936462
708264736539614701936146564880823663856871274311814628024296708416
102187589986096496923503129824468202286241015811833808695873215776
018223789618015767799892633527484089573104084610685377186963984413
185846115041482873050231180669598634179338953174698177066866352511
778784457585311149522528920915042815749224555913847113304271589135
534111423495667324043853073572462286846922284083793734484706029169
360139930409506545961900308183802527282563082209372697527529895207
663388958110039346705019550272152540707842220315869001599513082638
484197487776849916923922196537231665447574407641010099650617126279
590178098819258884779270810165459373325352841280200757510482591165
251375783401008951847685249101862211735553486369698415740199556028
651242236187405834090844768789715421259530355913270720760615089738
197994811856383567450625729469513436916789182630665079524311558263
643719976037744462018791016135424477006273647656549196267616444680
772190922929080866228112959884924781148671534257155367603119524547
561565236967703853704768866046980213852480305611763820687575067581
364128628780137765140619072670023309784587766918228333916512907870
760866937517116891741627690809551757279698052016821027162621360697
060416091561788705305407209797684930266376389686627360501139889883
673359135139603154742889747138541696939293183086800212328963295993
812716537221185295145016710843207469420490209766700742282746672221
568338634130232860911046024015612304984559140952392104609756543648
344461508550374997059553366602213462604839960122732736419106949439
922465245700052391362273333832849727557393356495943155230901233230
563434152351056962822867788210930846121568661144438707609582840671
518753498089714331027439670232072868301053055854740611207109035935
222453189737748928721223938874431654802090390081081088526692277246
193143825290972586311246157795502759822070616687080487695005424198
755620466992746639886393484868401422335287282144403954784228917454
765827194543973721716746000608536862066597343051524283327289391251
077992281717140591722360207385278926420803737929278643557802115835

757358359060207378135471081449504904591832117553015450198368358598
699232679963224360435624602028470500658345386692633431478982991 46
785381491643329373377064728561197775471780210572065624167806204924
960506810145055444000836915853863991964055830272863321210271592547
804919890670890668715121943494416990158181216296916452965667288822
617346469987438060666365012549193529951563426706148368629787919482
858285181278413770269402216755806828356762152712510027225321755011
753916761857082441000569786355201545717750960983903500128251287302
913885175974609715463747851581629121900284214937284094101651557061
686206462220736689554939314543518226625017737147979968920612065325
275418413914476113922152134614069286419355821654479670253694488544
417123943334998602980270985027141815954167802543245054515169222688
322276807058416857108824894008108059824363880693049401400360540588
251472745588258232852976555436600609761418817454395068351899430644
076770973415827497543346074110326888197363251029598689127139183273
239406546776520570126856021583738546587562512867371973052008354520
571151156379671524228543756123193671768567009006373542503094551019
185119622995140325974379209354374209118431551168349851325491586543
819283995573831259453672482506711473896296729908788858488825997010
742272525040354818049628522570520823634851226938315628067283671474
350559839279732606684295072635138275104197468897676725913690173065
123657317195324310128695610725805283363158900958266984173539638487
201652004906490640354985449237229427418378411033080777386940393450
466333220017422240536339407527505982935749108698818541052273129910
222333976612533246907300802700490258911978660954614122133412138919
626388266617865561463442108132633077176078663806363424319091832541
943227473241152377048775613809279604835764996953687485113088070624
761273257503766591078316587857685506619883979458302690860270196411
220225736605864073929143877225989074073598764473315527068466191670
819909421086085140332806017878514301499679947363423662229182069 03
244530152501557857601522884736715182349449265482885717976992631635
275384557655764406775448906682489966134559521800978142208672688630
564291540138296534685734899671972420171760414562999130251414447524
884878859521922268437236453296433880788511069933976961277826916254
630973235354514622641376521392975068099247765973472612690939787221
392807455856470999815894296685155627040336605876230534031098345932
290012698322692050220158143430280800022365801546328825529047068250
736300630781189568719809407671585126103344019382962718808600861169
707239670394318126130974247111717901985743012399075403547165062880
512649200956055134533362029263066498091540875817634021299600161949
111034937515765588752482721423767972642564269773681988908821722699
265553049872479047981000650468434265519305954864635159658341932560
199539427254246373171515176352439514038074244238577247656596152456
746699316286028264855040685912246340769663976942701366938829185607
675820086694383375631459329417145519966697728221814706209062034976
317120864660306056756149331091673596869262998033651598521300888024
046181683318608779296650027792213476011617481352000602598951502189
632641538513117752107146410707257793497182252726790059933701865961
579327800661769111831676147157964648741069522453761145688792311508
863036238407176919599890035658075472701071991664477592277306588961
119394275860246657291646477829595177925217284078607443295113156581
166361862904473631555156558233734102054109705539810551185877748472
553177016945619773274000762746629160155781498938499154800670015880
287271396477885553340501525788846719648020524672424493955186437255
844099960063063002898185781127353194003444827534004296065595089225
351860155422289176429463530972839057155251922006718225204483456957
283024858158777835437195126859654267081696079823466167925926073561
893989559028954229820069011350119992783370576175407125834805827806

812389428561931111413250493316426052802330398141690067988980373762
032099516438489058699369606830710123179615266924530464923746322846
279908242334798330374034450178644803988127559225902046536608469102
137948423688814526296968662135664767020142672532056443503131171444
137713514673671697265452265465106144038202743516102254601111621957
755376909012981481368832862481603668992773908339405147143384874769
201164261948192396449463200446384746884791115170448439193867653031
100824238704981524520405573962058183162486945032953509198580564589
890874425683122925434507473734151952742364764180281995487317377212
901252421446754042945855402423906471702257004464248759363876636706
774796202442768094377854825766512843776900931420986660710187029338
593310437732853034371688951848025477991273331396336672506976240044
439998428715482735400213622800363878357808619303281309994905965891
889375075383442692557033208903758324628467949947720161618492448067
072606642394323219232071760037252313600260079381178885240250457731
492535024973994254405139909972427789185118923895556724908833225321
079886270815640003225278031510976686622833467773834941746122589454
209800029093296907437726013869091027914062598515250597766818440110
781670669340007514934828540555614304755391105331437574642294862097
446790844758764683631789277308598851113550957947449542186298667749
669615947444409273113200669786113808590545317626494590167819697149
861399492269722338452705143809389513504644375512254861117089139668
089366777973099438488086193319091043578607208496730738259379239635
993284751004072727938626271670447540757192025039341919725451894831
172790246140496689551790799869102537648909648459816709017139351206
782839579961226311973314913377818343384193185236753866290190046334
332853283434182492107191609276730801323536947264848927644297987068
112565462806117260460733219176304741215064030201789320579568760510
277525053043612227096046758453127386616521424194086834083758914009
511413392954577009473174811688536409418352861099710341672355817860
282349820203149986794017404171914028393621051948060481901824518008
170137104219745921279840402426896300533553685494164008639217745676
953445865088328629821658460164507800003598985212901239590070420266
074498878506271760776222550606390745319477188925099058100936729891
954099576633538658378098044793537799187712142977461976409472721214
023532655177723616341966074376391538660194501776914006240684127316
295860806350676293775125293436759607342719967513520158008737395483
897343990123825656829078591144258828521366169201402990041314621368
063235265086541121808474998758441118362909790891983411875376649604
809362648932910439059772563295581388776744695461815797870419159755
833500037728672460735185350885495464202930971585636949785764048374
150555809918840209343333512819551455518573334888164916769442778240
555433697731192014372254154192745059760137984144373599420287605255
571893780411489261257983036183801077110050042392392644156922779005
702794740136494229184793369253543248404878023177744518417795835582
575476184254957539587854945834308480441737087984926741952893230314
895899160068542287957892948159000687353733333386381390842099487255
511726171087294408886844785081163455852692141540299672098869371588
623222033148558174268263595068936234481342247378693946092325660075
472125674141034218490131189639676547329167424718993424330280290926
985075229490797094430092386728774393625511131362876159197383241349
167226826122826312779381175080740895894397190025426412664947980176
442016454158813176018972659624614413913192067850352463830657764016
522973204708920114691713372287863037038453134194264414249664984474
404861773851352937310809985947281511173650719139881269307482690614
039286422766278494329558828593837214847507077835511989622491195588
237045064582005610170205324844490215022945617655618592145743990095
548229413751544136132915043046914179941224609338165136262790282788

421381220775894240388406229433172798925941836682936792596042241845
941770653019548643948170335524102874704473117150703477508343237333
116325876727636834685844486562539187239460827324713500448957086803
150325038676401735315079400345572855037001845602769961506156829141
611706104616174082484626703351891525518824821272600725957656788991
256787020149786679084710663107487646730989091471979879259859062573
364973498235031260983098734686162739358057907354908246830497284097
732381167082491516373468087051052191917620541698826254760544581771
117993776778696542169925785577204263442443042074495489703390434507
206119007697363517401952656322139282583120528240037466954590928045
259687984608140870719534254761388363353511911214414314850552012358
138062649231353833876758091889237975855157320365883176242411691675
238145885920751640352366843726791759063705392197981126597708113947
351671962999970520901709075898569163890664270423570074450277512010
490394814832945744380974626035105789819540763987060787581774072491
184506978842994138342060812428390148814872599854181480292949227878
324356105549156410917488770670678201198591095890983868839511718380
134914825499274914269855259517765361262421572662448896096148297970
308420210516027967147856159406461363824775890201102519923421532100
601752302574212237542107495918728675189552155329945322689425188409
422826757744222275528207615607277103947182566802471066067738312063
031466284744338620474325956850568928716265329083278784399650716724
220613829453291660046380872563059524153288120937009922989600698139
627968627956763519874182401856293198492262333643189930901704881995
258827338807953265885144939381241154327320589164578642294526841175
188081840509569046913813443077848902114879712116283977344305484614
980793562435496714129473332666422483050043644545470274917232078399
565086801876173270331906565795475395206929252015307064350471306881
289032053854556398799210109566729828730479466100543163584623448744
155540713381432344131178842419601903702868696242276246502400420813
713504640159934720532755679503533906712721617790603021623997804185
763348100544838870317307162552564799599998693531968130101562970978
711673069649402749166572674943432081323146138869255334949294118316
387788990325940109341115727429801331814578281687913514243955787342
059156145877368988080791707114551154762661682081777574724842787971
298154823852036895916524054373052466728731303019258295249876393209
857312091610561357220176392716995986861088676061338436496169518654
848604616848482407438381180741734222448394789399827801473422230578
654102177292648209140948503501322415155999016307147814852705161444
322582547801434401095336602393626424931385229405475364168364670415
743005583729847040181455710029696234698505997788557369980218223035
492383187976298941569772123638440889178927527752847144776218920574
254531510374799777068623624218990630309628221177321970823034774545
506023858591140389896916662207787683260123961741999762365506363266
016990661168908868774133378091951614977054921417118191101495434786
938206209808278789573132789906002086339112784715368430438149815097
030477088688163597419253413412219139628745781099342483271410917827
631765420963125071366926365882135751199875401157902802850780669833
384183382667394553683196565717031946293364109272667274244992747322
913800983574450913407993085683604846778665354210586680052115426486
749723953793642156653587208185541259819454742751611841693662556324
193591585695088905353141218336623998780221150424566001502364691081
599741920254437873610226712941228298888053367943950329404804961677
110728788103161467730371502183528902414648175430954905944887541042
501680406999981967193798208277334648558158591077617882849750546856
791399040113845624360403574402461912376944022007604214335733872603
925372466408966491471465473007219529273741767961356523306780426820
002423074121866748669462805878878738299523278050107066101385402880

119138557309113805727558936819403093807188637937582154443516265599
120308784820523937493518955971581551828048148078313105350968235671
612355096316152866585785901952871775464250641298494510573035332433
489097997209457291700958784083221404405014741143817095657712526715
779817097676384786420648761367939457844238586035649664463140131048
667056134627690174928322900432711208192295641589481528265584221926
189025305138159118034231088514912266539179448174339043227005068742
888199670619606885006733283536146804985960814028791196311274825466
404370384782884007981579134452321016686737411212944968134510122379
842940219479469166481503124318739691196021567121142279430682010408
767568093738982054684214785601427768820188071717414127493674678175
132277109046730015309467773069223290953495694491601169046818182723
477816197512572785153556986162922029188474530269830156638749177021
947605094174866841197276604440854832975049890782580615390200301035
218027179209950817839121233970070027832906193713368218336961540826
520392716151802279152864076831503299740609280483105062551572262871
313276818247801169154937862483224017238370727160886419294211817185
646748118569400870529961131410594582938703305286297918826493247420
179551244232947629034883495272292830804012226372276110275203800120
937520885624064531125037264015399643712337079037439512032028535025
474827798093030202196632509329094488190574510298521728306996224219
547101651939329769370654576333432433256433136391221083529135557700
812650539793164681494513412075121883505473968595553878091947374656
627903186816171364161521554077453543804147984545840634474745085680
280862324126969399311229640560327310927938491691879333596148535170
018149698261648003440278228139858790148628521286746175384850348068
386520168074767440211476655696438739675796644295886409577559154192
615071966537341081706497482220419135035223932036927792390736955880
577997552601903081435508492475981857797980298126412519269808310756
574659469351128562797590578003419323476001369611447290131137293265
188773146721410741275221010515155865713493912768855730064565566635
935253545094573789688358002770721080754685197902156776635596085899
524492722049775299815458625359295568858655219086203488574294435488
340176616830553266024583599445433095297362056428294745396641017593
878078757904014319567264450256538964052007148706853706562711746676
157191810642280464382686780641781701710145579987859492981028674877
471679017435503993169683528592116481487861286289116375992768471088
743661617762393098294982340641378280711217412378342224144069984863
288239240656377438980846317043398141901704177045856467859934941137
671018510482883458475911298599113261806311665209771093260254577766
404781139798120273962790521780796565933483031963863726360548172617
544026268435726451474111436197474885328135254323132104065537475356
091405133576448451265031599499398475912338676258998862826742421410
656317028268420274277535478527849806255477967309562135010751357010
814376712097361065693389811287700643950868226352419579899875244531
513268435771252695511656842606249795300564134236794032686943202212
771552845229455906013166300233297861842729369705425640895722473147
960842279960643121155722898884134069015706079169855397062706949869
226131550595458162406654276265360989882469228424022170510920884522
641938004050647235141065446293973629500626870691857435034689078702
526688990310590307332020997707019556267080078421417500402487764369
382704966754069144895634000929852354991296996046540283212995128817
182931733802589343360888639561881453054936332102246702371234907974
068125687248833528018208796868376748932865951563500158437345427148
615342497125226151780864718350199049932178199155353873785623455087
143139050384328696513876903290534015238428243963913256867292722487
084779696809273656042151690402300750173661949935854471440415221410
647497242161523037246612553017965334543540889800869016307189158808

455522198431157422312945196703758626058776263336684850752752847903
442501653764916331792832073572043631496293326217198412041496939002
157898758405506883631318978280327068182062011470970782464776027629
635132451352251922427860447660178455463586895641834408259553254334
088710161510703674682452922046780848531852956030260393369987912529
602307327858469371627037491448388916337500526216973323321524007087
580984856462897791809848253427212024812725655515050305700418119895
186092167878092709372414804711203209275216382196930053667750846835
893865445281522353475575590763802378521519460182724568236395490568
125932747289545698934471141809813153649597485154106854609786330324
864298777018635531367450811076763860473346402957688211528898135664
493094931043589036451413175524147809925350591481309531471226234135
406460905980393007295409692136945644985787813148498347106185885802
583353800866348220745875475776920674100675717787021480249068099354
034049780869747523182136878867192069420344186215299004686685357497
787101475284582152153509827098635277035756293849313931604238871689
907398239561898872218842061190277094299032275736687186079770707721
559232641813601697172777407028841345717732062530298816172640649801
780192696415503872932862487451156694400901469103714665895280769075
063099165801904028249123963456798392695655715947613669438739333006
837383081812309420441592959517187192492679792971295029491581185269
446544888160644890910857473849722526625613598337391783194959915065
439983793983509005174338788171829551618447325706942293753652551534
711785761849774779733167716071499421452638170475227272955705001733
744013560297812168718861307295720086827047281509944936557613530068
382030072487238918604555823405676318762623625890434079602395529144
081868873503916372993471628157658628534654447671897796640094342024
901643410848630536408494091288771199650698045788071379779515912042
538025709198604559673886019016344148290099729713922827786387126308
017669213704934976904380819565618801297909146507544068453180516940
496248454952518633632422222614586995009297904562688659703930983969
560235643889686329923323765908114248603580354516978206769488924866
383306879176594263820092637789031658270502911378572975097400493867
535805163232169847363385841973520487633945539464273928852497451674
899046514594738849778743657589470771779435627074325722855361014929
296756502658510060032181461536129285910693432148786863743429770205
982545109580541744556250811648819563443233555509154070228734506176
634898004928111484860336673588052242980276245065806416616460933096
583227741201687004618647087354607594600235537343636685846300082989
115932174654321735117553157555101863096588560274473288174361764536
559262461764387774037997688608018414576447643035712983995556915655
942311745306594836250206036438221735440965535395090457436938651043
129342716658105214647736285604104271523729397812487106059859320178
075773843670656311350685930588860836202347780649879758531784016729
059787657028138758026032333379883203038968539991452029129125226423
489661976339718268106023709597596629809185348190125040315899625204
717076766399868353153196608687529125423115375034755005153674068684
699567677330577510792350490345778778263396473876429056401644333168
224565090395700668568009825182733058190424322611289485184918822142
539686131003371372243018683955964417346708417832040075715124168083
755413148247924004524308125367729689704440679643179742681593592031
358215146396789238533146916848019237856136657063609422132962567182
221408537658070683461685834963756556481521327880020825169618079848
404748675428648124718923245727848809951784920023813214950559768609
780439227213656679143255838729396288555753189851339047123780845594
071844832836115336157802570583643859337129234780510580916071579237
173588327126164560656794503480990799720508422730547419077811064940
216016666629509459324696808278858738268960848774456118379844865730

454320536892684034914234508815351016785757066687495637606663072936
364979840910528400845462332617818149797614121818728612686534607116
650955135477832306274118114820717164665265427099148478807776892823
457954861106715049473647399736480665299537367799934023532749648010
344600085639769736337322294714697302776409407753962883901258246629
396938987701586208937991855169330846381599653959133926292450056440
676097423169030780271559700744418327838639200591504192666472875278
979549277034150826391547896926883308166047956420911237931251245238
262891323911425925924887235313534140371673379930617119964803740766
444036108277815093719892667899588375052554760929420018430744438964
934968142535424422875482634672599399787959797254722441172876007712
608840939504811921596748873955586267106549030213240224770792652497
769449552051080564104822068021729276356152237738510636020972031378
948389087608773044443391070371387133163749092821549212966094355995
765425050659702392398001339719254398835335819326042355334640709072
858048652141533903522297060471220130494123455357789547124863000470
556288945504460862226648523847310733172230385641865311802179351207
883678931632431925806448500511452822769747028812975567230419389696
691151684034252191266686056130038452053893369112100812050252410879
522034623660139062940958680638104861567034918952889959443421807536
331295702241260765662837857763105841332497008021096395134916086420
519405513798037382651472598999021247513597282764870480302409576924
624369255111392235317869482928421855572437473618265779560035189015
876875981442665762658258377735379880116858111611885458719903282377
257659747925555015170352252043295945668950459779859313354603058046
287121050992002653919990220328029855134990558952962035081185191162
393317684154510610655162478821980694146953140216385429484650993057
590184254772587685764147409170800507658641633769358146607818722445
058379424375762870854540335011053709267102160168897425190172664771
091187734019658761319238244016778669447316036433235910530463926925
486473405762211664842699383303050481743973119350053676164505084476
304732563764216661398181721671531112891024558516207319700700311254
050520823829089312875756421875416786156917970097009380501429846668
521200159151132720592321337362451009271857273494135289481232311599
234615896909789674454130627626872563304682335313715965130004061533
200562642138432231482703547358635091988446533835684004841684653835
147529872668521047590091051673159246264853570708194303523598755049
297733307531722035573470236517079315932787444549816445932473938377
943652780031092775354373492970112030888508113851103629856205948834
696156362320828310842502873761364128467264503679574982163446945992
440232038663609579652731250489120123599962848087517002713559088212
824459212365972636236261931717499306027800672703852044386944202396
265943071733875448439649357394016544694254389183159395967500993298
475531832019709586811147403411464307749078732347323831843169775800
695104759625347803929053122789720283171030770441520409620275360661
687750953874913564865489935285108137546623023010441644060787433799
388659434319739727628098872553596421836886252528684292094076693531
017467606143615040224106100657518385781845943097381666527422822359
523265631560043634606066179929543426812782469802311078319221754882
824844348942155905096249680955168161417840810580681730090026315272
179607472345797447534390082662384080704887676620125669754226324768
414436029961270632036424767255662395841363346696342018339637825422
544743063280535146449208028401169350717192513434491759163446481547
381600589645676084057980165902557489767118772253952700017838346180
768335742506196377939729071866416577565549223989511605044361810205
941685054187658471590296259403076771869386054624379389529177382974
148463926922111853547905990950972997607473872962647276052104595112
827435631270960595023755731806195452598765319888191583707806094607

301046153140796722525757786092730919024691929884517398716315076857
663518649379762973806603165031986181720651187867335919765239567443
188879533652290886520080140499811495657664501357148699739221946031
665890119202261370639549215000080968080493651561790444139936547321
194227076271680662326799317221812574576091524551670509765492798757
928496616737619368534354686042600792782105050293510602699110772925
833223313894239808716540504631465861786721994997496972905979157918
464803703518738864382027060980493400455678609307653826192652631811
202186385925165692586594832744831643461459403696172358059717958882
905838152369788533168230133145229299915405923022740585503740628535
904597657859645738065216002631591774288812480183076664637707719985
469471432460415430871067788218769070796941313678460577018382839283
094791609157762283512933081599452697606840265942339288587110544086
830636638581020473632094496534988776275786269905353915532791188119
263351069193127339766674091486139745614525452756883805182285266087
134963283884183684848013626458289540073065079156074102889083454659
625380869625526263711592359482638454560872583686037619817067263374
481287637268299887772294454040770678994016438125002755420169830230
695451162998313431071889909862541016002935524415486259137615974787
091851534025688881301989218844566111649106326888324234483096288804
965797427103408223736673828255772507677153051860616483063355980193
407109897163098840874672591044598887595960530329248899363852119394
909682837108645713523456015804386429710353846683059689879680835873
497857626785565888383649516272354520708779533467658672049161720879
813149141698913756646454569742166633880288709311596552908909685832
466209423687259552095876586763964525810627609539552914917651817376
729189876677984251192168075525816004692672892837696502864920879878
222442107140464673074851974371397605307432234515525321212358708516
162852145380083686124629515480394533296837736424862964570243581892
316862264994992186349832468121494673751199481950450744965970432717
604013383552223608153400047959819368522309258814575459795669694756
238949572473252484866070654288256287673592883976141961905913290839
946916384100532188918694325657206677652323824731643807976960444572
969446502834306166522169870802617575705124164111092073577262039704
116888594037610584331302931444829065955684296487296377161321950732
199716414417025654689461012703082459723841828360403116603489153716
076232863966616561856000946715549151974673084232558676741097334029
846169675642107632231536789799885806831989228308597633750929464903
879107437077400846349362362400830084950073558164966916759777865808
111823047707424696436047327935182038471988961170359008300455685125
280509551066769960237387823537663727822677405205106153201899838061
149198595287500266294554227352605814894880997940795238178862244330
133236455432574103883704232398092141614963275539566361768675243817
655662510133743046480820755285156491167182834541304723879598488899
915627591908215219594535894398739198788544787855883930195317034712
402500720728681966631889154092375629824773736358962930373027486492
513691978890676358246153690757238118890003868340893039793751993065
381722873667753851141716182146406300539934079367210950941233283505
714265283194967469485021225058627404548110939587405372888096652279
949131350411485737581899491522734135204022017717426975610324053900
259385589578491549406875910851090044566987790296358999578580439959
472228433554403913845515754590510122367709490042907251632725064389
114149403143929338160492141148298696151260756811930048851604289256
533437306862399621812008375911431730894419899802933772305065225027
098497205959635360606930155405423580935622219990176155133694756828
973901009351898868910230562033456067478159556373737243150513104638
843684616602221058076605501633844395133927539050471981112677893784
252742930717142873760554379041535781461948559847060404010163365311

893068369659397595008458513327284730880048487740393171839649232121
257591559863968310050334405605278988766114724308920739067717133448
999590499556528668704313897455638641950652562633820072560532502544
979125893728660693665274756954585549379365494669059562648221660110
907222716643362714785994645999926749049549615126423085297640404369
503078388756766522322785359375567721700544176152275789006884922246
472581006854831676168705370471402932937942957597129456134524552754
565682494397143313858049509737206075394125741562910110232605185216
108762961134662379674361123425177993400596014944329479641646003108
837766887059200488620180196337697614505829217689137668547554193811
607470643646183550950427836763844727462071613819711917704444879751
377889228995884738200829650704622312728062105126991039445893607890
442906612891216488673293277124505955764999531645688888539374049657
657168809130392423485693764450199912028784002735463631080280488390
398644628663159407840102917796877741898286222270285329191455035549
524356674461195336689703928030763361342278481300805902452322986453
653475937924700977123725148559789622335055037140823738855989996357
257681528249857323029102096633655298097518691642928076192749832296
521944816960437154884759085527235864404802017581425754005293438794
619643196734902934702696186973283056021266285835941429141764327043
883997140379848629650744072265264163463428982919154069458223840053
228719781301203428465115000214159693875605989584674337883215104458
173349102931408492619625443010694755863267361339601315493600270288
226537550169131948172167727277488797907130864591251768962270234348
267173744476254036044361238342663685290979347302622177434409361004
734383259470556179624247014795775993839612828218670404659473671346
740038028868906734350560654196291543982323319166341577275777262556
671272875677647403587825186324014293512309926524186105365262390689
805767959009378267212512205511131583478239251417189326668486967410
493386288058933141258364873085596852364870957352273551564173160005
704356833751285864867987776268789119701741989262975037390169668114
553105639638913349194784911260028872720224362955411330210328883125
217903860915341167857762338801385636434712302531331275834921496268
168434618056550643768648443216916009932979358869713263459480476580
167302876236263754046417737117123337555290761684576098412031481490
671265447881308749669265249507283763395824831279554241699844491415
609082123420144663561151439878692836764403819999607361303565876406
833411029087823685230451771621814804943262467842034037569121810020
485713346838603164091890493318702825651133422596513951836285179232
653400316253031177685830439058553031434700094995404289931062006909
384298598494625764243642747550200929598219970537138567540242399582
249361468818357839052925627662578147628254905215311884516726792585
106299641121890474290326898290300195391955649073481812466843389938
765229124423119814574662009378080796511082080920250037221765646622
654628516780697561651229846904028763819653602313563613649766389022
197923617233885491819809573522970142320454913947600203098311826513
470831957908274217797322911001498110420929199727079391310541688430
564766806828814326393122924847528035155632827952732656083410881616
979610239630665231004370268230915339990110185178408534796664576963
995643862325907256630756039470551358975851270259376353252345454607
115787656777517132314071984930870727417259903834799083775222662385
016189000136966533102252957386519369099362589633379104117758605187
819207087776560999410980515517605243383814985788441580214830037738
072942220576114221873419120173809741916039089694570812869196186877
182344133397780597593991704085257402459527953589551683671046122414
148488235070420989494640533461029814818149838496287485469500407432
484301004237023770269359497780909390095551642812541683795122263408
103402400601731856228367117879128635057651221052346487547089228654

7579799198663491343367457317808000101592280324580415606204058814973
3690546883830568311188564269274466682562826056931827114971107908022
2167427372520560852398097051928761728182949170796810849563420922
1568005222657659050270810105964689600419159413971344320797487520061
970362143653107681949231466574683330194256177258029656287105746356
6793619642933509041759154760564372284823152054488326426763087111474
6063603473283432300941904208674812321963355980892981390137432172
3190294403380213835037846658247849282371602845709018509399863087689
7463612139222874464372822230689814427108087673984745198685973565683
2475850237902290438743386565016354309204949751394651712042399638
0485152404382714004998736609221595167943665520971624671902878497234
3068794146220933065403515146533353438176212140974995290813343668874
21960854630056294187185719897965490389226099400645813457698989029
37525260315947935205172873837166700721547961956375740701285592607
563335804361178569570327151086097332660021100882712319916375934099
712303202613810988996422209731755274142553634130615972948437204856
730536334535661832196025697236774376640419855390845866547731334424
306172126008629760529672078593225622923558462628463331892928319836
100025921871137039971505577493204875978319767136286320184428233653
934716363794571277158840984176877705511446946524042208853820282498
9296395084670465701934759746010821090022379813893938891736619958702
35322219629414991396048429927934116586704877857111337265349285665
368928875089626084058604973081180779650060100681097962304559548011
89768566196762423893127565241655983194566147167582205720515073318
743864995920772921715411527070251573742932740603038899701817639249
284449614498921199600874145592011971106191781356008311793943700218
567880801261181157994641471730354790676703916264480369152300163809
750954986478577153019332075870767280738241627299798565721981668444
519311512147215394626062415474788449253147452223428981444393566896
520817283520184910446850123635346659443910177128605907892320369817
781441406576579531416934275662744275647924957225617711224415082059
057260848147140459239530795970604271724592752165506005137153967772
426318259343164959994084881348962944398019280504469903074332958696
420234779944060855529802601687627459481652470013204750956993158251
056595655451124681216741044189201497129170591132046486929952189360
787257198954332687027642112079711125806343717907028536568681881306
493152903586491231366257190281160804499253317309331886197389137211
839634887920380921481773975151215550607142123784782495146494932858
500089951732313341794491957195493738102251620433558509640442999506
549482181749182982098028501613576469799306791322247849851295874198
976234020482642870620205113048454320745643630546829858997903588405
564610067458997230028648958952414533037922186076757219603000199840
534610423696739484659819890332264471364177828858318615641089981063
332566958801490875489118552944475487777070332542816540487011452902
591625351980123874140104360363092052813996081857821750591992044407
167990153384458640154205422766039495098525317146476566588668145884
224627681547177489770602016193969477385529779517920639342593819635
523803884248395810392756705640051774584154536477920170734268207895
378424193187181125134001913549075223018853664245615188640302650415
206225875269746448305960600086648084673975874694031004430507110080
328575618526003370683321783410069730737036403212982503954591566451
625180163585488925018864904176713086234741645120009780526716591409
458666952971943266261362755466627639747988313242076600247082565551
686346541573227303798463349743541760533911805549584851710030650471
465709174000822022033342084271621025506998295166962322792052067901
249991059790172386421758536906421361877237109133881499560018522736
296536092063731683978473714984116231404795457163102850830964929465
614289077736030956891413757372422008431953676737207516723402117432

726467077757072056644653817086104304913019933047479859635164887964
619449295289058257309790409506277206668356944279880549619837512732
746507601821215205334922080851917626617083613572333425100696831041
782079718606502797325776115716507240273651199086137670389886650327
557743894227581566173076972183643680194219584768114175757657802259
412293693762854837615011371209444777267883907263765061893274494741
092640586931989590585599696009720463567109558642594222507134228108
316107878545620838655268622496877558956742840096517078974399836452
194028789699296768855225099245481131679811960958449994512830174395
658645542534924673319276992922114986442884274282189623437770714914
385776080580848564273037171353764245306793786945790575233507643881
521810656607927075157252883991851571052600561883391085362212756884
261536867998180736017087567364217013332426353061934635454360172603
885525567745806213146382054889977943799486592528073247712770153356
672430512874551113052270769226215106526147981396130120548259553984
982903822165854000278718281794527593457598160626825669353591091970
253175878958507864251238169022868440885550462338617680937438712550
030627677911446042717502369048260534968220034680627979397528237235
197310708566623247325548156691201846615653934486047540357405735326
497356594491899073616081176313352491857472402949142219105534669423
028663739063083795795265229165823026099765074091873340013305234310
103037778507875206351658346277844835368366559027055834947961298350
561239433902475739693050522951308110467091324988204831537894226132
235566343098114183098668476810282571550286827483833975719258734740
115366467168262937015526822442127711052390917969319239127737977176
880154224688349620140099732275759329391772408646348928364644435387
618113834143988823534353062371783603772709222067901491315749461235
248649803680609448488571989384119196847152483689960814722241875431
733533374236949008176264368931636788534545428571396360327509922895
712580324146305634287293186997117594565836310361683515735192411529
174585686548397470920699815889999713360367051251783221658704694156
520627129418830367286169732775954748983270093376170910380551593860
779751831647911315109183280127525159365150488607939884858609930811
934938012778967091084784084709431548945577055045503618218115460265
023335986626211907545728433334065465592365267796790097240225590376
662770157178745301067836845974624241194146092702551816535744964080
370763592182048070317078004950305152791980468414802926237591441093
882117584542230025871050989743405380196080967060017294461315943539
319021552498639177030603251093959660382062356444137941788427652406
438997872152898548739082541948112684974003413805877029331600234975
233143783250683468323902948629612891771070449877062308584555830072
538581936999562749473165520895617944627647768827751115761470499019
482284323868027347074673862814752675333720121010716466861961600633
771352137821138857212155477397216933044019471665348302682850980553
100367604179853359108399121356009460421182548283826080594511633337
459993309854009095675531282250406764600734126635440624826166056357
691252799813016159259426864869950047247753327706499938464441139295
589691465220429908122439062860004650357381269521702256392168217733
673202091586750450047889816870364609615184600485574143988172609738
447212746441954501983355932317688295578306458020898295276315490735
500546541111137194763043626173297859028465131916146665516766350641
193345758667137818109841564659296236576908362716072233814788508550
638863177490787234919198767989674636254547126044095416819779922071
042764025154843319647739055810496100016470598951578529439919179851
978060893956786471973638900985242234000544223763116403151864279380
217952656842997814288383213895982439395887701557990783944892067091
706552034967699860541559021057651477157873786751371254730446603349
942396378779627117683623689624459677644019969786930278957043516744

091001681738748983584562893527517079117054900054362196217064294043
278062798737866767302228520183704159213524224275962837503729973495
914312511292396853190129753561162967230848617432317638929881754024
703915228062546847910309796282934304213394022884129310475940434608
356683432324956136475425798625445544988963516713644459379357350070
277317673488451748870996682508104389915567709984887401788299074948
442331678830180605973715188996429503241991398806477821402001041117
774789688395551080742111648042275079877931031511511084338317728872
525585366801323192752339690786930159258085241915908837680453626089
750834211044419117285563214646123720291167406146603573964319013326
735113831808685442753061156822006420507762762431630907633957468027
315291779908101253549436125914719887523866639937485253979788879307
769281320874570272961219939081254625546254109008343052040387108383
857226421767563029049301769295115288579085478954134272967908455152
662497710743422153765699549216931190035675588797196959360894440560
279853228297459230057292902250191869049960515196346658738457662861
655837092622954905255871509887801915359854417167213573070063569746
236645069994385958653045816955766814151886375167295828551947025681
486752161213458539326242521744695255070076927008328053505369516383
702482620563452119086436094550878637455568841651474777446606868027
320905274568175395157132037546217069826261484759071069250556887184
036300053051386810957538874806334632999487695537434810824541310981
083736907075104526381461571373620350563741206115436983956431136731
850971749634340583229072358340161829060871658710160130421591917139
023645897059027348156810722898671953025796837645639897187692251759
331558679733407737330432167556539107557542389161607021227173759207
005630713684374782121352889897986682177628325725333451832303927476
186540391920879656525892860819307040357396420809893210894322008519
941676117175273539788820266187445622820220531528314278483033868602
799651751797558198833310256254382304160189423391703863378357921032
037786682180618908632648674196620083453772672138030048353496696010
662241386427252925300053729177370611623486543873603335731512271393
004752577030921799026910568336986814700700466807131900900638769068
942035418676510749733938247585986386289190118498516056762720635681
836412579224498280289965361861777668677814102903988473476271790234
863835367198841815377642843579068363587905670058581213427848601734
299273114985258294263813065849793217196027188839888813672103025293
737306872842842526768184950327932957470704537523032086408891882046
920257811437470677421043931464527865896380704085307448240780539431
479682354362054098652425446130960956897299253200046937227651926452
523471043868060162844150263272906125719272577676314022420185623514 8
905903061312244081943896347125981067222308073624874038823446428048
391759971189069304948686931188157335894643337313783064630758220603 6
072722165120394083173607712722910894553099727713041310614156773024
875503030331197459367823569692069290086840655178435506990979613012
085913420825236538319721640025784386531552685180455950652762182978
072927001626238075398453178749321457176744284224049806304873363455
955714192165559906931601168545680475785694458258146525410569094213
041518607874243640505075700116978451599285143636331419198562203928
363376980736964248023175052955013733589796035987444259036369071 72
462752084031212380308926131880232071030140461195590396734783221472
762103942851915451503469923928686060878948272040131551785488924211
858750326012076622758794866107806199431669460231246670036640694569
880337894191692790930563800771255696111385511106823071862635087813
016591596657199561663926952401328193712268673839971023130237186061
442402816750002785237013297428005731412337710767305263002918854321

844114794561403613372359291002402880016954282527637001726055817408
445343080511456901070920068537510749800562299567939370360942165585
240126826086201398966399798182668381464268876294549868698695601835
872133246870517519561716040850360702192659349072702510994757252219
108424455952078308514342748979583140906113813687321865624780519973
309894011004757071818985229378443814143454127508270984979196457929
208023550364363535096012517177168056554998278836687203057953323354
892273581439095605122292942545110596156659899801588064005422943187
694927076222111028476180826159644660270430972905492918095775775902
696247824342719684252106673708953912879269571039131701552419895665
937962888409428690519523491907549396833743385108678868311297484077
425614288802420254564707508574033953987674644706472412244050841574
598773169274280657938451080829334713697457317170780121504655607730
879875787050244201825130662513284579379669342674659176754473212872
979925532293956541582868358625639296227016169581436104796463370168
169038370025573649440139581902290259043029179330141319851960060539
395930151180348550630214863817390059278659377962839746005016600245
619242505593516523938993857858329225917116476018331586739589226917
986771992642677072280445165745545012821713074850073677934454702487
148837188476688243781859855683230039182925072221247233954381450812
495920572738582163867174121455444072000774625667799993033885814383
952246841860609946505511749493126247475454114900899298880244751171
439414717156204843166161483490190460011909296255685162877604368512
092217653725203066326102792604571211234638430897691005760760382050
626948945183133681297570084946503627830445034242988558033635619845
445054185239483989082514808667971595308723718732952461752619649890
592054696908404522519746654776340655367050839612952694377982971982
255074008724675796073755929885119227004014089922309976925072908243
725293025365545849629334370195169448316009981695382175397508939318
308818349025619452689772640110604134908045313145010713937697105463
476654293873327884965077889115704433899875968620677669411702353257
219697006335598012767963217354057017373873456002788461555755391888
881901579378055441731521247110485252795976660872618979299145615755
209797040480867560544694283122745402563321911571044554296315225230
436082636308442210335683371003474828646197343120322024724439322933
802978839273166096657341966481397117329057631860775941914189828747
832911366843302535292524964792110196464906521782428021658444801194
148375608706264682217052788555866609397308492117248898807055295004
138866907636839943008188779348037755411804546951933743693074015004
386915629027469361458814570456637297627949440616093119931117418045
204492835456146997128768235017514453738928383768007204168069639536
492505786930809325864323795839791850836678526870939433960987834815
131423664526254152492755877990653256163320812718635364405049101813
148648797280648360849666502489207356975362171904757205540792696638
443542621309943524325188309713530823407873141549480966574879196207
042981321762437810817966000036278059616864585833110248343312510293
526638577653971473510052211816165672635135384136775558921388335613
754616690150611753776496372976412757297618546006530588154337466466
375024676618736828601359389979661048870373212998988077189303455828
424798632586508329716478813687836934043158699441584056073105170080
740906242322983421483645937540666898879902452967750380630886424623
540006792050169402576684226733377623470073850861041947110699435804
534455704891685726842546490009371256247610593669638899731278465862
443799144139951569466810839263252223181911728640074558763410456288
751255056808152522939279259781486174754526947763804476995955050607
030405723074802347057423446724103122965349505065170116543132428521
975922325039634918024654316346612918022256977140890212339328788604
173941013526311520369111453920038754109001217400800370764068065824

627805087557204105425815704573154793095847220829190461794537539554
705589553490462381008166373194550235848101992227192912016177520244
466860590096640142655724768436533186703522065580145891365214214884
279565586627059338874918786393912310949856121262994129219570550982
159041459138611252665621879456917858641408418466291812312771701856
076429841403475924859795364139295700295399600044765241741180636090
891070329935760123567450289489667243683113234827356381370078481809
220454448786971363944071406838108537502379067925491990743545391118
754416978797744588070612794679102610597267850687549686191026642898
726604155400035593706209614687774802119559012974434756190369015032
298765074421165940749484642116358360774214267964398310275681555461
210571679153107221777930254573228745372929891949546444430214743494
433991526199764617560500446876141445197137848176211789772414355467
354670242507347836422189788430240648952846314388163435075296533334
782074398744784424394834378621580005295941201695844997595662195314
594638517065744940644541377088385322747539546226747217816410785971
506263948291174373316912308779475584016245243467316076527923038336
108403393227859230437069614233618515120201630037106473392360159409
348761413931578014737552099930571439690936141780868692008127299950
126894063381545942618042524870705529369512012093385006182670089611
255740262928428939779819953658533467689013525133165447848621138975
793953376384770437896000543746315267922612655403500132944795933203
899504403691234100895651264183332718963511651306511767120225793729
061014842352467437854740469612582211063998326045158125475467792932
216010039689183516686733683039313629853299887287731336514009920077
181159495852996478186606788956873041297968193338686403654299825024
897608389872787996832247994861290621538119527811750653500768127034
638069985321345239607040850382203113837312467354408540035498055988
976284821825502984175192421381995295183553440312578431076669819823
628904985956993976147201950407415341828849716367955572998115769289
902945365175441526532860972531124706097743100642810215723199143928
724718284093996741383269759137019206039345242446281820941620536688
358917458615199461028929758611633262539363485791650895497475561088
704308265783353395487206043619313816874211841752981800450246240132
138892154135406316330238216156546453399121918866588028283036730994
712897808165487889720625876819014748657643264745543586479660505441
931400783305815766575393391766014969017251101351930327641254379081
724668999810787074329882860632105428729230684618965421625785755560
161255234030058019732943125553442148865206998097004301429834507905
380923445824367943874916281483401529428782991908462113223908762350
246378918770122675476308791945324149114109125348773470452902228567
697375702167027235152036832281449865303019336247358246400226530178
178623061318267347574556271918313859370386942322406078341585617375
014810320379510019132622268591620758399992841425836075502370367514
373472500610845652123178206902693232717071701807175007667107024383
008821349562114209666627925762947535365612231196303487297992396129
561001441176843099144359806556684733183771111489825166860528824315
829668157736742930175053096196635158767553864128836864640168032901
020983641453901361820123106000007757966076771445323744903666453819
033035659640537748486377056891952143606796432870265114198205871409
962188495412759055289662717437163787015463118234995808051672638789
869011697057433252523306688114102309063716096539338291754247405722
813182485286562201408429483589030543317686693886272529901374301298
828533204149259647248708428870868121562155659055520118766992099534
928858695129349256205885299759459814735055255033151457349813024963
566600571562129412882016229998481452915802727452943367134613398954
992519726142954671341106550874868735678990505261841864946700362615
455651785900747434683022033530621263319828205481038832943772784801

543529854234069099151220711408191653284216286828263663561083432356
662184583496584243511408252713418446704388546056408915331118311239
118605398781167627676137036584284818073139200562490471623282329910
806606066470626403542859190079697347301887058565429921291410650831
050249211811525610063742292432939047324285859043691099780873687016
127156576086252725630413794618685327159089484682802764387819688303
469013986309935348904399396141313859646288438544854517861469894836
359302502333888613866057882534938204297542053481966053943726434779
791355298709044097516714256549173418657402833105637868993037656181
566235182047554994194349715215489121425206470792904376212845403635
654004264438858799154465265045884415022083481899808252376263783493
919299087741863311950233355350795095501985044294646003792077032647
231653837167797632255104615083447057278649843104069195521190290898
190955066437540036659891571296690373826058688622094315856762141694
627700344614655883709614541838367816532646717988704751624200276984
105681049934797417474275556623834482187239492559872480817902886979
178082113078266738822717569953681581006834944177014110353141107098
279679604574215573467737293131835742900988172510707001027056816105
909256197738725262954967064382697487631527099633231538978021419584
628370235855228029775802140619125039390009549720689692293515487099
579621681865172941430023651007141027555894312623499945262174251834
073149518226654136707112054359504770529840551641440823160400941485
932276718335610410319523348484607952364661069986963176588502819331
709092775438097502969129282902871827186867656056560103883625974768
993319181508684635102020444319141590772364683025539168088913774267
233213994712884977597953977347922967893621930992511200663011566173
906575736837676365937189211422368499120214755719722305574100572545
257245385556505686460537112268226795019296310394584417455806521223
889346737778117487711085358563651048040070235138414083943449149587
336317033745247442076903104658940280706520404901026101806307144095
089011836528291001042631126121723016073039142782998260548259374871
970789455908633421705653769115599133716964152529125865507192748345
937113075626822331441930750599400673536365037293550076798042151214
351972532560562264394541411316747298778368714099675896563070199695
608246280145049119912407121857480537069396403892123464697871225506
696069651615068130060629410740080470757013091617350773347556487725
473691225215098135033192993383343821078855438332361873569071608054
558098436700574350850753199409765965336872104349332228061883499200
482787381125275030492088881748464283190165396006304665029156208905
322947319996678910429991774534128768910309099767188148030932716269
812365720604331596496493420535930974995546353415998438238620494250
693932014266637836894812021997614179860830589838293900673951455175
735499971707538924803152941090118514926140875944737211593932892637
050695832991810086208404102864899234632603456423704248242570639091
293808017046596536307241449281837553304794862335063366363758865610
589066033531629111765979002365301217285753662018901016498159777247
290540226707862683387773011170094318514035776219454816762064375319
687841485184258934481405295839963849702539597621263597859456691106
370601332279533419105303795404259783041376675167497687874652996491
783423364270007454754819494713598668069156666458532914903082320598
902812058781268157674309307210176135822631899928879732309320101492
321263732671517964955489689247766118431545671617574939384016165229
078440894231508766870546652757932388055491266616937775989539255610
804069381182248000513508093720466860200390550091411653944819959417
321871903409718825590726295842837197700858179418507010239247599882
896135737787564358854963425611467807833405187109513125759999318051
909219022665615695039282619479016017023122433057544312606544646250
086866227480438667441944201539426038115578275400004040207121819013

1574640109342733574833610146940368545125644320347075284483893175 45
6176463157220928781027991202902189223828247035054633830941944779 74
6913088292457205029220624985552355105654164304576298864176806201 74
1134808923422728834745432011980726606895692588229327024544775347 21
7655283908061743002542828159828767471359831392486775453184684460 83
8048150815663541946256581329635955059482907633010681564496651788 03
2837772344264957347620939757595579306386710047774393400649834055 07
2362181198948448212717277785138995684904476270086131269781577572 29
5464760335927355634301851525659582920138209150226200880031749728 53
8998215039065355933128282835302291042484991010409600808129227371 34
5158145829596251438171754663416763065800124676193027393974968282 71
6096494372311355756290781019612424029811176798261867140483954668 52
3855807275940560932973124055553730447992932564932798114831833202 13
7311156232404092544419362171293573215519555826930703620869391030 95
6262579288494465718175291633610508440138556367920711571888579885 13
4962980427561385500828166953039086989620802903035538006255632140 36
6817790818021164843877948278512937817115195312920602534993978761 75
4640800188549112650947737794033808138516582177365474212231932820 82
6980804014905348395503964389278202924723734827169643746781354156 23
0792934930724033675169925218892240566961469968147387885186031428 97
6353797624324269819101596656181862939220488095891535233677225687 36
4384668169767417735744924027732442718521724582490379444028830673 84
4564001853656074654175605371082546817925693646964418020722076795 01
5297467508384135231135616254995291763514168838458188793842769207 70
0363308166744134057171504307644550557086201962187509021388493215 58
8493146361185166819219333310687104157219222584513623221990026708 38
0222348732157579711911396826803848840402814592695923283959641886 62
6796149305159820371186283109450197691335579388158951141532725042 24
9535888645052952518856976647677540641989646125632782259701702333 75
5852088622218260028535928144630861090706741716126123002253062294 32
6943978032683573008816248684496380618338129063172066698253927339 74
0049475856958735139345476951134818732264175263119057768859219802 37
3209352298817249598232180205141646542331746026771047957385695174 67
2602807096815250437335898205504740080301330175581522716953096751 97
2016120092056623087754287106964586347137428066751678319373513256 52
2148383317367230811981653422398026247437655894767569216343687766 91
5649479989449089533354826086379823253949166726899416749983477046 99
0214640899683758249505829081453314220062263702658890856758926305 06
2177250459027499099932791976237866665291918639558768793566387776 47
4276695160789608393162352529278325036415584678024618159880514292 69
1440698965247194819339632313685463418650909284138271725216953836 20
0632300921099620624942506081118148675129816086548637849168389142 02
4407461253734991180744446800456578072347621011306844607797942213 20
4417518481616010190843118577837369230285339399275611116062550093 83
8015931151113590785216256048538691432381224590429977294696432227 37
1518952580297336604535600707534380412669670586791436932809218304 11
3925179378607725904330105369386056453122825753943173722335852211 68
1543044363584974277208363422787961783015362502801858578484425971 98
6713283424851276948108214828989987454317220977924036083261987325 36
1585956014193841365176258823231664971366187149882091304281551016 22
4391160452496338425654078620540396847598413729509315814877737124 01
8717977838047898994943654297767325701570538126378522274697246787 42
4130079364232849781847838418709500092027232765464981769718563115 94
6830120997154727353173557025264097429486253814814007859420937562 63
8394686737163244650066947567053159947268781125617046006407444558 07
4290199970126210536944280714092216615182130793486989728370011952 93
0810736667285487985787148223531687967478735752661938542362400071 39
1130567555033885290142377185416489229415567163845886141110633338 31

2041108527645828402610255558454722593759618723463643099396380l2344
4892556529898272029900367843915104186874958244295462621261555251 98
6745094446529022196496329554000070385210563219676582484226507331 25
3620546260268684522660968880438380474372623233161666015912793688 19
9594050257199993291767339310271001259536094694379663859680832664 31
9316496584963397729194414518373166757368853645182023023814826377 06
5301139112891369134624913279126032535345919916323452775814247666 02
7954795407043305095730585710512198291209443165335943240846813807 48
8753872970853754416828066098849350666996372897944316567926876367 06
6059226649550352104308343571403351838775617420963086662250481525 11
3784858825700250405158386183616450763591975664564712150619898520 62
7461072077885340900897413337888452905606432467837237844238116245 39
6078973114333370609605259730199650937439545624668662812432752778 81
7857645097868654913892339614539387201609954727731688759973321677 11
8441995894848614261038915187575363531390084472167159313610565905 52
3068822701639006046465423719340430985082507735015485199317471835 73
7304491506949757248007908426935982914383093138985548754942322744 94
9162792191174416817622585153292309062702886262732717271373674728 85
3638621232215215659814014346174420864122382487018421762113798480 01
2891846115029134072351040099682816355155884962593470228424529656 44
6314581220877964148139740521318982878542842696578224348921862124 53
4021422918738248798312836552083881102235045871328965119752972393 95
6001142529640219606769679575814793597999314184981204111154282216 51
3239255907098748163233710191611397919813113343628756085191368235 62
6377581119520159283010980245617011022257367211180753783939970561 97
8347858998010395864273294684313310109468796463429787772268219444 80
5699924553685460167533539940861811762259294277647153412400002769 01
0274117619239922487172129321988282132145815583308103276766568215 76
2232861023578888664955757065519767142309504205762010616009270364 20
0234045097208356485771816682415119269448506815186293519056117128 03
6592678213698507508774982207319334861979007258977582664142806319 75
9559863145370970654776647680072587850063099401087894554709720638 03
8948369303697002697458292541889356827697675923286776710169803509 73
2769360272883121039870404729507335731757223271391288682308969775 88
1257914549379342985359447300616559315055902450692901802949396531 93
7276413075979435030186195182849400682794556592773381062149016449 88
3428475015304083572143224589265200027521484688613520268320201235 65
0451901911872323559167037232979034597875506946927721571147182379 96
6807463907229451229908534653261746797338216839409164611176212107 56
8264733826146148780221208854610941607249501461647180175574523157 57
2564916552016964627970618347370461961194707193166112753777075198 94
4648689161124640677993406222196506865913705310362339187592819657 41
6107839829879513864770740645142244930292524076236866433594986611 39
3309682004315015611717440284570589293425048899723028035902438855 79
7151315458832846190466314888130054461650106191633925396324995668 91
8855812076832961375023195402633096304694051247986849565872764528 76
9709915657361774733919053124868449462587226912095502070721707162 18
7967918776070558048101519229507432633782002791831155368264376788 75
7991198116796887396317781452995442566577309275671432912484702302 27
8647522453354025140790949182546253529101439352289666482429845599 34
2557559177758575824235233897374748446334004725931357292624483984 87
7701765481318806697061502288930961065688203005792824776845650575 91
6401812945586968803264581170511288126050608222011029481378867623 47
8883248252825039466886560479147243495936240857533931494128559506 58
0257628000890635688149232095192164003188097299919802546703286321 85
6744877047667974008052444941230192577061686079333772616679523723 1
0225603045607349457206356031317070375962721464901942508795425966 83
6567349678477964017862775009417083925965179211371888502696058085 92

871546564185389115112448280957513693327438297878402591292425270152
380183942827764230778659980071499001127186267256977974935585858827
620784419210115012275557417797638627058527065459889378332964951877
011526441421644875208632942410854193186111409682762851290143784802
343912480472466682815277274311548964625567080291229410470366544127
961945872451600591100008959934764826857346824604969511368019113143
213023072466341637342509019213619905879348610301168491367031251593
211223143289163231551486390389204907304675379060333848146223790476
336372022317683354114329733331141893942473793198755133365951923692
060556418153629274571724953820445747470974338111998252069390559257
390707774360691544834954645439455506517513046593883516308482474634
109905199496236797103589927135620109785056162499523189050215581468
446772463615546978316832482756963135717558314707897091728778337180
485198954598438285966997768692505943828099531163809644381799776035
011308707394485192851625490589031108723329631488255920742740143440
238814712545595907001193665470967389125872702735685273077751939688
620387064305373419963678594085215789772443560955383209737122234993
636234092989901253196033994721429578747514768554321732167127462276
803323300056382327072275452949421725751941251053916721864335859445
349268769220823873314300392613546630057296367331556490979598048992
472669386456437462022496833800005055789826856676967114280187672621
778655766491577003524611009327946825636771615396243792771294838137
972344729096852205312331820157895844795748404966542625020395674682
354733532589667464821188622077302441641396400574395899344512407081
002310766672834298600575549363747498649085556304880457349808974578
492631977514473595857773563077254678529823332546870200957197489619
002324974468772535045331727309242308890578830620727285549856746790
484861180492665365738111300318254729987787422632245052354172183013
256341795439649838679382945641196752277218079707950556444508380435
789200441015991008710562086545753586198813375442552127302894241653
075803033080796360397960666424281579324486052872495607487409036181
364062430201765226239558683682892096274924609118894291902212698768
197461064344788994633549049367584304851434998064609544967328877300
152215440295681234534489469325923497985578607921934790424873864420
679228419251327300496390538838157937479116299593373471044258657372
191835913423118924681215100564175537835673127527720339420845773309
322939337741474629120414336423758453227800104181991754841646907988
966638034903140420857975127670234369730901782041202312016332370681
609809619376238353166428146780856607220894937814058508265825612064
156690803913301450378737470086191634207868965841313273363314363382
372598692478567010819887206149431011600932643020534519436686730988
893658736185274628304684865086589316628441742815813439912058343184
793314382513612576865309377574773719660808241387978909568631924278
782136534145351715120124156321455772612571711542375752304356111855
891966314442308693667051199138153403226210621594397427120765465231
765198966424475262047151909698917445511843743312560411206000481062
834177449518999061022134126445340065348576158006336404668827392202
261921447571115941454757056524353886708217499556288908901677082397
310205194838871835498343288088860711036176903323107813796305572738
981129127703868279932165904053146896325986391942915207644121837405
335668958199426520915206072234787011150784945992637942582738100456
093734037384700526043240047651510339744262915897859421619027654612
437100731415331339560670120992357055893555586437332466239368273381
376660988561386081756085525751881822982365620593039848026846892564
831572720381963427502449053813871227283653813817411890381862937066
879655274018351601106777215442748763186693851665268919096693241241
457652517547713861679070746876905028630836414931896559562354278624
551587599337404868880593635094764046404226690237394346938131980094

553983059637952785004028188017151896873158310682541473547504832093
766879887868201624977207965462296599869709286344427118784043263484
658241672441603790048629063352273962632643689695218637458554773792
770810832099800656037854978617928168238018443737773965967582913220
601255392869706131183075749736498210147313999513780119583630465784
586218819441497849370098402756006854980263357350276903501334915999
410634061547078733290603707231161240341870550788379000898176947025
974081263663402332052347594946286477202796252113686448377842127754
708906075102553014546480725459627611374316793083227144449518251553
202530688993058531948318061909762818016677230361216606781369517176
487597906417672078274900447456671143903026184811709882218889166496
393868513193469112598949862477341922210392931856537346282447189895
787586796112815110996593701003783460366990836605499539314209994234
926019517200560034985940409635399105472773946979904135787022632069
202545498416572677427946396129743913685217948503406180772850987428
122769011413681640284675288754276648347172854512389859157160621106
210386115041453249787913012138068893780887857411356940236010861172
747490768634586796259736100199117029869639675360563303585985059024
044857052314430822798558580421066760697846685856139920532524870355
545853701010773500886495196354119145399547557065526277584357657458
461258415631074749536770190450884604517896103563009644983256798030
526371100903483368327380092585578396397697887574550409776166609902
795429188589308917950986812311563053892116814233795813108294191301
853876649838432666048909688018607869899694639519523554122354444951
091490427932659316891653845246476121029773238049491026972063197314
457687223018886454723103520336088032734518925146042837827235737381
379904451396461458772415780099182945276911489256449554457820811258
549746299122723264649039750796423060026194526098123860389417285256
976447427429289378164410225978278080809381188489828665645075046990
355036823640457028786192640078460831062566896721051510787599806260
533102100122358089908786452552773927454483894942460977309768103288
181048377558490535754369854782487209796239602856463370367224447256
026531392813243226027749446017801180064353804656177241828844153793
279733031656426254964457524785225151403332921802994363668920045028
087930145836265592745303693208581992368582405824492867970470048929
341036743249680787167805287155715269795387317616139356132593003098
518745852607460428071453029533444248363527863091514581116770338434
981090347138086879895128909270471036033367710778140834844624686061
472549883654767143507897165014452769938600227922887312461611888686
514548757124587583194866819607135129780973428900934829960916137217
184080568891284832030737804399029801197767445952608724501787884827
582203141607966468453384013730036339352239132617464228397146920127
293410803224014862978907413563620435519583015124060878235799054599
370559833996396427254288844432193508373960444126803498854986780142
412013984799469474267251345750432841611528318934336257825555762486
044698920810829515812130807472484883173797144065755237092962046872
292297505739043255293584811980266329093403989497358092902735650265
209686282787669265416618779365146499963351891859198281123717705808
512908891479895022969802277536978183264365803065946080924008017689
072325841441297192824247203800486590762483298432666125302685103349
902657657857996330572857815804481506534147291034011865107527575914
621385957555798465331746524212479703562965764849716996877845984143
281364150626588032373537267205100203918720865488940491633392384480
571756388725093012313790697303446402836948914532187479689036890800
919107696487522679319688397308313632855164428485454389551192351626
705524166101161919395623212154044745884493177603326458648332699995
327613115608198673031798777096885473206973611106968735230086661325
732823554513288927073584038158089593409540047766863373882209899947

758887252513928947102576115811305813730792569508911106036337471430
048180745447071235856967018761630440248592810294124481653304300197
678125184887324291465465374765308407856299090453737847639163410697
134948316414943177259257359898774141042585256506469922575333866702
293799929902483384497963616582601770376024042854352514689293826677
378840100504077814980106515655297476502302530481848290223516634907
089449876811196121510650807088452894029830319134369478860719723091
087906545455929044269621024126508962080023775486379219005822137541
799451367737550293442139478343845408824968300559698072274271475234
618638449633003720110584884442628228471835669510578364180709087118
119763413467923048279991929423344434337452657118519582758172429556
975365547653558571537158788673155823731791203581943368590246116895
493584548438102182098056456671152278111152866314879791224104671503
454122635917402310507267578156516907949753545695466102343506351789
262820073876715784184485323312647481696410450093295263939107255140
263809723450742145266346791248799219504249506398350575631670024222
097888845014235493326447708542073295642006479990456728288273089736
342401635598151271655857087602631934760357479711367322844254495461
412410314411213653295907318446626781110627401990080057408513360217
113291910191481723858110359763233621594459068606885817458272106636
078324721105539882285311162308307722493207187603142835549723999909
543867479119041206409581635030692153653952559398291083644405778947
686559064269590785558892780148115312960528297396378305223965687979
329134158295623155010561975005747882584350683894802720131680054492
4176554134155367229678186722262975219357296762159745729299827845769
995818570124704106527552347093167602108874620189498309990562680354
732391917438803528523945196855902262991234055623668168420261406594
601661426898163648963226705671454527615520840319977552181122228306
946402827545830907880095255382670011748089440085824422344840901443
930995076045874992960929194486787284244655294260355040153663048572
784575067890334206375485651802605864654535049231546366067214955979
231996128389256606862453821966213790956141809481962680234837370486
445390985796041171310701243314722770694381624261607791747358930604
032216117319023629010812414634682390910147478453481264736939378034
508669020117854092269189072113908427362640084022560952795366222687
803631074992951896334937247807842462077385456462974698567818779464
133277559594959198587419166844783018668875825985063358095304730592
627958788266281356695741856635183662967796335366162569000658834528
478932612307942533321209343130986170013940224515939863015330027588
574482615552011632165583054011090247991317443186094788894424719176
565857393833615293891646315787775206926098992237784272107622675471
362967726528338260458031548818429183457905620558523138467486880987
475741829951436089527571148969698465913305515718249017264063469473
831622484570295688213478321860050219757949327957537140194691987103
391921503460117489549350057648311848003832107045510003197299460626
690817032846523263802422485817403542298510813693111624820378156824
026705402650609276295741300394590674452791979966472993327500323878
534455218203096952772541183313194929837454299674345083494569207945
583308957219416697425544682533864656106651416194740273233068918055
488714423970148248814130243332211602822892880315913275460406393144
756504510247078478270031008306239833790350592085387823674384874690
715910716613835270975585158434913210350122255865928574296051076914
689252902235938279127110071779878737897902060104053792712655426027
857235385066544466670386663145087099165571537120692078442628725803
234558565314378543259039018168386005327549657925726069037995759888
134730688880747447547111932240148505713258006452814851064945062178
797173665367581241855617818186509256591508967858838537346006935149
766940200854142411958615773819902618019957555810622548361640410620

7630901758560745738847326450713338575654604173253371130280137894O7
9385793643399533627809231210839700233028364694233862029910331376 97
450725192819304481060536148898127237275534536451517984386997375729
723447961754122638543250152317808397255726878502256871216813534253
902834781019191542034324259346091331136425107319399433703843238861
375880568271561648549365380237850903934833622482273182074099684532
966420066268037268782026486705782985119284503022158150530356417568
937754210631387943008169086192587428698754675509234944739615670759
250191683691129303774804955199097039823363826364891350584303996481
957236211574077257682336990702374463935429270205346044803831023144
481145913795384749239157294312780741204406080309758729669731071819
844106693340826049698415157716592955194865581648675779423393689882
775900357894212761604704734211672187920488769423319638405449947511
159325479617623938498538240196477081852094864855924228773249668730
030102462610446676908525456097605252767542609722175804231532157673
532907534815364111375640334493764477315701074417762358231865170035
472053759404317824599133598339181781909632989815139125754319138983
372544052815081545703102289680299577227508331211659770422199826896
583246941224511283294989872970025288692782008856127159949775135621
524144621201185726015252469722909151404640753192199281734280327618
599645702580211560174620906358907518617575300004249196720092148567
640159697615970405736942590356827057621726980826125845805961670618
866332840263385648286426103859307490073261464768345904157879793049
982050590432130023886393302978738736196275533218595801266221632584
664077546476579363380854496495709655491946816120511556943910807968
135393846059750067179506310256122479527199604657086253843127219194
520031050931912369479545856122379295846744832208038385192663153238
284507062223554163557520160484585957877279643566407928977315177892
543246016374195165332485486621259981179724635762805617849167583923
257722366843923139046362506364023028317230013785538397844059603568
332684279926654616067211095960638834242950430023363651087339459142
466082468786791422410972599551213534391932339520885700159038044698
266311069545853638464299547406804067960963340089659647612335011815
471822156520259380056259062150272901222426040414726158403427816260
545987338572456371477792160474827443344111296731237676119862978669
863080241795004730990548680786839290099752578360568423252734512493
884646426117437845901765446954096515947087299765071127626661175786
960190328974286263834806460258351836519591472871025416090341710762
667758150465169462544957993828996178666609857273572606855066958690
375723054995572090626217139470396850395236312944870874152476422506
065216697079552830412810868604973922953659243154172599393270748607
956674145938706841232750041669086025761487588798762359367347546293
045968347057818243392374739992499318213776303837529903851510492138
626855948622828980290986040984011665073222899953224056223484778426
908827773333002520957071638789137572836111316150828240586774806128
378099765881224500975070984791439925240141622405328598517840602677
533819228267849527886672456893337594516073195686216856063512035024
630992837418507807604204773843053254838763592412557531636452293163
511786522282328964448319102692474163633999280535309366592617986442
495494884617481328995681373139483095949958112473555488373871125219
033700515468040236778326154424065669062240436357919958919161873198
530482800619742223209883669364708401812321417476276732239473306004
051726285935485411882461399122545993604628969714134165203093502573
559361261145139647646649021062754457374977144715508058882686031359
661575978888082513408417512408423142188759570263961666676842178850
151166502959559786184159605479201785548116465418583113141209127828
459690444808181998063143890380522074970999644592680426847344541557
810334493205950156320196305397621418073990862708480643032178002476

090143977156423120072263543493273799157315518591100652472774851997
101984779766556894491967164861686707903757170664835628080325964867
673404584205686501819370242695253648169558145909093659078076868124
386399342565535330465056785065548655571282118381739659983633576780
308871040572972063848194704823330793522090608439862801838670895297
949455903398039750568234493536783744414698853888045281008136062558
329519311211934751776284520665192222527386696926300256656057137987
467747226321911039544639384612184558577569269211274469656540715714
141819794922961446403902152393652171319701682379101625390563079628
676903703669998720101055197275883904902666193971648348114974239117
387042271429504980391844092035353505646482647090883162717270439141
214238422921600927129123600670102952490482688952858136084943203538
041356964515930924338277301069829074003637819107214220109190692418
721602775558045932649015215059414233531352862778826912685057008770
944176708211117803391613231077884547695892356286062676906368115480
861944886505985634982420785224821027355717116828604374279300190532
421473269807603593366348559246819120239471480921233284186050062585
379108555395006935143257214182173424169658289168710477383110597050
998134940969399553351724534645472530194525002197042367256339419925
959139030043726038976533972333101892731998036678696654613980753800
150074994058963965696044446341048889101877254017581364829927018989
318743514757195119016432624565142006231074765452195530622094090762
265639677031822328595936090281625230627976852910671388077127464190
257056024461237830789026485486006490087858777224582811024344938706
614774453567937320714533408083249610368600316487116045576385296400
092889905659038807872015381686860578453602623953761584365876413156
895420184516680425311250906128057586051851261261263727019604236219
016699129075528914842550002031663943368032040571141160423757980358
565956312474994695698495135867617171614861342118764921998574023604
525815531400876092991964744628813382907321688226913157355514901842
172781677852130383660376356129007685445315638810905871806940817781
907895745143359593839035139959232215815478747867262033422730377952
548101343348870112698825270694572970442925473858639589403162524181
768010338837389259782856379513349072256643982136040806076951229905
479422077455869779933034443904419997357607975028020352529286365622
716637530104348154145433270653391650995946735871134933382166989485
670557380782200217312093915300782652612868451424290726599905709582
548604303545232999965155578146176895285778053665489485926317291170
626203829798016084121593188188696290282871579787179368849040257669
196947891197016642307944711630047969166029633665411656714129266583
864167655842583466265066904169951079819904156389709616578360678034
580385887549043983668110794625922809484398900347357457657221278020
507663612424745302728327791975368709460269211026412850520464771511
319590975978475200954067737217411843995901350727293059963625355275
183651258426047228081305017016364582988795296387473523944228984041
274230766926575389191293792702273587158906780074048592164839630908
392340398840095128732229830618530714944412455289744543189585611874
000180952752096285137117256845726219543878785925627372240011892859
210973591774453090991376018571051336552689960979827966913016647136
456697073237081484653493878889810099483298222045461002017204361275
120315381165829910151186943049114475937441511992148277698846672390
919809215085158244561052008871854606987301372553634632995644554638
726445232695578243816895531108965313406984204121863850069053799029
011296556029730496460548219601849771499587160160186321879285776555
148260986889146641786743554863988576721640798099307846447004158925
298121200807754694044486902285347709395086821323407338738152117640
444760483455529305299958930207165021318455760851637824162907159794
086617952632265090870625000097852945988034121108255776072014188772

749071012838936324373447553276610994629820292478457279595995866906
046000101825538486745656582757507158587296867716751114872652122065
893051470298169911459357597062405238204272016760908971936440310471
427018901122919727832816021135597563255486999922433885045198582826
291038182261226559925915970829197862331931668810629752541068411710
862598703071440860838890816173401839452178676169102178400070572151
153318183463981090485505984190806537984039319676526182549014496282
638131368683019053727762902214082282532050315758144911052149693400
800591718428867474232818892082210987075100960942227179606118687523
108651305487940730324755855158572712956851667115026585753721524970
207356655428748048188381689438092947193924970783426911981621047130
830957027914404488752944652228809164269322120891437353263434701894
141865498758085443897837542761116048219449857339919464163451058827
596415074970868687065333087674685517086367220041335506908982270336
380872483247736238427671272998279000472375033656913596485058948711
797177358953752351727045329880897687060371716893813114926117990763
868175722778250303799481053099714133210679111769390824978575950173
473432748633566843154997425554342879813872888588408286655506303116
923034262400651924618208512940513841094383383099433732356118408341
561095254744542667874415394527438085478157481718055429230172577082
542155145523116898159841576303310410248678593445139563372056885889
069057119083999697995213344839522875922839778656931697054502755249
860375938138006589525705654871539925942499971731716797625183078071
800651730515295633826327932127180626959480634164969102111602364643
235890477243477665172504370188262115843764422666620475127825235188
702178980272672802771182277315210303587264208245289884331683157916
664992568216114499034246695153246391357047807685268989984893205253
372101246744855945116869825165086315065590233506089461332596475857
404743071645102198896466685765093083300456818727572416807670592160
435546810092925593956489533295027971567218920273257081506277090717
387103132384249600991967666572621248871349920569723586933240738111
770557323460902157984722345213827715795803472205402589664415390534
121374470050890338715383493907836511458079202721014790629798992357
303403249969762406078897321843108824493082661908026220499901128597
492955627721746461146899392940105949420973381759438287802504440996
875340190388601771701245912276768846757152369655212200577242450365
397376969028517858272010843359833984544568178406257490431917087743
824104031867141420417203681142989503677256546071050128601831884330
590267041359144689996884717833470408291468124154743093009981538425
850563276047390876890492793224924095349919034162115446082138983645
436273452554213715789152132060376564346128773346423265279490788381
923864447557705950091194441422576047487273346424607959246298170464
153460651330840957147648715521286086578470735295517681275823209533
816373144290474179840268935951138362336190365221936869489539292986
105606543505797027155112426086430859272820573203824528633753600405
3715977663220991349122262229821552464413415294565515770151918986924
072986258052706984831289548079625435319741712858425766114028200953
496918021677875090640963889389104845046640865750372940522104112800
095473196600465004394421684859422090144524292722255849059366382440
277332836264503030533539930969877703430907123325762494149601610792
428731668526388452751267060300570764109526142989664881812039606387
340849855500941732198779667532749962011869772569041446902062477656
553506566423759146917333072748915434058300807072214809831677481930
217368453700212286491519944808643918607136506738526628029015686800
738658482695676254210529864116593386924825383378787521349222969889
549770334204786174098071621018441516177221481911004373337116936743
087732669730859901037197397794961028429161868412757569913986492114
247878143531088307012871037747664524240630432837083773152561194473

055755237910984617735312906442410447089451669048806950914607318268
944927878884930939500099507022903534353921332558947652503280710528
542412429468783066310827995140399264853819433461802956014311243285
648469653171701739125387897466609714539422772741011442393879589598
789105456641042681478911626857280359967830398709786663444404744950
237805374940891697917385372097470734527397946057249275908249278258
335068256908378083545693636681739559150054891171229458934250193970
208963987204233608131092995278188504027712873857443547058197144905
713041591925075571511856871245138617084613762199481388155508293884
837569145259023563970063111260122118004370384777070672157882914472
504703857119508588093475506374838293531777538088558941174192632937
495430008507222483922287543944226262695981911896956305252790993462
946153609361635494478791750377713741590706231759672455836853259984
215588103162562444276554990253275017631362943127796011916430509716
959532112992045271774496546336026515365658288419363879337753552210
007523306249938917088603055134982454203328601498822518924449789713
654388787607325839686941239838808204334626611722043799813307423513
678499084586481666628620111016787728549322725897317962900838938093
783688622430928160362691807321356465371535050488691647787791858427
737908188613147543573704383964161086684544778605185698836435455394
146744883622569075822976566805177777484874297315217407172224089962
520271344638028938433145251698363807311799214678365333421747888059
772842616020358430828924451439079548741920599467031093767769273460
011572635478653303787050882558626169169547527360426276524262803824
771112903565310920613755010415351635002947736028030426515606870445
034565865327374888313887136360128083391312685040569730582623675060
661853976157041742174275940940477766295856099304644453696935713123
533139018447310167028536689418035023616326193267872577312149724982
450873030342047625852256587138441719279864813496990033682366359258
820601075142130599354051182398221393595842421288680155694960131634
265883922903517029638549314312026857517209243416771675953673854595
891563041515434088850041606705334665408281307008328977614882496962
892401604211416151036179777932102971347626579979402154699780387784
794015988160852142135353041497943485318919528301157500655585264236
187716442360830789734582505232932436402643759245918005632908723050
762970364843550869676213310828770177163493776400157407706192909977
311832715705372092053654029707767867266532779330859965800109221736
238904135842795193350888221454365191134340143428094289321478884830
738725770497425332464660948353516657817670744653783648028470969261
013186450293122961659943735558202928820234450728277138137353108738
681566684680584165539524063151157367921549112632477501265298070868
653207871804612867924288077555706529278259419430075884278012095910
166377504140143144565160939204213995507274987872445710941355550450
157228970947059287154785784237549555530688277162786068182464764719
997298003673328772074752265359103975496408864254765241431378861906
622713920337483035553828434447644598243634065327813631957808343899
184020729563312274240406329113697622685654000106238986047816686297
777840312794899917073967230675221629509652878963150378284676791610
967455848873593475491086691961722460379433265367175326679642757594
399604997668066120400165330285049499728607042754250937207738586370
041544102605952449370758500363784138186077956760668064617234295007
523977651459432948967190018394508535071152508262845453452737577969
122483337192342761582030801462801767562825024071460250971605630931
767245930760486882487900538471580520750744820305264966898199940251
579494232115902784104325425835337177788899239517463665743261372851
811005292757346641848185181569944340408818037687535197406636304783
372844055818192994312492820699574561423826653050981344664300969883
579889993336482496472266767866094838218200691764477995154100648314

950923402313298555020760129765976548251432214277265806665755782452
114351183573372377304924371425957029585515871156388807799567099240
416783945078541847459790898158041308296754681520346603969878439383
081839237477162789771382844434051340951168427734092176907252476310
892244466478911687943251322200736515345623118550138742152335445030
893885398610146110690748910956635966294815046854675622928941510892
372943193195614918231825611290602583642878172194929175345388235279
936285889636367951806699435539473796067883863943299218278556841255
867799549795461330287862146465157139916591402845873487027052610698
170625680635866012816449797246668164161969879886773684747096351963
496757677241171938898911618410167572490652812301639681693150030825
201489745797388900152682402377798309724004631085064997410591209501
570775841414891805283246730618351409517491370788512597534824769265
004236854623584360159707296182804430916802637005785921511372201216
585722336541374444765941395496439553132516160965801809920877908352
579037577267506223225185883060756932318956337473318071536686998849
863869143703937913796902175096258692842571691290542002081554697777
269088469155248117449993657072604850439508091894328530447493989630
060318518772745821052465577410812254858746139841024552315469728605
615990049197301338745304316023246449926512501704259895981825565238
875877954291090967219088355357772498686136615280495876215441732619
304117924969971423648039456363099307208351692249539620200838815119
183724445227526366889209929576677095667383518896619596160537535008
602323364828721267279457182592717871773248430958282735739819410794
665642841537352097892288900210373172027586642901183835024010382622
694185545174623600553905336844479903780231493500910433515559836118
650126049569602591345769365876222655770632075180259102970974492951
788854202119297169126631041420914010786425238613208134763454729775
317440856732210322546073930167741895602027299203162479753768627807
845971052132685726453736242255319250951861718760134511243376299053
555461959175254804303890441737617913169627434487853115501952536997
807793447126384248899343581917195672961221652053462230855341045461
753516028530353170601408812136373334586374744900712868341362130746
643149919626458766238324794568554952649366465017373025269911908734
889383808220490313999050562754439889044634722325658585987911886688
740806970103202682039916459653939827631397675758675630661191248528
924856994955452747582595838059266980566909315279118773020597897063
137870893388322070950897673353008555615294578493640663912452296012
669596009626249236235435006717356575093246211009787638785940003781
112255481862865919130265466121607498035325602344356367631241181589
646973856150829519318006571541263062289942986318094470882999697782
360083731932759312453730293080319708391679211238220377593869128205
704012577244486157978289703237209824314199521913569339907013764865
723794090189731026773146477049154312463353149231164982846119122320
443297987988654550488550281722412719641296214255244081433833184835
548531648915655594728454941595183299317885179785723333498219716745
220939822794703894830938700688809500950017601865107568261901821225
987911067357657410237188662724699351075758553148758506068892653011
018387189835211172154353512759382643296902223394938045976378558090
731716349568750327089865704509163882052480794235222753568708256026
031828129257727003799185301263523799010619425957035219797869460379
340909076816903937424032942224485649062429240690413555579781762809
939784678087508737889272173505415721686106252435312971752870170021
713892652686261966868712382220018081998105567112872178866928665140
868573073142030260294175800614754339699614454195780173471307453251
846650248739186631226591260156602363786266520161741609131567212837
043023831858011836035956376114402196341988517773587153802798083923
962306959289445028812707711326597922053836475249006386505384673526

547127960139270574253357239928439256682119007592076122827272386085
621382765349933921722883132892944445541855911852332513183003512791
379252076038822932907683177572843476710568432799744762007968785459
609023037060154744372426114673976376141025089181934278782304188311
556795804541243119210344135149026909550179999604207232260994274936
196212594458470157994267384082691680703420037334281735882422048034
577619180553844186706109278647544687110904276564909613367896693206
186509904811679659860550340492059617415840318373662444987889103504
706165000925499423394566216562460486362752367571958462127097101035
867837376285945519035694807841844204611642743268607874808440542518
712273222200262143479895429528267493240381500981848046570078416191
838634925747769264605691767553183675482278920331802726653109312711
643679338196228009595413304787068427586157839815966786353122126491
074162834036375848986732387826613769035930136232429615603116634579
503348069604146809599985141673315641663661063813593337029105580951
861002596596933585381337026672093038574257421678642906364254627773
712451614250089638653859527580132228918653790607932844115312395719
443330597111674626649023894326671796113079458231044119190079783881
715223000670094400240063875922694362285108398908252287558339675045
354327424126564539480106648018793370264779678643563015419592850256
243936980244407004148981012850384761255766532196799894997511651665
525680354364260473149980694157671149571790870619144799725227147952
932796307353587611207582906805645927075287717611731266294567603663
357133534836225749231609088499733950008239080227083401844218602397
156226894697590112111641763528196178800201355089860699803555039550
769601752355717349136765805036634809891766237472777944388920986795
169389361595325081576104264980771878858319166602587545585580207124
960203569014360241160676509441751163995525117604635545835455868373
332737835385586657651755653238466588986398386295793497986580783912
028837322165418055101504518391046892095056442926187130727188975971
985224621314295193673784547251284191391728084319502081487214486818
091226214195079624858367872512008495905249253663525575397130128252
266631931267472711708740168740198182053009569012107994936932404638
634100522606283359784777509936197411021609915334124174563222729142
898453154604467932463165850820599008444304452923568356130595857276
984976510035763738094889043789814971338944112155340478285383410251
586273452209298137895801512526795905016883811079779373785231407545
364238326775372591139723168587926879041195675255583255894326247463
169595932379328077397215234131651023326955510042567134781656878944
325901020307612993187061171311147244684738098761661329821589523733
673235671671545614175551623880797120001523453111994541783387246086
759196531569649459754343526241471737702735162352186908458816943326
495436096306903278636782721734337872342122012779145307361528442917
645937575452766551173616624937762583412326694703866180109952641614
144694368666554121465355725108823132604443020594782957064696420840
506639720639317365073513173003880445262259603654848636596942695148
372191940537488664822469551380014166175344131094397669630489259497
232083175309962614307845299331801650845208032363582644210162743101
366129855870542291846918585358000259657093610688867074083687084907
612579947164130996388859772806057254383677775945949487903959265587
631921103227650583873043998704354154077217081533712281697253470848
046067848153033982742535460673613310625536480782351719373197638292
319760687931863997393970932923584652592525487847560669587367260250
749646865932053476033302670225878939971559384670734638222108716543
115987110832363590150691704625433514220561500233993163621226931614
164516346722487008908575597842867209087050350595391102855612564991
351759644938784685362718801645021502346566085754400621293280842456
354780667137045656851925933592247464344496813893964095728803007741

566740902382614245016397148992671505880621404736388771776520916125
774301134751954522037490896163122104904353465180190009099352151964
047145886477356021465122479417805900423518207667965078580026385108
261666788485589507491586451266193098383058348262969071317067310717
197280807768484805796587544413799294491070474718297280178725901703
403330009230392538061041675484669889573135068038686174263992443690
005077832346982406924270331223442533816869554568542413174368857672
121852339245514519543662928877490400439455946655074084270251388154
225801277993624948276125530109575941070155752457017401978850976999
525956172086894340915972589634928359294917641877073511887557335239
097533861390461174419754910858237894535948817340978863945061700922
735583569004058562272378064911621799257526358767717337825450819281
108102159023883118495295970302890454635523927760376697180464550093
663893952712937566183649877213557432275835211796176974028937456777
117102096019708609988891974822781834064514963240191341985954666751
863043821782298053647680859641348784882765133590607031616600807385
645023674086326510479565555523905693241365906176731915774394061383
800816228694214783660581227966013786881036422393449289901558277560
476721452887526965184984097572531141269697999683583131116362301667
967637406820847352207648198751988229863052801822887374008463358398
398351191797700578244819958671237440728854255425882067330386593512
348849979702623134280932584612061873144905739334295261855568315864
723465082335631116899039298074623730185896175127462011024135025301
650473598739129966877316578879331468114620930068822230887206158330
968583647938117175872322118804407563604562665101608344367723134148
461871266939435580944729922205809540034174771169249485722497611316
668715094906013042427878611700218095472391317944418416770245763014
022975018319380448844816047770140408137418939545475965527353961460
351911339301223133299133292565079896523080905985857406951069192032
793347339726614955417056166160815429521787924208501892615890842770
899908154102051635179645981575455968806942601800130355785547027159
876006807331451816194078367221223851733177317836585699996221593848
758839224899291106871277741693137472569565133430063065227379245510
5626692188808078118325867313696011209077146657944366288330014408158
630986317264190633304404450805822811281548891619832293273181879995
885127378387293850969905006909581505599686237771294231761972940408
139418136027240514820820737951363148433542279393342776536861035254
077258399965486187460681108544159937826344218383938448584852903530
456971510991250442313825249252954008960027765387386635741563882743
141102359953392227466142808950302357630184738996956968785796476148
133852746260001728400766103533599700110742958861790934485688238614
022120262810098784194778556695767060208297297284932172835947885074
785626883541396045205032733900115976899759234882378960037047435094
401262181010948157151330500526967347505159579302414192467718837799
078970967419856206206033657762606620039347514589512123138896768052
158583214858756210049398274337792910008563459608478727443346055618
482163033872291055643309346721926478665631468981442003818721109343
897389863038172129579092991127430767197772915783767485462892852180
820396714378507933637369500772766426675585419995885697258889417188
547324570412654538727929383555423041284137013213116316861676495031
820781233528078670360575316926111114131927428779044645590745310830
304595611994778892976495070646397168145146294531242943635554886863
612488265612400644093270462720991938024993538960597443351284906593
630428301040581746061232043944596786199437588182107887055290972403
150885180446976720704932193264273274749474657352706314008460022706
069559676968769935816122080876142890824382822508271626545101507650
711225487767831271048985812589031480801153234822354857562501636134
324457554897084018525624878558411615590629157064001861938267871916

```
3065940874726155832002885605643054367547301269759923912140930782055
1634730568394435766080505495391291458645641895198421266755703261404
0165264647181916825561780500043662639868126326051567703737576322855
2572372240337326563439926101072777491347627176169780024202417454219
6335423683421089982050983996930900966833517291556051800329123750279
7413833634655705452783485374761577072978602826423169895661621536415
0611916911869580576300887739841331443323522161086480757362775629353
7326192768075059179197538220805553608592143706425588204079459456805
3209984521619529034874717291587880433789802719917575528412495164553
8973669036679774585106049237316474933485138913107670478168931509280
0565250807400336026255879746587338958865045549105944187115107898947
6427723914132008590273134823051471147108625684424430556126457313179
7594669083041340349698418952641280813039237953542339551451643685832
9717672703387996607050275380884466661324952084705956236035132887635
5692831040423514230736898302257618939822104514711649750864957931331
0230516400702740418722925525312555050751962630901419042881583786080
1872268853303342603241887419947430948709809007796896221521278204036
3871849869115568928526788514369208111989669919917593944222765985852
9607316110272771930392287503735276746031364987286014859087801089027
9648101784192505137683639443277985897834996707510645916080799781498
6746212959429929495103455612395285189966483893151341663425625387229
0420785700310050576672387824994229304331876293310759921633410348818
5307894578621304384469902259715109537842483964045195613952649940910
5419348027833381410505497503863252134416652532578346241054313724205
1343656100709750264749750601879329897004952635108003848800122170811
3442311592330182506552329000805329609505296747617827764990338046492
7564246078440062247751022984090446457608824008438396537572516330908
6936595011316680590317606539941304678318604500629809546411793222760
6439454509746403119000404610136363756652514763392205640853948811565
6651564144995469997989615517085192389641735048616440196028228370231
0531954944557157777919567588597530366609882202496886496240237376974
1490896227826267171647090545458964342513581072155099312787001810947
9337800194288116971384559781303437591174498305039727200508838510020
4864527669175972913171516690414586990948299639788578847124158215181
9151311936465926550224699522520658612371086170777529724800962855738
1117264350439772399159135333727943712149397486397926247924574630926
8806016126833281080137307606789638853035247624795772962330363206079
7531311700321563157739611414172670968457919889678233174980250090409
2769839889486946584061970883711871716267895059803799281868790835967
6744386739411089537196986182976738458959543924411217443260728143441
5683337783965800933241626531197588638631457596991628226060779033373
9537806861549954334396662534041373293360583761357136631484960228465
6547531883339597694060242580533822826967160251486243661503046556944
0359099795880865306477527207618121415394106818553151735880056317978
9508106998207123263571467145311627582525008062000983530299859876534
5615877265590577677705235068007347211604498064883912344314969945662
9688381099231793759615151983265012059878144161793289005863320956390
4496732740967756156579216951256957573430650111772794467699754846396
3676474095197874698574050025666304304969303782864673088907406762172
0816291000927084108802198036646903200660489829526544197361650807089
9913997268065256227964180494645415644876701280058394443003877713557
0780465643299132187060632208151427506224046357396332303341712920530
7955186159422189134174393464269860039630400508002042870415729593735
3099577166776152516245240929006876484094452448708296372347946634083
4676788924290451598405086071695341689272768570261007189617528775251
6051145757923487582896882725005829643483257250194887049167329327414
```

146938616110082120730148945094220915611353196601506009515300421518
796800500925670277461697526917366335884789531048751627925415229157
364741848801293707600826420065741850241704070543171728758494534967
559692560581068000917355323392465539765805611506266857145888623136
877825027075077938929852212898376757339444061238960194625480149686
695371667569202794500414657284567068530994311959103987237024927090
104353255259805011571212263284707450745903405964385528208729191184
983955166804404872154753233912561994593028105975250334168546567651
849453998885243277578350372083477741274638187405458639292544711505
554322799818208681879698647010465613411868005396513134672323812494
414047414596471311007233355202812338522140507238615369316608946059
307567586266790750537016110407034090926282674887750285975237510238
545267796205530602752777436781940577953384076593719123390416122993
896587754050420933165780695965341258929276900210583534723405610651
455542932009170745043947019956698366277022470228830426650690455121
943837056846708382357301181632285434001843049497985826217684156199
179199351241530949158010079374774040392763688826938414993758057175
171344755815931032745774187576371187757132246892272242493377771139
575584843469314320561935228175997645633844635236679326941663652938
292994731554160262278104340459712825822426701976096925423356800253
834894656311249517014689940046400911880740870773299643030944043581
834084141758018478995026247389569075478192665221993496740541728819
110503325279002158545943175734588732769586397741338302654225157488
127285818892732849776969110852593856521723918404258258231447031064
760464814502022210286533308656374402259588047132186945014436681753
853073158920051906575480158234491548443177708176712520461196493945
018862567637414537686456006014215126129242464596730642846488510497
611191884672049866454759999268017756643379003400568470375447562878
249348984087140077023629650417075010454030069683112353172934030263
172213170861762029604344548209543486763695249394153847667538095773
551214637499080503829564670902052304673985564355823807120013405179
535817759266117929746014869213202630600393179956484944444731611937
253694206234756374504782675939732127033728057813431266211173872224
928063336398345193036803892593096889952092520352785719809923805762
361228496065232769709707647903913942314021333686903948086876847433
080122688699462034708237163097247047250281060331416470144741205421
382403081283471910093086202697490915360722527512865977495195070144
683595703426614602530281533848272776449790077688980744583010805807
625802826525398467121849904439425891880218806297599563142816384851
120947134995242272909662562131057200278602521302716524357302013219
950531711204193338562432181159685353143642809886601095854368526010
852934463784118853718262721650745414154290924125763428168346424803
581833399866073577309409498606570661584407016735064684551083448304
103407143306886135064816123133500844233624141744252203847162068581
577800344074422408997737952507722722425221632527079828648342393603
199360077015781454854997327914957152420477718325986255747117217600
047591868613465766800747913882388825260595036961456375550948364552
494331849375795834470825651120297713054890598588936049041984366155
680280639101693150794123984426141463145272074906342167466656220820
733874054410275552773264257266626032543414164518064646201359107463
690481409139488173491501909058878658865756080546331911539579519922
691510175261135535365480449153805235243092095683913392366418602184
865740565313865858918038882901004954410539504281908521709568970985
222660491497703442563415201133543595601620804833457056785828475182
886943264572847094155070663900420218474284148150729321142555253848
768239840596403694675585587065077805288811843705437076685394020436
679087947275769976742717439082411422515170231955326014929205324852
331644303361410581348081211878445401680049841863719537750791225098

168839594398437434905009276907430497219732166826907868189429886554
326039347996027915286663167424520499357683219759829614090609600277
500754116118234136595195436476826597435387250343355756076030942645
937176844352565845618941330436265854933396448813667814189940353781
251863618098614310938046860842389417657170919837530273529460531821
774849806764100727806893539487785620807660764628864349757813849167
549694801333616458553592342187444390487282202327482790004380974372
224090556657982792540792018271917640731568687889908698234433324298
461713480779415792607894378693787932181927623832088562198256262053
706157053365099866604373352780043029785385825777258208143493068950
997469028421232148054362521441099265845132086983569503793412192787
701548376684625884851480353282100576225953123843954998455977183661
774696948771337360294902432014008936954352464284670390628296129253
331756090545801524153362697046334156094771470387833114080455327202
669851691655054580885262347227009185683811529132516075575147329831
048174511690118544738905002707966281357373328915219473513171507142
118196223950640360662507937661011410909757198751950627907671992084
156183831262327870536779495872093480853103370037079676981443339865
319473005539550361373716903540441227441848250897255434113291414099
210105027940607813434527454574891039787039592557677197303266292058
500218412430561101471172665885887585811098013622427191783601505630
084506126095803511462218689703926560067677522712781150869182409312
639832388951433073092080618183670533272221996712213824392494133845
030855374291395100606121406401527729920472162474690601996875361612
930959443531967021387026224274713425295198346411502152585217190733
435287605089694898596618737246471484457540042632111691306610800075
901992768527723122943384537537344556192707318420770035251888198510
000910588697024636141339463736403629167648502431395922261089143124
584310802491853376636547379542810198006394675491656738822093720679
550224539749527936043218760416737125294186890416787560507158191846
283974995177618984701201417284922531659976424913961553463045596737
003669827139244746172240537756388452506057383117220633099522461382
751543842624841830545461614239767958075757928929553582315837910599
881000136355835802538193049608748484062324421301967312857279756963
808915891141510626829367132533904432204439351225627134258357193758
075555470995280586509274891485606150149058672218630181453955271543
377441157484301460454210472371065909374589455601850747065551250496
207976349526296285857419106119868433743595990276751351242842162787
382704073617901822642116190565452355982134650098443468474989342551
941906125685394712155093835778399733985359799208044091401401301969
205884816241657792074385061659298842954287592676532576461278658136
538693023064324951487229156834194893397732577238311018607285992138
274399553167071797786946310241209656299251563690707097976574622302
248111836993379540037289040035528138387916696451463050174466783026
773391565848513133733221345559120169428199446355869119901046702572
488630391431318926902723427880765595671435508100852332873676188833
143306258402854461381790916491511389868861020424410969493039946188
156434692259238227154237325561865763731911138393508298447375787630
908881906697487503462616077362010561476491958588895261757302986600
028449208833098693563896669357365831543219805114630291803035328239
125122651457960451129136204814160742634836890487253477414604979289
686654718043110963702069366180038156492646017632282354675257725100
634810833246049639745818948562507159279851400688234565876643286169
649944702876172786074016278793760031340305365372001633610673119735
402165742745526613772464146797016342322109483927352053992161846684
431657805148578748738995611735615742292710799789378304117463423317
231236870629887993956379693234381409307730316673855875833658948792
192292991702925321953131063137516995648955562792077326334439956069

913401234556618660027238644091295776817456745562042696966397914924
854631761555583478849112031329808109382020016670821426195383939135
194943324558741577258369481163492140484733546704852527626699095914
448570908316042386634335523479952419332174312708264275715015373811
447046489614779234323918836597604173276069183964760678663226749291
933231611318776739191327131564491305693705855153395058229226297936
692800988901273744011072991307584839194837251638752515268120935615
506896612815652785043774385673706596865712904074504021396786409805
016287163242664226733761382152956236522040211884309169244962016703
983772402290077519119901733887256549451670248842314466733016979564
931389538586123981116683082572022333724698573778725176730167468852
701156427758200593935709812258690125889277275347751245969545250389
826116680212877573805636856315644421994581874028106568017531855645
652955822886169552862742002819640306143910590032153797953969302185
194232688079468527914072587719484690411764169152274721106824390965
834936817407243912567260141392055375044387785097186906128308954214
450904545348523815226121360913632796256187136414316494221393554422
060053382734515307986723066880135629301316501765553770471630091506
247921237391937057541387260377844094217302591125049238615493819007
039732269827084459330938371631480611283411794863130846199594307896
849641116870843379334405700795264780255324299882702122395890715726
215449188311989117964922556872579518737643726521041961886235970807
637082519160197482234482099443332365004015103308033450987342082127
921412004204801805010797985617232164735044013698114855410588119726
087395394254908628737839037468098832081722147968307431302693815436
368453784576155752019947911774123367085935717692095928402888000062
917208716273797741535129580502970994424191380762872308506357855702
200290134327092777298737515761662474840473928515508631642153028208
352650155756311959083682307340304392715181050027526500337086894984
288132356849650072493988474010569473433805637338402382436072538873
309111373884076450003773447847091001864804554117110025614054087836
992886952741392619379085138522929789581062839805904499241387273763
985719482912844834765974014041432598188502453910670591768324622694
839736115718148985287706502379213217892695361163793044646771025399
165684431444320202981168593661665925520655979568482691676299777340
317373780308748208117848767254586837334246334524199414107801536921
241598778339974669973954295186818200906496976103678215280989529876
069957103569096028371008599789898946334872095708021792336657487071
467773673609724639922157532194023698804256015000075840411757319574
984167403330352603948097378856539028866590990135840319648037329389
860459420398689661622275489436550760002345814107089615093946926452
773942351146940111277191491369254378585394835272648575521602087204
812491691018527179951837560732844266771847679540711162180153539879
182477498161858356464522803086323997713849928359209270555307866515
564527689591621869203470156455573402876692638413257703601605964596
904032255123102029559970989164180907129145748986244302977071138941
462836484848715767856779487814312512432935502476872684689469338789
309686193721724524216873927185950326411718319813570892707560107929
918514562750286621924369798417811414045403196091642478439214390278
338683726417115549439630419405888067523818774291085517880020788117
679147787731136388027728566969401182727554750200093592740494837691
964174174743764309935882811490241585671721186545181954032744630850
849001867213652717887423627513368543825843572609765763398593536487
906202036888810270158500580830280052905250059204099540095599338281
560551578080569277503764059324228538216945892473083726933480345155
440890818030000971359031387603695064629525581193233191042017397696
851107854510205884117474295667426408627622606677216076333832609244
311716326610442814595860625069986860747754290412444646328607846279

2080416316171887736540720514693121214325169234294535433247282417296043324790290635405744264355101657618868875755289249058307322668857940636107263471314512803909679409368497933681487571329869781321192473059949601402778186483195060983999490292484822645196907669493668091852924654025480077219908917729196093156605486780271484478376604390600314714645229672337382101872503917315524586995542388613649792989340573091186898229687569221987835267888481656201217993235571811933947899729411316250382377160747601225078909139136007369816164496155077115624751848671864274187522209836992625110794587674427102605491018377141493953846017308993344936047697330192877913802703135070732109818820990421774795824467599020083571168178448830478739147534493519381401175208813705984398465495570550101894746331785135447806050024532228986959561098127445372003405068433161951983630318170374789862663270724406537943927775061173843773910037015604508542441718294622323098741599261379231030639797519629062149549367514934829553425573267340573662545638782087724780170125191080613807632941910483768620615503990171057754937943719814986022745743276873586654721193724736483688847386369550472304586168757789926241799192428880868466327656215489453775846426936601705490410696791396585650472314269792668892085692962878423316515400879707948404460626602059142071572641491184272577254451851792252299228998925524179313295561629333876104160849199666317442308779405887350839478730728309916977398349794368477463448015790691042083874953549261759171854122935975189900892226217219144593067886197603562218812780638813224563555606806941525529789749690521025558711666880394625316581559026470171799000399024833401918474186111779142508795783220037431338999438878790174932178328570913622593452398799211941358296486042387182406741709862281219018573336815685703235943091990841310032742416308560704018396764542197565982152583422886796490617533288038337592518802135578893511916333635912565307619634468859555679031931342675338330663164340607697250337461204565785754274944560577318094734904463642115299853304353698384660984613766934308461700054908361012391654874021552018074383204637467839049999351856781079228972587469036774957913385350351336075507321325600291910315513107283177116250772551531257304962327220860225214648402202842798593292836688799077140819239883785035622052945336156802820375313954033176431589405632531123586823947289210428147770904529570491287593524120513586876032845142582734694488545643193344560610392771958212921941310874666676659245749138539157944635523988690676799695355659403400839266309489156370515518332939400050322641708848899460174496659076686847228668101834234733580748167860825699263614579414341577379697276195362514816047504644571393528525921757839527856441202769631859515199253706473854375073484198045852310995245664026394525671148305359686783111380040816192340692340326098970410456803793483601054492920692731323910928948941612252877254172588071576980029023276929926539672222312595423789781879617297369231269862904132940693047792690340796085936969553082870633498853589806317037906551023455503598110472263078432437887502680866528666197012358431075487193446996135110246538230763263859894685743530656058275300173591298306951563954194337812121118044070153615314384579873166672036184046659229140007286157047092318388883712590113775110154672815683126173135845444955569340401606251591221445202630400731678624212347398415663606157002295757951259606789094918071487219920115133860334201249033432690190315247111117705376749037925070477108000780724379721999748678051270543865809773836608455714159223131125076994385057495720844706161764642897016734248531307265311314113296922449973258986937933620538231043046881167270296781793531514792526227715087317841347201175082919918140024385165187293321922367139330218393718018284319234620686122487712481614464372623846672273871692704

432266762821335957208657592529933570469301671377062937231618402812
814665403393929976479945083556352931340448149974608215731436782770
602090362000236561479481796166953389820330364315197605893882403131
045864624539039457563768870592794191446351316565638530341380551160
738301152349650160262898331175266540895142764548739436429000310464
975634186467063383937026404327199934445017653621194183980914054354
312046611071022949147957282230048881509892880052029822219560313715
553191807236758087351573949415863863465924264929827124687428737887
107211779609342852822185075176120316328786505095678597846969439706
685743639441733419056383504407294386217526100606739944657144525650
821851108112347349770923533060731560343833666812802897283552949783
468612079306536662298488684458973023686950573180717092710488020936
988600525942346855824214663237148360334708879050834675141631859978
731821377113609337967922671304808400635712740447619016990212805036
249551464937562857255534722015529162736004842071745078807950761845
913606957788248618747949602794821468325536369648055794751958042265
961817575455372575780063850078828942880154045147286645076936364292
616561464092376099194423025307177606647646097489013007128320670095
834441486795964288442656533480626985008900143585979343204954700739
790197624021831864502251873557681953846695713070062888292637561603
278764215965593125292492422161329574504065382201672623986186616485
814342988831519203159057960073366305926447801768242805793775192549
911613874081655519618008061914487729485116625699772451508213714632
819036518087733473753221390179569455469855894609950994481976623160
582738973924308531049444078708472699064031783593529046138482240608
140130673921194035926589229119690168304343622797153048445980731590
371353990852305554135850303305994263075300844749731213292281390041
493729257315810390367157343929813723660967707037239975647113138365
036061040548871609861903961849936105390323056035908952716521688244
120760002937772054200549345059959328476579144513894804179203585085
252993874450064907705032042624603929134791860529641815140735430834
587703620162613635647536716676059747800859031429361333792933191750
084855096043897181104098877520049212223706955858124962571331447606
650693367882768887310324454243982890773510726773326678439323322019
262193656446632205485521963628598869821245905948414538983552431719
342491299006181463959235638285940616902975647631126229146674653662
658235758329722166170996892121526322495409725300927609096890929815
398547781047546039704805640970194386112437832161330172173520169538
667789380518472999961133017530953639202007972943843344271324196298
010719595118940802076078542247534707659726795708604373801337156148
493023257101981328749438961494854878373297094278163362544363330826
939905579688104948920375875078545376405053657575719555719402413921
082687609088769052263686540301870822701883547584723999228940340790
795136796379635072634422455419128095770590315877255589220778969111
186865369280723830240389236271049807288831287553507175113342480969
287697201586675189142171434255763904895946985807500609279810043909
249452408176625095274172741560161454315945185221718557495526847527
727167389139029014720235059895496157417316989042855399440283027838
377623650108590936919201540276948681440882683921593425318596862968
867707355681783603419705184191167919396618337297000193822980150272
483083508010971773091110587089489467230542892715118246415440910664
532953239354505193052789909317281149964927253060347021598719613456
500206310713361365178616544164970357368209762583749438630392248773
435717559667110116693127919204620889828753887107153631368815546679
590155314546236533727710941929274840416913145419700177540191661941
927952326018882537303377523360121961711612555388978356176708377387
852607563134205865681970192980420505034509501358373027020583244706
964639692236906385933110811220139677100067477245536145173982465787

3343641455185212468580407288018781059764871776307978933254645800093
2156135484194842177917559330935785967194699195056191293167993441119
4597294243460001028579758994049694365666190975793295668054246701138
3744999094886569721752978624999444038632564887331676004076776907117
6911021713324616671369335776775179668748237981197416413893296651109
8321913188912306128832130617534730645943263123690219424876504426566
8006403723735652001243194823731911164158611994016234561070555880366
6626031622166878471348964997725714436357031875007532986006333082899
9705127914581977792060780694205054949268204440463425629715753409499
2743557972516015720797972660701996918700986122241889692247833122300
6988351930268001333515823480974691137836974486763207815466488126399
2408284186516024914954979864427209866051820357176456894993560204300
7104892958158650639519173056385593822175143013143480759866933265044
8747990380564683509565616398416023155104796598859316847452297845322
7115630225679638245805708738335984861594269992353031747205620220377
2615270897606684602608874471195186752852500561782972888719675246600
4895413107910513002573103926269088847423715651830991635467374572399
9438350009251653919181018423706418278446319964905528885699393252822
6562628242869076046959122933888826367890525889629799260043658351299
2859166016816271158503859509920045023828805257871607999148577951177
4107145878927892859445542297574892663914906061225590184674242048988
3069603260992416053731980399309580318458741875611965516941030258844
2746132677152860416756259971668900919744620570788760619382471443066
8200786999515821869152348094620599473367284161874380183762446843111
4622719971835404199085721267006752205570817790802076422387700830233
1459632437697248430228137474801499459678297652451111647562844948822
3911265802232072339396724873534837968347090372172780718219930457966
1708748466832260754831194646363162955046142891818703440255160661099
9604396822797600851051090393629109199421193882665513643110459377399
8233754702234897098938238334960622241445881815717448698580076801766
8098313544889080730980104598298840671012861381855597791311265857944
6279763440209325404642565232144878549983704521267864862959677235999
3867028758906282699279492148880889025975297177478906720299367123677
8763451986595708011874771798648451089825195533914045264040275752866
2201560983909743678839234336869029379490238806297699255692023125044
2770510894350978320237026090787722102888386917306520297074268705922
3543037688984749133115084572724089272768529320256830382290269854988
3092664279816962154826437896461283683804207320924463481062482376288
6784819581085473378891721650337093171623008278354460953550001570877
3258537182960697550817043583499220434782397372708582326963623701766
0976748450030410906040116877481312364157224925413673506596999974355
1468311300047904376338946838115075044798362249775689187063312159822
5736569343060981010581075128320284644444669225835877645410765424555
4461602377782788423014324148376077527228666458631768687572843620344
6387264647033781045580384086990018470013294906720162910673208665600
1527200357037700877236393370846191528320488231140350582535412867188
4976918987401837011197112474081661540189701457760230237450381123111
0997102646614140418572610895636960583124466251033442176986955312300
6918353300561885184871146756537471284253783753602700254047815267699
6386800281490670826847143662704493834208868297956055915309143195922
3053793897090912350168317523156938922081723667794771797136271924144
5558888086019038040749460151095181573219926316308153672778639533988
2650376935093196217493071036054682746385192384010858938048215379600
5703755413634194531791021440277244002285950510525078853005636254877
9620396305141675053890281548398938260561846059692543092305011844822
0244405735333994864623246986164271525520414183945458833906450700211
1202816269037643723678632709238480835704928556654225657004171664344
6794270566931694659035590025009811020461599706922269604303931341500

```
1123852020843307834927685212802322911525197379137517432840567171 95
7865482009683483949354987063341045109115580239978965553728671671 99
8063562188229089859713594565995658390071990841135290070321279848 68
1737876376919750159650347621049267928002979882816144262470454993 18
7358737961144612207504177376803842330889951248927382171911705995 17
7134534299457297252152140383430460734029321292971835990233671675 51
9020483679889348578542813074091711212749135188966503880595036488 66
0019305177479737720060385947944314665010410771923517224472581698 76
1357315539473606361022608171954947748733465753076976238292997066 59
3811303556669828383083276695476109668865318112204113255088889820 62
7979806480301121722792334181960798412999720720088418393872211390 34
7247310851532774836698378248679654485396054673324517828218373712 90
6848843226097503190635530179412648238953511473878639950548602246 00
0335769136031838256952239304394162767786502263671590542608217941 62
6062786534137817872842038156593007446364067398966754926876493957 18
4903132121136562023902615229840628730954812816930301869685037131
0954723293767472224785729641708198589405216965981052537889235303 1
9872584948893728640768329664533840034139733684656642998129620745 32
5565166107548253738167396922771695636536682581378539295928048639 34
6240680045612897367776564992644452755616222399431748911099786814 13
0034087861609640689091934466010685781739964966919294071519797706 19
6735563278083037486762428189532993799350174377033041267846394107 42
3900040751986059181564659477586096999941255968962286988157890242 65
7789777920452894572685978801512039187864497160499936222646124958 68
7721434771817782721540331579087438333180945029353819157281454183 26
8211248197232597214322349402628954695747621010807874258476571478 01
6088334049625546506324147444237115967899237058982776565685897719 457
9259926929040833893272332479302542032741208279443693635479139795 79
7203963663957821049584163144039654240938604752833259243435007813 06
5949179499077038816705671448564759014708193628846063438708308730 13
9992537252983668910410331359439434771325452511125521110927987277 85
9513827089163665909520741709272525129902620455410308034810322700 046
2081993134977410339699357052008134906942080378722037904643828902 49
9024012213804633979802421821068393468508849301679182896621304254 93
3387996548743861093208514910533275877322690267241292966256736459 06
8755914896312050174243945263893979024423237032649035968525782137 80
9651504544989318660685590975663239442261822295196564215725418090 32
0998044966138139531209853479671146993426949382451496674751598529 00
4751805276612260071977572249670815150580391434618240125218092935 63
4426976907759366545209082025060278046714933632677872058168159702 50
4818400764542843654950033719423356553170611002844230300420530529 52
9330867637602864565461103827537415474849100931287295964420569761 09
9930208810403531866948926239609953565722335747574315878618459048 31
9664295632227261052716481983985463379437083657296409394888203974 86
9606943333151457916397207480723433442706976286568508894730949815 97
1906831106120010286752309520111063997859704188194278438731917954 83
7460367190355693083899401548373818626248926181581754672309466283 62
0556512169174703278872845703153454448585223696069798688922493453 28
2096934356761679260108472382625897599052637923259164150327636255 94
6077462741704332544934574448894772168762718277270729479679929407 03
7256821061295089993702462171998898944678766862734579413522640334 98
1775308339939067029665213303469837278905324760606193925458206592
8970141526129507274262922793246579170483716269320753650111960155 7
5259469406191818184873777134268344442405293066005733445888690588 80
3931774845117741023973587752822843859586720282398743743529592115 56
2432243892896300272910568728872681616117730356952723169774369592 91
4248446211898945750116903129574251482845174419871337174865767463 53
```

189448975073053855804909207963214830893523184803790906668134527178
233532246612521949926765291427590890922621175108174670500059568093
135195284008043900757261786657751257452884330535531748414291753374
248775094899335437358359554578827060373973912922693703012439656897
123773945167318596704193173930742310205394492793725566951434978805
545703305908340113045524208837745301823471483685405703839803008349
014666157562782972454384494736538947399853428754328227478538137311
632469929383670295832152967629316901577016376459707311545568276634
901948250420327162355437616062896101031778920813007133134968338654
865407252699942438188746548272867272278105469899924363853838917100
915927170823049066727659616237816864044108587574547936667543859698
670955474999659120236471863025134234286603123083288725426148465049
133391408455713489742121332627956375141588593834370232883676361427
210910916326438119930711805813705320521818716890340408422833149561
397141000910171509937355016250498698021280775524186200458708968444
383063445989495551440098765192200443482688701270198303940692242853
914434437692525690603785633163635916959756362855116855631624525027
754537619629428904366194589680239185158067914386065545763066653855
089791671967227739752076358291905766479671803885645388188660350645
431683355124532052382832782772109259629794743650827334190694841174
771358669416235515154189766502736524377829273750109051093082586789
846241868494722187945092867305656472937265721055693249737530172030
034298462939903576040553269480197521260030669803843503995362245864
967571735364486622960867665022114761971906683670007878619525727724
960774953015827040191563034896276315535912935217020929815099957118
977712469085446761448508350541333397848261343953149537191502413214
925257011457627010326835919788551641036147376475962509762230288111
889540471348042461791153854163232840054371084642690950362186833728
775644555584451321260709365088896899626104066097271490158259265168
477635062365733527671953832978920009067298565323545222748165410194
800740749839188230279326391449423695735288995277041099533947552827
010819433573369718431950816575181773136178920372220046232022502571
019599757952402244477732146208376008348553877306273902091886835251
019639767078418503278393034561640141876545693800041666721878598301
833249780430668413708789977806097087513122453733179210477653219632
269292062444118602403823939582693848769438647992158275076780160750
653635601924781632884895067393170475081966462719511896879259504855
981422537353491882022522322545270640311004505864934019626832439967
50270941977257999621126315498180662935407155836102749719065184274
065659372545125747421356527406125514208736831953589153401830056437
614755600059018875594324899873423544185362988977246414291129851849
531060905307036852909517474704662175927821270284427202763242218865
003628293273448127781908247371971787331228262452933903310566123136
943767215970190562786251023149650838505178495474625792863354846764
750561938956048712724763153542713060257324619707305889144995762866
108051940160877383993594759687934206306164976101628938474378762708
398093652868909362413539742230974044012337734528350622583007681949
535057372712472914630242934201182055942858754096729982477433299525
328938891028826238500291868606622306007695414534401415437802743546
527798112480595010881568865390958105517925178926168594761989018512
885485330019719136580509343086513733915671442531069334558535936906
805731112135220901489843226163964326307761140249595727575518017958
941940131977457342289223309973919624542378153163739920532476645534
806101436730683257957605166743647366202346212054883257962067779465
896153466662849622512559988373663561545738099423982234139778573181
185266945092193334002783956605221904343907952187695286295362583451
142883374188013897668334834519923543727595099724884754998534821287
541602121420007167425273228186584713024374038012472127577155173543

806869321781709846930477213869343623935185177209438091902476791912
350163419749830019434925143922732839989527528454309800613975570079
141708167825793398258034505303504355997163018455281682926422796379
517399826256972139310348886952365033887672353459179213883115787976
624404445856862661187618660778544234578255621751391515121750699702
826712148235376167533902997247943869400984398033723926082575914971
225249699909162516822418830277064831538112236871275612260858402325
217728238991975461696687100468066683951394054683014706632437280971
730852617500405405846357996438713060250466532450985137113504784066
696740812062280849524708273677848967506686806656952046159359064032
782602281023655208379777490999881339305724930668665438786938362894
312535175161385304765696084834268921637953176445418916273051752216
789720804110223722838862096566304326937505381260580743571556442520
301536065982737244631942002726368400072903913523216097806820898002
503971154135638074184333838437755945688993432757328763589953934333
013215225900120838600512520109318668826735672604998799535122658646
076878454488418338341366254221969714632518921172825009741219838894
379964774246111828756492740080109580681071631909055544066376841992
483030382445386120476391807877747840955329367731266650623046349154
209455030131869928385870404977694987623086816011982250603784077829
331371486931955769041248096002894285901471563003599521187511349609
284646438827763668164442908723542365626241849131109707115881107599
568488241862765942931155326435533657810786249366068097352567283282
438804714953365163044632204199356237763659235469498248612224043693
305064454708669838194561327173167062872112922330882782287685661129
367040431097366815821566525309531927357606575536633813081141261504
182742591979158468609756617111553592650472452890139797307483656845
667637660750300080388682744809256019525018228778677511683518513900
923990735105970327066961915407328728911684660750520909200607145646
383935659156554266871106258607999663404577588827698230347449917712
787416589237797961170443306654990881949703719928121853092042455010
101872809707443290433948270288632007292968230071606130096726729562
697918386254192392746603900071210997349610532335584725675941583353
650389695788836122277161020990818078499423562309210965202007496819
097023368204647962109385232105587621508865676768483543211634698215
738765508378320373381431990074202634978428101168489575401021897545
070983266542762146933390803990467531115202415024832005665615806359
792361819323007628827294668878379078853974048969533931470031131324
493093227705326130028850543429077890063403191002285599374199590545
341982948388657641910838216642991461141910541040721718375771550615
135153927714008775520012284097818725816627089312739645464772598088
949046387444120338340398470547482648434216601506040978703871976045
332968159456039741792770386611691754030605604275947492737317580608
753112966079880717230218830918163103554699678999676814113223840584
872150451109777541309273348085613994313891956459179137461296122743
264902894505828696018397663266768184863672978442961082457532735323
785581012799169576075366163284457154796775700222592039479012456471
885952733235380132049867071615509158782889567274613439615495902481
052675789916395615629228002473414729092945654241442384279751348945
730605833955546620662702100141027670794584352116489088168624369796
568234197708223331301580282187684116710285191137349625550441565003
220132187807208363215672758328941194293009420176277343107493222163
016969037110211968178145961129850803567824717557225952337646404023
992449941171332270648140922089039340677416590793358224796176127195
757906232160753334804425925247216637653281249173787913554545318283
886538707564763973408162444498793361431231856965340138644220930574
391287276338158138712550673712242883009985818632101563534940227831
056311703310767124990400513200129348970272130995215492391507859042

140268931300460986561523614052530392725431314097867037236715981350
870414441556847409342428580682669188705870133146463205081915056248
476004435207080754087821149494621150927923356416767368335016422842
786529339283279284533215152892040943012000817086185841075044157621
681026060833568283697384319713651082936212468002579767991155399907
648403804992817180375653459518384595099340093392603110508797537641
335490529395708765991342899729770181614294760801328372843715905906
287968664004706149178465951433808979790174722888221305314151452675
047969517343623347261533030009304974265653945794747407885636678194
708758120346048621221197326839850319839880675123556072123142248397
682069335797025454114267856882868576218146166824675502952377526614
089492621794102342151635411775702690729443076957090896064414965881
671742121668318114963709144779139340867791720363604718337590738200
996909450123084402978243198983074299124747509659505402432113462983
344156393868466638451130417168804680082835017990969544557428581327
744030140033636837871236116275322391852009315108695478540406038514
296753370514492285816823175467578599332489704331947481163136568762
244209211961639847807493990632550658610472649946278570911848293076
400523023957169404530229774843375344969347910427880464975509168928
481102733559380944046934895784831966191619156876786647442091767696
146015961430071187637115981843570948763419739913808562861781819516
833566051319780945322585426552516534052564189836041918098775647547
009333545646386374588183707193089927774747776519400712100160212924
290428843775188556969379841374619594878664049528851797029944034170
922571269836434779234200044508972401956427768435740004691806885788
296382555685769552433481059235369632377665412136136594165896489360
126908303912189079669346387826994625689894323842694790019544917649
079925967283332015020405505639582288322986542152012739038571255115
833894601478826796130705936844627140731766358507487735153608785471
059945081573750376872175758668974763714204508585934755203715928941
449038455518882477822488860567681794844885054248271176560204127562
510816987302947899169290417780732082029453912387288505780471150279
434067819720679870667734689917596857017096422149884386213172333014
036484090622963366139731205126785480197514010687614978678223829515
530144775438800942091958111908455931728419128447542459302404344156
046849603652232207039181979547390237479288943062875587989550434633
272922942658198189938496433903901748591900745498519432437746889715
163506178404476581726383698089797509331606867090273629067967365282
770315463201164237555377992984746403323273985355161097777810752126
229689498605135171656024102871037724129408278075558919925350758497
154770214909146833655432231086547487771386228875760810079271785897
902598759188635196306045666633363192174079445303345927730124049043
232891698863107254908590395013066665927301170260376629810683291888
015400774006822293021385957645423568417243649753033910342475954667
977697080027375943580064715248683506681994620785001781035428128258
352865340395212327966035363240822318089825447710520475037042522647
972286991591452243000708332000742959773227257950376529937676872026
591893146678879839618766508408972120716214708050532965530683823375
864780997017362177525182662259448897555479107900294328073777695412
037888193857533624535557555386215137215790485645195524782723839043
922555586085459837832420422489960586622158423688887828188750328772
057840977870899101239796223592813041542814620670946907294304427637
357079519463824063853539753893251455320403986581318766650671801285
529209290281388464944991314896215110965735382736711051946125607048
321120628812596874969053325465166098551532847050207218448979151303
859961827075525350830941788153307371333348324728777479051810609940
065062184695791416090258633376537026950336325159012400610772655185
040857437205040286964190345060154341482587482135948966805169716220

412921890901365194266163349101517770935418782341159443425730184584
604796749773411239673674607693758490635429993974545300717430914967
401452185883758079084100939528251239939418878000980008529832501179
715524696629805239359426053342566834841710659646899602406759318187
300760771656964600274918478453928375977395610105436229722833079671
242759581913381790783409621408277302609459830241168133924254021024
79082958427192272091231049877743600822820404793982383576317324431
719483156971330010828525340178091758465229417473591973493722187334
865037756637694545173480584127419296480623788474696003236329645607
187500856199400629019636181432769610791010248477449948177473036975
189922813557693575014468454703817964356041927482029664114842647553
884616092843173667326095117141455146642087759372116066400571318718
319403782493035660264211455546550963815617175988326306606354150291
012138175740705460347589437776573433474433136145706954995855206815
968719207955264670235032289325886926552115837404571767926978698309
365844160521753983979691416469213052887124734821526840486335441603
666716454520572878920653903968965700988330392782283124639883225936
818488973007620295019139221746963919008129822447578103017840712413
711813334742069153806373196342037227001353128231561282092730787333
606573188224330352677536168514401284814216046927928006261490423726
475295528980672386898011246352617089223609419514298318505493877642
205598397842354960683843080444630918738982811032326174942490245929
687054290959893277188672788181490220518594249649786437220191508458
725241513773308659016343738990369096183941846604804764128577485733
960243288848561594816539130995078432615276124274373041981321913109
713823323536521962565656841321009977934658671253098091631236945456
552408670990257957373786907357079576233304152045776015138834558474
196237479266731639431708110461614928063588389189301292776504366428
922229524865961964742501563936513045554221841369811550560226456922
426884427092190824913879746046884262153522223215969529720460035628
448018051430923514649064831558147073373909909403306351628476364524
307039902289069103226906335776036486055194090278268031593780882659
283867885892833398144312107432421057444077972553048758075438271808
973816058294605104830293838632112044063237985310181200968004784013
121041931723115880198941289995094490518235202855017478454727620598
636970700962150536736780071040186618081413859627807691530330859735
979222742977968064432368930843822216161344502909244442413428682045
989239144100586494855598206028492271624778702699558974228142701436
725836202019104692411143248113656782388531661678230591013029577237
394942182206285322925329662810562789429374661505175320710232540395
606954202499821431539771325543297586855252724801325259204962363918
642824022950565291717349820738772748645347449926663833468080472843
102113780927195036693983708889807928735328153398474260645007408084
432945021048660235279258531331296531322453687733089541670661483631
106827794190105286544385525475882138943087838755469743892676454966
213807228842572393450520834542156644577393590327231975817176591609
149922300536477728381273341666225338414722426299424119246224009785
447297982912784403926399816969824983199881028202402018949606067126
365661007469397089206468940335704923809270107051053509385611794273
021697988253541628015272720389796835160423690238188359887210402920
190710560875100167903711110517939171375466236838325414471785938653
029705646262609481596059731112820725571828111332460761042177477596
454839111797136188734748786825398458668974921061770350321736670657
082169855598660531527270236429296210603327629512934921752142974883
617498973053872979523177137651695656076009410257209652641367228047
189019484536577305232471845768564345313341809126025754013903941163
886109277630735614771042982037148558809988280770118207688604358180
555697428951349392085027036099852913656672420004093406815626648000

4775036926701006718756549830267840394977902849840804112864904273731878732357492335157779266546405875527019733174152553436934333587817743947667698653410903424182885600468244271735589195629962507979082737456067765384902422624243413944410514766832862609811698696299573291894803130938365787720305440635688808739217336431685630197495882407785656469110058448506322174856027869870825449234350094624287811424792557092875881603713337049894447935541357861767774259300350199487881893546457069989802388428594340235293957359036387799554858018444355982316908424883535500656784002486228853977905919021010083266439140423785834714153921125211966441271967985001403777378161139546945562639343963722916983970344316180226380185350772747903835838776038137578386470121363152065502850438209268547160480494467248770511152956399198461966070480199092254387591936048994177642432378782451275097624575285549009194966343005632504215824785039438656573470302650698002722496538162275541258123830022466850559270192809527663208055323913607488598549525769899507927614076643457646429097181815270408434116864875195291524250698686972091219727276416639989480352938155720610365285429980422793399098463092628786791888447458228183849154137902575761730557372190989173358087060952118139139228370173304768818085099171090504440130249073627223745299812479422165881185896380881296789286027173502479061364226666970965563306010179050522655425750442591979884309629281030657817347163705682133316954675385257041275570407558625868324966666399960771750714245424347638073499350925572652501928764086492761849717304589762514871648859159895125537522822295535780927551577177349425449406463653286438873433421753070279721078884393578408051947775698254173932129280352819043283042266081152776150337280932222161462728022551727594025891404956780402766853683560064837512119565560379290890174970568828924953768000551170445147090402764770952661261789116435270849476633363303750476266818134938316846983860367178618075709038409994416681888508857156733750859452381605828850594971914115773519469163826632910009363196937626265633997143885908306240500221068562483937545166453897795264550145438489917421968312193140137299511841009750979419923746884095425013212364767075328956871649059351028968444317033151388148484475429115655149549323986047347014309945309592965732996406961790565718915571395922522237799661633492924092690021217235135430809375801321612313775451234872903356146091481642758104995200236087135985496480127596933372611487822048271416542861097287385314981069732753696855913268893124658863973778605279647314577443870375512914383631014808801025549750185056343837423264942884038079814325557800163924992969085284589836391329251793706254015022668173596465759859416332244151575736726621923042569556660186221382619088019258120372388154600776003859044903886176642120015712766244087630529283978600293770835310900439186588120008743195074102899806565937988870123097310069701795579926044738696361160786515983764865016794328593343196281286555160845291905914267792049399041189678877878667430392997576167646554681342473563258183134122662690037483369850318720011746032135725115548751022769220143344417041475936503892095454997699002142804930356954589200600890881228980848054976146423981241706535745430437630406316824995543589767797872906794047694811267909513557437820175241646753461329758000981295779099727071190393638094497139562962587268238358947185416563847329242758695098427673777219233217527653968660617728237159657082113035581036381518279132569446119309158813353194273406654288407415997081949240291838849762574831533937525466736508284396554011389663016763904630178354753694786523936554798320943468147633789840657847245614207599449897741287517449458145346573837899796025595862692807395026992775699077608492022155897434673926377917530366643870530714852519470220957089282863695049358558489320909558686670233

7554319254514425181288007849915141850033013801828358291679510151
3506325891573568794876741918382330615913768592234983322508319995552
7848625505908486745504695160242052152523566763826664320624401134462
4618483553823191903078979048915649288167569495129760602261851595755
0909144277224633135683572235099411255280916228242406941132232568999
5320003903020219696821738613098796980072131163862039032729719199555
5770591894651777093108673343597019086754637507774941816421366245877
1228362571309690425221424092741444428629467641111573339026068992883
9959142073445918210129878938271304750373833166795787273900628348811
7212343279316790078017927782057627942474196395117775094573952744388
9586335334736796607055052701214254429947908049134644735738109236899
3078603566236646115074415941923079912263580537526359624935883885499
9453578608882349064790166624557209478823103870099917210865162700177
4127764789314306360703172683961166817967359974052437218012062348199
4046773515110115013575303935536135039838765404636303172924040044399
3454222886650437559521679638559904741481073663576222693266433064222
4662251361964755994793944751674283039388495387860866631365660876088
6740168682542438595093098390095082474146115182797151479253878236322
3116039101597908370532676133565305010892559736947825669352228981544
2080144662048550197653232102181616621953034657184128801650264483177
7753037857572107572167270373519222403141487053328125560252379096522
8551710604474479118067319707100206860334909523369289943517346016999
9107184507049095286957774131794120530566393158259808009454159456844
7401673376199344464418595248458578067918467182672145793302716286488
4233254970208681474069158570524830142625134913013179318973838245255
4931717540343510630594415185851799332289184639856876612867808298211
4112906685003925620774769600566532434851625148542827048561423339711
0633961326931405248211840228020876493282460070792951867118770746411
6457364563422226184718124284383548826605654175590498795369362959566
6497254118371933569798469939982669708232831209910934125599480819877
3220386864574976150073150130803594050406734056012325709787469629188
8299464967009553229932888316237622770234460841617862958418100330599
5177229060066089581303058313139558858804827622596251755183942649800
6312004512718100192221949705766974884459659269299769162079726642344
1433969809608501454529911686784527877225801508574285976431805040711
6225494651215268950797614098356924309417467654518171956674720444988
4286032696803718410825938273355849743868558051359522845528759636133
8602758981945310017076089442733247468724295891164782188536209845822
9468303040751100830546676612131694944656386623369731490536304878900
7883287404207267338339692583482813533324626119663972767295769874444
0365471360016591674771423861819916453062722898155773566229226610899
7177977150083462794644093605843157320637834761951700001658106021000
9287840443568206520194527028568264322176071358160175222197346722333
2778027439894359711559780381276278065260467038575085560255608105166
6677818382636916211202759547635750927356103375651797699465779495966
1144911621311679160046072342568134822109174704161025408424839924044
2350962196912636812091643490379266934922254635101740341015465437522
8316207061053908212266935397141446701638713391957245620135139204055
0918614732321829361951891236454920882390497224882879142572991339777
8222478186521013391414371601077808100071612966209368046726337703011
9405918478585965573088936450785793600087862868663380797636890280788
0619857010023224477251403933032211957205696718642388023218633477611
1259943548649924474751661178360316695263754160436006356632387105922
7935792175687119998228111108914246461015465535659621270420999440311
9758735153439833319560989417309386547454651409993893976935539224588
6430358876231576156258615874628728187813371123513458788378558042166
9776439852599278904996242906538896215821222189788781162583829659077
3632484969779876120071306818833725190037037740348722504297349609933

476607284864000163199296066954370714271183192140992442394695825665
420541914512045536142364884148160260537748661498641102837595162909
996220912329819216783392396425613907277536570773639372511982297369
678692468915137165264869759651344357671228268583753144012628041840
462286935889735824637378784844748164121073381675877592282230791549
822108047959043483193387634330733994992519424337213632119153658108
725527549390349701179531056527436888894408547466647272707809098080
469973094052028161295824460656291655076968023561451397859995356144
918568579496089902281540993189622735210747582448567245261951004826
725615270172300934394302906880530192645535304297709997828918871272
977567312250801215690898921046206103032654195355883084346331184232
432667935024698940574739104932355738728624978680751440498738143435
118385255825859650848619765348305165577465354849251847062358463611
289612310634756134878291962080802902154188956716954705784126360651
280509700448653394569212676107464890218260151360021764042075934304
251549604712165606382690726688106032814204741220088866734941517997
246444314878523028195667730092434525278035472371332498112111447527
196051728526390932400074100850410135349534043775070868269290958964
505677769755051816997654774249074987137689708054222310307369986214
421344497048083338862903619246209941700659527552349945608408883527
711995125641178874877554269069537890166583717050597606881779084911
618570218746070392004112101273866342584584512364593848810490498891
712468573926962218064811341034779991028533450433635293559946999107
974838607309383222418565543650497649458385709575475892418107954937
764316896260734364589130152744656521145786141557709921049334371327
654855020781975775613019026316273124397224267414660169404353621126
647620668955718147149979122203905936045317129530955540122855108112
121026572969756546972317828275358932526338331110696576581556117359
486795475942419029558201244575090753733704603536226154971039544311
293340778323908570180904613579628848552716340269965063129190994269
695966225589497975628721287757198542419673475301587463470733050737
845796145354542520644230878240841408078462796736873887889585204470
927908101071211987798727835924233933557556193713249979593876098414
779898213818265034322401306385745448854935897070486251427872429652
100656649787070156838838645399590365080171114364104266278696724141
998326547334111284879330934395808667890754340702026793784774345535
265665129290134337234628978010782011552218872393731201609370970406
863255367709261919003989418841145926153701166686589563673150522163
304107798488677211494160046621106586052485041079547948381425140499
636824970739689414424666717487367463168517045635775424231051505809
864764641739106287459904737924888107274308142629524890931985495762
898324171908961998384100181546644378082375332840225441604148959931
908730084724728557488183769758189534793908801984580628942112361133
354632540040466535717316673238652078450308016619577080902713241189
165462019807089554609187238543726113874966299766658252314714818800
323795003806670118598942195994821840489226045885487688802337594014
215791542952871235682271939416595000624391183593426781640483541623
849679242080877099637573387339727138570051017216390102814006199640
295555510920514823727888939586225063582005762355860460670033354514
888308803206990796123184492379733390033115058829483802068767102409
164859588883944324656344667574633584953224141651880328254882089918
241029406831822065117122105635810950805267608243003755064833828150
365835077439401630802437480979848295664872770213417005589113966218
009352304353954888255902667090006571188533595233013070087619042828
655767907621690733940856070615889361930134878025952499670315914362
442199778519830043798444145486551659082820414113937673181040877195
965789360890060092985718784824543162180245386116837858586234510782
849182169145864330972476771131495903082734618524789221595763617619

415803413031319184160775944101215743941774531300010795056442252510
930673752322368854324279121353055759262100311221186203829750275381
784254173117337551711968165240928436766301402843312361032923452318
918370223034445497414188639739865083728606861694362513165598064294
041596095582815279474489407960067040156066354233056608882241092462
255818735880428279346966420627411245975215220276131826209673609273
703418630680429620948015190112156463233546199496049740834443545314
211500672682516687795066725418216964868333082337244585492412921673
190981802291517170276395390765960845015945874597607378223637182049
117249030372172847521476818938282858524186640140709140991778837139
569216170769790100601923052662978421951009921840123220629140060482
178194901561947902646631312875029083244483715546624849403861524853
532736635483810240007269503753672481333156495813195298329582489456
994701540983863588375105274450387542374163505345484390591298010160
337024065091419654406408930992874563503612249158486023751328733731
306177764383349114167974279563535826432656509343897438769712540489
099543976103094470122301704959677916134546024092072149464971044874
803825509647557923378408501570469654092609937529297506575632445478
640178182086899760355012046052898752372284096564154010430487466276
071995929003518139082264656278006917499381889757550412455262811876
953735750525012728015922292778267737194613719216516400180711103260
665465764572568399041112655103269898477620049452573207633661187946
685468075655577881616833650598047997528938859344621827274827697573
5348116703851100088120600268451219316134987292227973543514525162066
264467550435130099588759986591144348733027605783273860610819603256
783353625213985372146327137143733668525942897840927984778786647466
449578835896680820451477672719070312628807710771780875944659287949
105828127513178451193902192043621739992710417481747355300551496807
137461376682610245822978890268640706208716918408059066840210711797
550731636097123210429979833192165994641187673904738358239727102066
913868675822234076837140251287860243360137545695612101211186685695
275760983876242013180600973001510948770418647501460347190956001644
591325816011088700342409510860065966824661861370034045403056501560
321089611195643279401133232062441296852718973390029304387526826413
252372811818734183772668317079822366819849255173111404842922636004
972983664647174703203589251119031455263782036974827376444746579634
703436277452109556074920999706859663108318811970526754076084185209
907452641268309014347673429855065555504958367106871903849244388501
716241895924159706438955179197612480105118326621083951809191931686
471500762285491546332100294923138434804057097770139827445193483922
197130608237016546337256801393503481012122660511497687829462858623
206434741211262663352782157747324524831241354286604414019190563714
456193361673409966378271496009780576875416953445440599479396181489
149477288393444938623785445710715026668290496140379849969979904177
314534985205251546802967974626481966870228616423121477629241242965
831276366150132159956875563032138349578504195023669392820440838149
111506865684280960304485296973825380024172676987094580558823876675
088046999123811364649817803232713886275399981430647461240477541717
786963338613965617885964831756351902365389428609879813243105964107
502022049060393598709159547794213809864179132509391514302291823591
945291150294344931341236215198182158312032029486003940276690126632
207066257065165460052747522401019923873469029975061163166192818509
7677922609310051354456493611645965660284911284552622247857268747061
159462603273049260387319098427270223022793637175671192685836471857
815155133520340200942557325702413567497929832606616892374772379092
315644836993982219062239598351292774874049218848651486180067646836
474766349043546852270441815032947346688059802599704514719086726975
420909271463724793987242908001465960306126644166022286040507213318

150829460988850709805662279815499892792431405323990348617380802149
334102633155051103722338875148162043988162961449937118414730654397
256358037320605374193767157215516252026287912434751769566274504051
833871608859843704646720497692571862630681746111178965071273389413
431264004219002284268846322192496026992853768096158931948902304258
339012860852902135855253522872936927725731124839805870962208930684
664626363416474317898670004740615668576378571075842749474296485796
676254876979594106944926811657656925791063739128091743329342766008
234744512264681724479134541128328974457551465078695658990528246653
499093871511116967951536432826151198278986688971093101695910417845
024882873523192384486242263629734975768439282263112020004713218720
170178371097575668553713938247585338118980568055245617589253290124
104460613862625297201449551884592486107615753201952379210693182246
551772858642266047786938398442151913918849477120401248032226146171
138594797565685911745747244418275564366755031861737105312840615915
488212497193717302126743920082041117549854683639841356774608838167
338674171004703304512237269963296753350355683742559327885052848439
709953415176590329384028506921996426091089906842903712845293454949
079349840370395015943656344631399512982458133353138964830395468537
774238675823879959507312791666339121357622930082381374995088042414
367706341186794579020222525197702359954488429230463287549234967220
873007422618532623944158468650826153186560577857699729253351807440
463828973043612131248741664955773205830319852649228400338212296198
294000357218896092227607689217373213756128171401278947680860184617
347613358304779955557011084658991846526471602906243268230972137919
999502600200767792842812801021890465503686449406851637469474009796
705228717466533465347668328529930583917529192568422946133303350299
266147490355309970592944301756633438343223044154343703476466404927
395026581091264882415178063849558473212925984863914337804053563760
280606861278168889215248356454361916584590511206537019448479242546
205587901558333354325586591018391532755634325304791374070246658885
855173264155785108271162140919115020187616175817025131170079441408
634863148319055296155411318677747603095539999860807938437497622730
370371697491622920021830001353391812999182960402359294816224037573
499647898156526226824692246618226600334656315544406919091944613359
229476476175309840144569649498541787417212313607795570046231575740
701647612827340963897976977401987616019415760152910910925312276183
834786795185241937060797916559070575151805129542831018535917318636
529202913084205780739267574115631345641060048148590659777277558923
397304760176118610946668393831705136767659806086454653275384417193
324821035003461438700642574999821742521821218920424698345969460171
454108096717354784790289649005709369565850736027996266961684643111
823719819499540175555079372487801969337650451347412370232785817170
527287540676808677865733191918065414650702346930410760260438076498
398418776243353883813581831396332297830451927354200012443477014391
410202805835137615248683465400922414855558907372029219460949678309
380472522542715397156446139320717562051057479473825563034044998409
055189312252581517064469135994549410789766052839379950219126026120
477205145883687727742039393492746617423169661272448881420286391420
227812419353329742169450946795452067395697738280628009153419557209
296208702178162359573153985804940599130964597843674616880332762471
313321871636394671809366747344033575552621977254844202499963393174
861668116858002416101935718158763913937159153107634250433840677491
100623988859795461135553975583965305392425113385151972957150725671
949315914454388674809412700929742577221170197678077755411531464441
588881354640461003436754633954381365737941755202298370463378124045
276311595878742915414204162066248912616238500777039286348472623334
350644174622655488896432896084716921233108446333350533714717333033

190172115307748181597531874032065206546663038340247240443641929585
156620772019573193519488591629815330055105279952540010923346567985
970645405142910657105043026284793759359388054120084681207165725958
968277943629931119229046287498933815161612418075418388765972717000
319388966533656527359657047298370555651002680279238516795033652066
530917843546800532221181178456832436800443266590254628594599103854
757965209123438950325105958302845145019453098923277198489280787845
546749643627564616966261836486662036715578498138398682528761956385
736085204193320257641086585308207834609373542674417458791798165097
760674803423587943788166111998959566679446483821547715844522357513
096308613259832304456646819209725029344903578658988884052055288786
406593898270941098662152737161752449269126472228562674431790670660
513250331457206783440463795142601733349592062646181273287377940013
015357157366237612685283211039112016619481155877955040394086512236
041949757935718979740811556373467207427424157737740444109123848555
845196736483504888683099313898234421248549562023391981900603548985
180440671358033140872413265815580856335529235650556243406221738635
871005910916669020110600851921062061521729889870836133794588258419
892972813593784638640814620973215748876458545452905648569347625409
289910669192256024642545291001498200945147538869058507821574990164
238325826612333030842360171331330197402643659281051697460061129741
349954361141507789015523161636275821160734521938551112723008960033
993710873634008474458582811692917284014715929271271973822535539639
876459247383693261128037429940138758081761750693604738088259160765
499660285494158397942913044178937413498128512994331756758244076734
176951024244331173208054198122503155476482578701650865766702309785
271213432609608282808472227355161279802179432486389890602925191544
808852944924913238575321489641284456505833651534398394353706322369
086818474891634082930268567197857966041601215228834098644950300613
637984752788790195341774348444672748004363482761808233991610087405
112235165967744517830819216702751251435494699668726687370827849191
991682025903711477659826068464297168288715196482705657937945663048
499534258282710020287457514253134878869888080123918252754748629359
202579750028182197510099462917837851103466716091576507689424057643
488232454106255171455768523557408154552326601776743209479015645205
466875325651052514797632011122660267524833213955089931268326349301
368519242840309426023879053232047876793848815781799120758839989511
824201625492439937529250292683380891296512472328149902698230237888
614435318987992150720017872694765932161051852405076863648912351751
671937073774216243588562940623594770431200526606256976250968421781
211488829880026616044059222329331624176122908743379022287804561701
357723750619521603426862806290537864968871393385712562416964079324
475831369885918272999275782949295751304825043666028532371402054964
473380738245775558257092700759153582136224878739519806447536509228
338732197378945098948812224316665010698739616672989920964420596817
569619231839866191793408474257868154615941458938640602296132950038
120038389767450208633855782667988106569036999081567827785163782934
599361943366980652979221522153666288839940268038621838784138954997
920072289371169507757061724002344872898683808889469325821863378234
356912074028956871885668709606127386219349873263224096065960699176
020054536038165896602143871771287553098370990207133084712303917975
574483810050683280951189329272191231654940906640214568359874463216
265575739792837308702860612939772368385814919939258415742549146335
154820414128505256116414384738621579485025909591694016701922227152
051592446384786736844034101976022548505859620374520210340195867216
081712701926460704607959928713121074803511882506823353044969812655
209567080884541941022535199131368352911597222819779651917510914125
749066752719879928439903727410645886716981183509101205683498176773

```
10095484699100664217037510129640279526680792620134649082657083 7312
73088770349853816883018041591073508780278978144452516540770812 7488
47379655033182989360261051517090092011022207100166947994860649 8841
60992557782103329254222023824316169379455244507711661278195720 2899
49059237178763071379162077328051900435950630278372052428607168 6319
75683199449535965461758237933195492218355714063821726217011899 0624
36301646834987994306732489748402990062675636866393449866870295 7463
29559276358227417390476366468327098725400825658274079373013421 250
15787224693593360232027880213344754711692547244278823478857202 0478
48109668249573259465069381839328994408562932954845234695473247 0720
95768155000313628681830587359766552445629233370980039202446539 7081
98088097516775900082945233933825387379751663668484819906191718 9230
53029327528205228873899777985775746443306673842846683423819777 8223
94151522436389874249517010656678530267666437085462406145075109 8238
26508232921923116946593605521624370127430023949246000846441913 7134
53739088171054329719962592178595836890022753546934419927056354 4994
46464846356921147395454234209303509561912596294276033231402838 1564
19581239921684357059261155218643672939081140498842995401350304 5826
16685615119192470424806777874883871318987189673826192473974908 9221
69656489981576717042890174496662059680086819901956648398712799 600
06066009833665085013172670506678073810536253324043615610980110 8476
75549487749423658516371946527932849799057701845104909170153356 8613
63244389487965903436759034956002166426558152444392827872271735 9459
48307893872024847342032229005213360684605312929409749759889320 1349
05015546399788069991700873715889175956776894727061811503019647 289
13257678481619091938845977305288817391379634191013912282881868 9576
66915817506640190664257591185763887548293436299211719127105497 7385
37301557783810188444186065783059241045272431966922764394688190 2302
93660368939291435279006783454920522889611788605187540831049180 8917
75926096257118288327086436346727818162747255714850253575093560 1944
53370570429727931651832436970736387485609272821695707559352179 8282
93176304039885438950572501794655341196643840611832817122580580 9313
86536690163234155042355394880397710071250704105678774162025859 9084
00717682098941448674622992276289202550858081621731508389755388 4053
49427919056734487660483097079108665922529310175974785382557147 1946
45296290878451956517409594789439633768568878413363340735390727 4376
63722052802244591605345737540661837169580524217180032186022858 3753
22597888350188042317887568940231975197437444613352597457897400 5546
62442432497593440495376268236401505734726953980110100256582513 1195
75389158493821251296799677253612764607639106722691844111059166 671
82317481206619477228053502579386189871073293114319621955835900 7573
25454926544053448587623799704869639821290623046526023715456974 6398
21850402406160647212247386285369145422758281589303257867192003 8152
31307030123145001620383558559708463628872856866182958803815141 2597
92427122280065807217537599560975302816812431908826758112139786 458
97799156707712334130061405072073787477542271334772231879138778 0604
11602833892807322946298610943899480426877630904138282008249327 6437
84456946668559691350972809296560296837842483190637664897589402 2974
76523373707075902957322967641074447790285422057108331864160628 3483
28403937671338148094153081003836346209867409231416257725926016 4241
31076838385360967743938964538812198718470878357602846575850066 2643
01318356377598343942325631956738892178647425115391464830610561 7618
52266148498621173529930039419626797835115432471979210099902359 9501
01850452263362136629541758790411552911630059509298870937205111 9953
20951919761191111785656856314582374252733634874426217894373425 5594
43882721099552583440480781533630123181725047611209858621391195 0381
27685768422270602288085792280227899270135450268982898128670846 9480
```

4561899656929407990430103180055883851549560037101536288339329630838
8861183757251344292962374362569028623990818089667478407215416482154
3646698525111835609376953883824779268205403556229310339823461721777
4992876114310711261869817671651010213281748432268604928999621374422
6489178747080052177899145979368325769082544704995573654607383329544
4550376054556169384629935269555982548143695227451351596350128443811
6576238782190283447784194348491675433220898865725107216380125745599
2050062613832353331001746356326967882997952231221335922955987771477
8425621528196600958248039796074188068648146222184673502384649962099
4829002372167471513017616183486648569009580445271292413610774485500
1645401658821009469531851670949653202836855634394275525867623093999
0262646880252100234839881081031395915672216752103640311682769802044
4708468275102160076529128596181232892391998983761546540152884764022
3895640080091117077716866325847151888652134181009630978924681427777
6744424909848194620720991186061783788272560602774892025077565496099
2241537218489819139994930435616869362159642917710952569509706990066
3231006108564855544823176816949189203353825938397977955201780600266
1504538466022348058428066808005409727242488709889918140301721083744
0851971684465550686866825957621317618534741443776409811689746202033
7121131861503180534816370992805100579393958183960538157279905317355
6462047256467564657337523249604428666275422833411947710115861482255
1329057472336964545935077863028139703502693355867202542065320191133
6456842785222711304993084740155083205534502207111518292503782458411
5159542385729092055909315552709371573043650713919706627072083660500
6535925780753879966242782962602719086367858420342617942729278387200
7422703925864789969888520172854373296389141798564954987123131742100
7811174584783471481011302200669188117139063322377463914427871350133
3910136146547582354731321638779785942292590928732660398061705145100
5189352613846356490075823486329152025865510311034138140491015735166
1788607576439646188341017565444148747743969587125931820683929218111
6808835597127226537119526746391354728890901025277774299066816800533
1987062324755696316479419943187728998958907371744791382210691683033
1443021088327187369372236371598248511277737658543952206649876302766
9812346134536976210473975484439806649685515001824287241396293002077
8423904077017175308740062110388698127516133110767104226560953420655
6279648030919591712222568605086606974919587195285117201863014572233
1112559580580300595904517801620915600339271581356193961524505829400
9423830675223104876027568331931395370615940569008254671634076885188
7380620283937649414169522478976448274159243938907385870336831106477
4829591656363194075978797738936856825365656798119405558294689075622
0974863970515862806160495538019929879010726988526406869489610033233
2728420340618845421289794321886897072738273627785013727011496387088
8466407860794265555255548566482536726238855301957008990944181411966
8192782252147436730237577264790630213627009294351935375202448246500
4450281774735303130558784042890529565361669557574603044684002013022
5834728578786034796429662285633839093850405959963382052013134597755
9495495915282491367582655773708630985343103165699186477493523558766
9856167640256943679496508265951649361856390677613408634494961092300
8528159575947324469299784378750695779652340662703489343992200393311
4202215964047530798772485471928990319033113606753874099262658761455
8295245462962744782530607137086100932565610910629015937032345784655
1094254156584863367597171410368246906821364595022593823586898034200
5214584241562109771259419886051741846098105218023309134915930553233
6402118013539938273907609601881270857002216149942023522853011978511
5148254092669518565676534104044454854881776606291533342393558830400
9770803826294318944385077023904084750171040932368683097904497849322
3892155985002143587007713428715847006930247167663122853028391120299
6158483322876218731270254402750988697775241867197258039845678345288

672337262681942591376892237327986963699571947508280574929908016092
963856488757743668130599330030106516567168643311600381784331809476
982492426608263925647221085630282122858435912911420360327206018252
323794629310925410251245417005166491749697850176586800128632544573
787252932671277451624334360127340397859302221599617751736148666793
967656332219514934298376790374930825170161852849338344242504183230
302641005578188318544289364087403203966008892343871002340926852388
496732284456687365704234315669893811311708549805563342411090390294
020698788366865009641636917052815658583564774755048831911648430645
989966327033980010970627143154871743704811217006209186084162459632
196275818916875947150823636892761717485163145845180705435637970723
278957450538758447100756455874737245671626075875582631624163830175
894812372734658328426433498421199067903327699506187886673063449032
827837648590916868065398440393171382569665959236573482235687609504
600206957367369539435373448928789454142994492422965419920687071717
990820275112322883020637409332430828407680236599622307450723954829
132510014623152283856699636464816199306108035011934098551287730815
954501549790983261007000843516322091409713166839050930777067825793
848915921320992865980751664776274042102287069580316910197690665849
294116301449041755241528407985584192045942224722408579545248996614
996312956745993178744978341350197476002485583093556978815369731363
214452511408292841812804502491992959845617297886498365296717374065
350275757846534187078421309805735758509870892321833860276680968786
744587673937042506105294559344800003379484411869103438489819927217
800456984088256180027740246971556963453537058177132496654317079548
795257766421120685694340740736941652110453017077751449502966420150
856734185613308793690799085988881954177426188031441417486935293012
862868769796349716441242177380019690974862799608946093642530679104
174593571283190402983113155059303861120619275400347429960129769845
672856800757478682568526558880550446502824723406212267230987650952
467955511675760755189736710818664873391355547303871771482599249820
906556362463688874428163547597380200927033727972357562058520194887
311757364152085688799839625539506720457656370866786849616739928990
516639547348064688416321612696232140430043034978793765895525591261
273349443137493187558515220350488771542061283232155425010369584201
177706058113108574067217688447392412151890674297679952843460462085
104229892955901538861717778596259659024537479964480573754259033955
717369017939751600199875836990940353460200600611457081297272864924
415558859750242749010199752785695834533449432500257802043444408682
890750774396173670553837615787863853870009535733359025946681197512
373898387266536879554300184150448072052764944570257994686803499492
941686747104745236313650471152698278105520599626500224544073428713
991949802538333850588539364994173363696431899803695321146317461717
020388070866349064786340422458469135590424245140281425972094336803
940426469576220351976052537466919686468405748522732221411263468200
732680991283596804337124898651248471333865999581557036241284311923
713805206985546305223962860016932609247618752321257009959416454501
759791304831585226900924405531186581531978459314051354967975019715
913056364079678742743886974318121596332024245369509081085401074867
453223366948874174475845601897763958449021749345971047703154197947
217559031048955150713033750922642894743661501146171128540489836287
823217755403355815130890086002311190892831719794615273396398147557
956104816547218228209282412622440866173161182953146270119621366199
594108793583564320932964189356289507521834160949562866054760820233
943903693829441070690737842159371100843550809934951258480556142602
794888117357782314109215630977556334356892809062401470430406809674
541428500105312911014407193181060055619529375994409816126454374436
773978928455823616806573056868188932905552483777388697883384821269

0033855255772963294850362572416179456068806250567439834358406885865
5279847132056820332680158401186122537107299459297219831403982495466
4301364141209379446478463296730804124031579167146810717215465797599
5084379065463926894416436702026717333214286872793000675256808959944
7246054007739214366237747036693706479892806834363066623573549188366
3067408969053454196254921859594829635299144250681242195784939762499
3626997668432011717830947897645342821592110544192539567389068025877
4295234702462536272058624452991614257874992015478349251604342385344
9243843410303807273770776570147437343580798451121498902138772611300
7493125184849712899174590950039321906256807723932545455469176735000
2114278251591537139224752151026195751812555899192237756055625186255
7778715204024235643008015440647378686471774548533756851330395773055
0554298410274520848824563801811743244115088666941720292251387140522
5933292189039302348495217837323533446532626937774732504109205508277
6270136010762688057349283410615014321179125841093281226749115294966
9194414057983354038200794920527262073123858332785887564779056721100
4165441661470761288361000624384130531050140081010798755577315225044
2463587242081346517078196813266473052652668700975390102354840054311
9103030585050732848566621892307361016097987301096045787862725969711
8214977745931912172420855203832230977437336272600910791708541590655
4006954047576946452693527389589089465660935532169427129426114018933
3681755821233660928809368683610841295931368976682546341618079733799
9311434919945061976741403836739591043325083789609765463216344296844
4750471808775387966011594691058469856934631546715196731054018943477
2532735101233055566492446253089798552598896344445382941448838257099
6712360533891982931364903499133212284219660873657571369432863633833
8749656941544771070613803436739395432994554889460443286211704228100
2975764901084613036258098852044456528892538656183554374669050757944
8211069811160943622727817194688442239013643555822522433014093711544
8401136621388410824798090054297249286908770786394338356975808913444
8554837537677719658959365875154327502949340116362862841933060481711
0939239987919008842034872943437214946123970170343033707916983162455
7660506136245905491835880524520307312984242588018709607849181635833
3763214177655264846608626744947775131163374609853261567716821601311
4781904456057708920308018521540881268882246108542068433312797584800
9219544443889667113144461789314741293665128198790259329194565273688
7344836398893389984361211680689865797567488516548886337690035375299
1987757693081057355173951443795272703800447049007285730925262631677
3099074000684904599758787132093533481479980729780306859252749215400
3250806206296793680290963657119655454749832657557609467247229240899
2061371056217009793399279320665670945892120839049846047586480114455
5232781360281453445795438733659918540295506010018789625823206446711
4509639808919967566146598237014128736646880385940326652224087508866
0528841067197999140854487007293022017220260304786380710886172631411
5313923748994778191781040775452555369360545903637816192863942002266
6964803975868226345375812853550796206064636202276341001562539119944
6325786788360870252437252628303301021044894326222627532207366765291
0916281998887191616786697698617210689500902364059291757218594584766
3068921247043650275363283506500433461831897030508283915358506052511
7223442293318962943725776163152226873950058135915637995009070457200
0726096898837387536982426218631495121398835716735638806300562903255
4751451996618172677820789627279916563774802991022509472440941980144
6869018862528510042336659066430765016710023678735189804175086476033
8055608827119847886391169660571257581116143322031623986195399608100
6448911512990383218924496711518198579850277044697851841628073295311
8752217073754278078367450606084368877693049830230414364689139837000
8269396406069456288164169529166544373908475728196961464119571580666
3688131248487829600519236538166969144131643782128080377458223191422

5066537273186601585557631000545881914608634151201340660568618305399109284222097222772766642007099875823159076294391295156349672083809897247230420387327835086014741160482852042007439596077939676657454574157343814376292956109531158482092000199683492227462223320349229787977209325937345893185303632200218523329226604326327773869929525440037460654804476294982724042291465600529045986614905305330341184232774713475287632417746860051035196802595048933541617738765023893191084066521380466746652964357719605228927287905823336267171801047870741509786532744155522815509790153143256991410913299525027699164121847989035341734180288807859437047470037468716698072913698781051913482317431997181756973247169341124042193238323758158340750503251210242327215999762255953608171639659554592152006263493832779387174509876695534287878977466443638551706502650448577147587895166390526126187671738754045938775924493697246870221984680515191826034341463515337517354398346584036650078008251337619581112539605958941718804721787463604668507755956087664614979975907257254669309501812266059756338320451208463864464994775567471104882581862451811482160241731135115533763941801619860829325848285029721564124935435493701472218371490931352746516140422988403472587358480471003049710367861996903966400318902701410218747143973930479667127146967452485821961590443588575047471161083558276186011598867992523277670077913488062706784305823037684432240837285577758508216273267775521575385493139188944331137097187976549309906304370838121079727316041468874104427329403072777436378288442397759487234629173289643236489504230303395254772328529225182097863294127922776106997648396446154980303910368747636700744207201348680420978344656707808500851248926091270815723789681795389714716653163541792741324937455510490677088305382910466098618013354927364715788231751702295292557476729438071843235282789387873058507172169878408709360848912758296731845035233009100824500894600016837289698553478156770898606637024379918187127137483594244296364320947275727110418044334601305107022177824729198840954442915592472967931476641686467990945637704603699887007961285734670508717635799267264190775864829795801509614971798643932311870590230974516834357125233587442571650251307838439678124895410287899686721555835181821976729237267508827191325928904570539216996231573413598101626063438419741115955987121948557079154040991260843153449472936182597146663520940304301794361263079707780953877079498286453666763263533422068945343030626974857296018808465649020974995525667340133802823078862980681870520541252040119990430428919912401546306489607552348001929454875288055706505545235487917895598744025091300741641918093997294682210357018981186767215490449502844625996837678251688727079295334993551398511723804911695566612441880493518211954231454579329549732901154976327970042572529288516760556706957888189166689264962782660684281788855135682201059864864426897310404206420388620214159313343356506079776483728611747854101318195388209632637397818675015670201035161382762235549907816708176258006326519090723097311326126453948061274615763974697290381991575806307458753851733483346860760889646227012140401657955979081551364317926971432781160003950929530530156640538544014468195674168914395005060129899532520624256402569997354056335685117062631293788209655763055783267561616292217039451858995939277954633373520501688984643648882073146139928560157646190608827005218388294490528350185640650433641753501393498570510063500442327755328051663250155536001685586063216180167882288985927773980708234430121876498298821950764879374527324977571646794376825978653807770909315826869893218567541022181137066289150507191692405571751537297985467954771694404608728583402007168255878503502580689802794309461846725801862557176999143666925677662478882563671023905089297579824895222709418467443414664411911976248630886752269173795608464341633

676355883085129548653711250744903732288271992572715199660002166693
895605038279519066233710710296452611253548201808162340593161238338
327872154509090544271980320064422532360012498934484363675937191423
227785159626576845253075848553735830841651524777849983556099679145
290553712899338041735803322331380434826019161910280534759866238541
512038956061132700556496689281316755512979967633665355404709079398
886689530685781017330266056885368956021180772162258921919924311730
489252249712553122821175882252826509235822912252041383700828638795
994087533142292025537883192590178881759789430772711316048915678508
678373881223628875585526611865733674460226117363328802255620686149
584672266053779357525558360934098916382963659980073078450013699458
821205197427166295188936366178872451938798839149850746367011614623
559091808914648782476983237759787096345593154570680055282070629464
310763848171836412428448832224163453064177764928060031678970019137
344145995290818130082735712024454178146061237267144018753822527453
515155242450079079753687991871156710284530318732656311281918735814
205042307746297225403696352335748306562048561086409342738331833463
432735127159264239390049127973028883783463744236046441965815843840
531910829032323939150063705253777010659429219076639318081087876557
596907078407317323973235506344135855674600292812281944826263869185
813672160413954633187907256169308154099694376002147684823666895958
338644584339139734195773954458947379964993965019731801875821543881
858049244015386570718876787890605894343970572439680676623307775021
542477708237790441269041207606617175145832906568124019088806520592
144597223687602261730245554637405620748081399377467009412522215327
344884170636815244358256118696626136383402928664497006603799675017
937631676391806943767843386214908962356182010564061401237788509883
566708473114371688891394268479485387646650984117195433702892158483
507258601976515246041543560676274641178103795805511058527509424472
965571294588954450204682256720106209862067718216674868855967781333
673048941388830396566121918933058714047784551328767280301432209927
052902106177139212773759052612468035783233612231672451314301832827
898769529904655069884850334348398335992274164810683359631797050404
801716357581169611089887526505039455450818904578203339888055273361
740589066076256788760234489965581821950713067982798437470147130567
036780400279088826260968751798433062161683649757733948961813441882
416886502289967808162797562569227498087402088768810359436929920395
726561305068128838761441195918624002236445248003947999424405825317
268246513520948959665872693663488401509928375984646475342403056155
906510544916912421860787811768003899307609069048350672795121403003
404594882920845356727163295012007112146837654494140706925962143328
567874574683801853930454624313698559873264207073373620952682533002
246595654211021883163555532210223258354198699266649135319296318782
349014915870014774992191089465012801726186584241199577478446380888
179235896723654961582275353549969841302939493205679621375776989665
420961561839357548510610187366314026732506199565814384095884354592
710427524744855320153629002887902736371511709761157510447444857500
232585814856078898512835095561221244135323878162333181656119292576
209991851687924286242308017058600765585346450992122213862919320062
916671040534441325309940503148420160032899923191032840172480360326
411739587737364393158054756344611667442195905341694665636800499746
089176326163936997268005671191900811164600029600999062976664950801
085080051586703858197181305523173246301735287693034978985330460761
267069151980521041879216938619991131368284102584387483086310227556
652408281412368889519506244729375224036690315921818643234026919293
237688715177080767495238989921492457417629918580433486296060893631
106258100141380630601231494362793068733268768771474454961182419660
371730117322672115488941447158076434636447645759070334085887937938

855117519467423534045394122524064570712144659046656734809242615388
417502636499676403979640395064536033058467106591608694936427670638
418752507639689615603173119292686338323867446335181133083074391303
543372279071410307952822716588411034443485788218090802820829554428
122003801952660218635952461056660631495761270251582303335224947078
904661050788518613260070870531252100924188140333110980343254472187
535643394322040465954026928504488556461425105265479584721663057299
456357882715607728021482175044787001124779365707056730989211387219
305929058086497839863194346632579224283402027520796201076674604694
071705609535133339937607492713011765060222240784478082149383963988
120054778938005665780359904311487101637277035214494728448065980210
246296286374329333750053421098402485860977159460346265070892757684
803201836190549885228928095376821315150435582517203728601695960958
764251395013822098404961222624228173404340280893997262257739330366
102986819922133775791637356034537807550181325596156935501032832994
238499747515243381001151950121317805013879646562849154332489194337
183269470192681676759606159187889763652603208651265852264245241199
590189818788450828769443767663184922384879924137400767229406807310
528039535402366035209984205504305921203882755655930480839116597306
245017725252787907988546850842585551738383833851994344288915912253
644118669644171242400135887960721910613494230833029789663443088271
107467000523629974326102318027142266222618750572543969077381474263
522155244832400804375696699071029472640517803015161891026800926358
769818418013034526647105519950731606432675504048745317728164797498
493168163518881132546149964031831401208499997545056544056651148358
387174381071044446819957363462868930027137176430696041478322732756
789030508095769143478308670354016162028184911441232008439992821318
184843322881342551248886686544852708423042840098831385549010037940
264844762363637536465110550081029406609915248147926317308774406420
709539199916755639316762475890832242707294829544328151229529031647
506098107151709493216616813022200848999073351928484090014332369886
937917599779238728056448586363560471696443660204452597048682215144
197815912322747578772163986575276098408998893737775089374040658456
115404534488982635679449628642247071613265874995955834400802449400
525647531127582945824555238339938861602167095409039509622984369753
605794438167782815663517171901560867850102297106693787721490988919
368454386725973007974938110342945358118921302910485720499673562211
673336635007643262740588295448615756961432047896330591725250029655
954680487653626281549757676580277875590236787348162504245315702741
835390649972371430862395364283337985258809365635808787587211416735
820002378585791714654161211202603427255738422155180158746305776779
395293678912532977700508226250081937416451141684737365725226777890
874579968237345277327360629946429241699671550862292806080316787715
019020416166302049350758837769066186746162964701677056344183089676
266188744032970517788145243401251222679410412177617218288850815972
164208384793333982956699403495937907282033978008796049701378070294
014570618322752960348530283719222610059395644991241509277875461366
823146128946998167258824711898544761466413597472404001167240640338
329990401502635271218579913818751838215422530479921538890284616537
929472363796333479312708466472272737043541076853791213190343311924
063452076733343809168129039267109298757177148028222733070858590225
913929052589740024375710216995532657613335185187663860276199700398
052393052742893370901691023675207451770169640472375386382876543190
430290357981930446828632045430189142160750516996685123364451883139
431581404652068503559767528406209686484001463298802638325495627213
258275734485355830002225513318596228864977249448196664152819040702
879710950567775583836470750892928012992146550898465270072696571688

660217787299739380432917832163467212887284336979031693485924557721
759863321692291013129964934565694568312672848095842925093551561535
868203373672201361285171957991790678887948977874155795078582804005
198795143793102409735137542445229106658730078654625141882080807307
192689839135049253775437442026570165148549039037849153357835239195
091842294100795817946261304621688184412174680622072287104625149387
649178333892585359415439913580058590242985408557250448942910311306
684106105252152943640589428225619515090298853496701185208964643320
418793215333668475009093794745862440500944197952593058084705730441
714228077856570371279475809345629087704798834697169323551696059155
129039465464919469769565801044772122115297178854242063014493599903
647048816869639454598739566495684468008279740648593976288861542063
449595204778764796022224814045187112205762128289512096424262439769
107779187598915091696748849690140417814624882189920472153978970100
410044519163746354849377767240489630561760857490190664199208564988
244166592591364114979721105709200483463562191125920531594952077285
728535022771786911343170950747417740461125977105440663928887571839
332360002445026038759995174213594979764940400041440939868093193286
423323138073107260523470222699550297533641333336376838307699122239
114777055859977842874256964525973045897989161844009118754738104698
043805595170062963032943375011243769165920722953015125432139405443
377891627819140621551682088473634534197999887951611726102841063233
698534566227140898250206912867044411690258204796576506806083389354
490862114387382565994643497880323272175829269451699863126735875109
548455878463140759717201962433708521996779288308204170836282188671
042940242600584400437735875331070418881422192092460714913350296369
058466448832031947410173461128786735179422094145466041853403015518
155623214316574733266610798980310906817008268873210193645956178585
173450547285898007872872115417256740244197902884322531541019214013
509123867111032321373145940511561470672128959326381967580376907231
303216158247304070138858933463663359767715470701977324954881451714
956158891597270403164434951218597470414671715097311329473848085021
070730048952123748421540389981859513224901441857291935709437524159
215545692963115014493847033948930762435538342354395078579177058758
873286872636137723131795763188119174939973645829559955961684714478
441518985430774145594300916272777064006784526222188606338106724847
269024402642674133907219353005842440622594642539483685654784505343
490529674305897486495643892935250696872825573073886534797956973796
373941631251221135723661242014026468319875234913753259196515806193
872666193916051049359265271321692209622463969924533949416814876975
945022756931601737297825225932113922797264469907870797211292701007
289316414132897554051129860713004542449721998255923017335593991966
625886284890280161029774147281472179960743046863683943583762096637
059217800358151699129476731548326243472252980038009595875555451363
524852923366036661334521578492026850615194920345290214617851420324
233104228486352089687974218454003873494172832011762737822647963978
467771365873511193020707222560037507494078103946338951998454416631
432297316080844049828135430303833631635314540529914831642560125106
820856569001603029729165846789183221058699489100407801076924778257
280672186586644935759237706601999726065952554332733642503894798336
601431993073084809345161508804807646366675290866716936206249287398
148879904365333871639691167273697027312653742840860973486972932552
788541993019041684282321395857966024873754065439260849531863413469
468678923583360680339445576185648701132596427755820263192568099715
894489345407354516693238449214991185549338282445770766882305254697
961282244041599668923715929509392373211954789450740806774448900380
624434575224611555723894226838593051527754976545431808349023872919
846748693162608871792151248292476158935141491415890423510507353496

796948749186334430479362520365105567215698882395203498052301531223
852125132616644947370461248186099014395654637271017556216112211047
224792650608818792187856456477020191870817409827426388517851782319
529341904819315715640400178260080474641545364258579688221314712021
950687073703931215333223942964710143388176399181150742155542260482
199024500820520315515880310767656881219857503845120447360279692388
489439850407766939191917803851311790463726457872800566499501595762
530276734247490355778730320694669762067937109531408787466090719090
054787150227573861562284031199979360148174018140726855934642470818
651372676127973427764124089407024122505759128332044876750838248233
549006224319625729282648056600967750928532573038883418242504410194
438374908292890770441518151343279012631862709344102805833319718393
808451124787757790528799614248096853758097666763701569484348743174
757489914638891633504338362739885110295590997268995590471511291794
5559126983594293067385743048698985594432619896425343492171171761
949868813811537360119252837634812218777109439259322057370956269816
464526459305254130817680476849179967094590975627099457464166873129
985177713155886207655433151026302360849223532018400246442694982220
093885619814174235294211012044888786517620477231007235577371175696
454026773786987829323848846586854824307251322459971819517637820651
677017349639072911973231521104508388963690034363456497713884180568
029841405323097836878788733235745843716778596231931182129965442642
274603311656218995807385709140748170907770720601258255372559881825
540001709679090974133855179150503462413627962943375279803921216124
494228573480554092996174221867552670663871540197164959258041982845
727233943587273849129806250522990823041441796420186323933597564085
626472114098710275684232847105442047692737227958693432551623728706
130624894831768300595031627353927222155596037191260927056320900168
844642239974599076283603861451560114679086719522744225341537356304
363680765820929448168157562440758354209445041481836940072478719937
160807471437048052724122720576200148265567384258527615204225756167
756634489083551590403475597055278114985130250874121655616058542729
230289933165473549907915612178664717813433928249941590501409236320
169840868059967723646311800323091723144906596018394433573246799472
136366714309332268725922769959786634219848604764038331215159824633
481575389136213747050626776094939156543444966503071575601905256149
343412398650086334976877258201426160358764218865753091740518241749
178412153032223830041880663938545588917876200687881404876692760597
626388508418767172390688215137534469074205279687593862965749865441
776294251870300911496135284438920514500715511087309466495949907089
979305234012957349386688178592724423081521590660649960755027237608
127238705851213727455288861773544544959385158956877519518026877985
648252026624094448618828672705420747504353679984584680211816124511
917916408388220977886418275681058507677565728648482836037024932871
581980604355587998037575747633172000054495984987251668856570630335
287606809308159018141059372137856078810315129253175041105096097516
542537103085517485489928079279216508267024775246374998378504723411
487224038878779685621658918415735659396870303193507502981382895299
683035730430607120754662998058479510773229041914306816287029509007
188141342145828415611632764589797794318524467033357220151830080677
300984342814598555943657389719903262861007167469115090265946427923
755624937423512174450803121349987410210504026254115763114123064033
738402302484473936132777143177832648722787200003132437991158454107
320083254717655335778841973881119878308116128253343500137910973264
580456753562692848345510253175697613783144368252477854306937063143
255096407622494270969727621061679816307458647731362102916913190193
505391736338772095930772880211384952253085233564200914758211321508
141634559373276638164620996415041814279261478485611225096974418073

9940121864957617087742985390839419901188858773363731131301710135777
7903347562044395262607677976568538504151780028622026017398315357895
4904544427165705596492052223188354474283111934696037119412186093966
4743696835216300841130921221376123619315550911877534644560429373799
2151668962024254716803778182746385907968207356409342994334271792088
0288752211254331790114149116004796389603318772204714551925930589488
6933504992233576520706393366578610808592005775957357706056346934577
6038849108050669551609381069436621287588273316132286483143147176722
1157046192356146500377405387217627411136601782358558451731002982077
7899936468177687598057719690442932656414928889506161743273954534822
3316663997917484984027478354053591200222609439905312070766019667277
4321466731325059919615374919120610926487819537779061425351892234666
1396095319606252617842571586992437826609161717464971634720477389611
3148671942948249029198941916758308889233973117415554172680947533100
2737797997098175650450547360227678621069754045059261438837781516177
9253790106064022916738026962573434304645300421104252766230305520722
4757393067927263937131887228801269585549042486632283070227740155522
8034220557317260915929275132872044337772363815466022426272279552422
6404790691285346647439567039015366644825118623402780402537808866611
1353566441069137697238823654053705720326485133071180018862177768055
9795321806543675321022250428000439940618518128895361407337239506633
1151707000571386315302132936855380184898696963028510893012021795066
4707248775032099948367568717247002905581456984051446746945071887177
3763680287347355619685317530756612015693057034430987614972306895288
6644415640748345880898652566166437972028958684422039218194317151277
5641117761475637140593686400010358802638912596923817062276371676288
7480628381602275941051146269228880912943302776649594724973844730933
3763274600371084359078599766718005586870287301832296672925665119599
2610059415810036508929062603999789107646931019522717446451994436166
9991555641564121510871438208088680752297850814802286234135318439200
5666397115246089048131844519231492910632815402792224893782282514577
6827162459611763956688646174239537158657446266439961554789051637322
5218257833325356445898929059519260586597986713448274478262666789844
1919627360593520221496681570436556904167082575274458817572811609566
1481857224369546475050830284430753170779235571329348761178390813022
9105991835522622374686711575705937749093797579381952473316322662355
9826956998047343344026168796547513042934616242661346074732526957033
1148814696916429336907194815454817908292910720694297318759719731011
5426199335646153283618228701515590331070614653042170066882533797011
3234495060714168352686098813122722054090309466460661858579999141533
9781448477415640822589035406449064635106154337194004013861603507144
5597360142786234514865734796217978467570218989951333364438192919055
3008577399504523493495718968461271137688957597933234953320895381455
3984677028512410913999962409428615356154952015641889962125930051266
4420968659725289941843503668188048075291059723360083654823570191988
6855092603500487657378829516292374183271323676865849464000596709500
6778345361003674425949188581955959269025123931107259512121156338244
1589606737480071832468778413078096938241482915189560427550175420655
1744208813401454360707135560267634995975759600410361609612137736211
8202235639801014559249360156897148979333658549918634973041034950077
9055097103732948921976405886995320189664933508204310048852305942988
4868017855565645389715296386871398239389278862831305388987044163388
7485323665505625430238286131768314743993446156093107653849475846488
9316215158358898933919567329443347903909096450020152545297422360933
3487377485709060186480705169912575593325182030441205733891169249499
7937444418172101800486952748158248607577122172413829852529767035266
8850421330346370320590111276927084231224737403990344676189570010255
9178589661470456118869055431800013574114545384809162384560193982144

576980154036744730933242141647275552190877396917417373506414595184
607851240181377454588376298517906609425179969503658723513291154069
411855800405756107804357919105154389529301786070568857811721742139
155095320721197089841522154253164791930461604981176009994043413190
991189215513651226115501813110735194067489641860940284869283055022
119924343866309661229983761658981274730669004071331315325819303202
814967455702892711980502308342949072461054910957995507893660263469
846566281880554901043878989574409314152964143377690260506436409832
682176336287098826272397430230055063851675289226483750950886137219
833353460698489068556859024446788863364396043781823649316075069795
253661777044807862852104682093268266828972207159109800819778019264
952538304724636079589391737003693289663580220506598028533870299680
922286754271291338699940266335773608637540472021149927333995596386
713941415995063555038162271317992876143298924595866321022805072720
176632829028139513624639259879408411977424214784974887928534813292
261758042969536056849641633358836124647760471766303398537726717373
232324351929797334237646067007259056978477822590102247186184955108
700414015527634922430585064979174699941224701667003101044327626530
993015284206842468595235910530969681058431185510376080853681033317
095349081348831311722359377438741462183926501715609032794028189935
612694496396714332078290473191666780851825519771728800627735453 99
159278903401078628896366115708075792637125157532125643458797675822
798605621785390463443878260224769831644730911677313769865439441397
481344800381829810375495058853983542914632275329122606239178293199
621398691881771111842441962771878992305735044724577538311943385179
322128576603521216877901140477658976778435696351365329151493096380
391047545011699675878027999795538955855005904553329793563702640770
333481120559679109660880465445819911756967335381794020297742084467
140554762538001665796195719912620080781668202885915862485723615599
401625547770791411160067640782608077107894734372899115676130685073
224963159123163419758846276472881920236762716375194766953325420490
891610249164837334965917270800147115271012908902961211040472462065
622820963283626670888972846484919450548524147558133923773626921276
628090107039603299462627250947141176912142913353975130151431774671
685840290596862221708011103666071463020620642207396736740275444511
153186803573711970612632143552346855544382456532551949622309244222
627616181076353271218486711038748633241670789046885223329211115019
790098723766740155479167534474858911628126868673604222994356076826
978331735176394113756818731185310939147331613471464295748025886612
098433336264478923277992171893811049025750898332957523113851163841
181019244991329300877847253627365880167927323911956677377346029231
167147252754387732395409644074174493088103356901689944732650629356
812407468591689254650921109142316433966434965355399905226030471148
711750195108603621437887928407449825270332425169177953432393380537
534154286334492005727579681918742184272188588466626602813391591226
508703295562974312100608476462382406120209740885851097134502445345
626967484521749379519983660135959959804421055339305799463541215659
260373954548130709002681616473580753090700579469859512185766928204
335931333658021043935801610790827942664487820352801574984777718753
666388687146928492233597970201859216375263706470723923280711774975
523653624170626315463270059026630402473980453353020409393130497397
130791718151488632385160351409187151727259632060397751818987737942
983354896214929883065168797261732334295186029197912354209146617618
580812065785097554051812624547853587142349872282450762802185554164
393735572873413177079533182641069580231812678272926217247904786733
132302602879014764854335809993244372349188499585994862583067600012
204733634466868003021774428308956732120657310909298521268530829353
520331626096123871927047491031694115164838847479745677123433557442

981268446143275337106037702381158730688628896939413236300606050428
996520045106037486769613649172511721417104539723698376574825092862
531991761037960505070047452751987069243830797208133651074580862533
987045295036577394794375194325536600142105564641482243606164677079
171658511765610859235634609485497644779621165511318700969902914073
151483903908991815918578332650277953957841825197056152467518107456
330457082959442889150666715929760412803354745155100439949339911357
400368108214520100371663337695212133753239590644515065233379074750
428578159695275696181784704238178420315992417112157281753138255289
908317222708031933401849974624661506864137178679359480593272851964
335736880274143158690076520872345466373639831869120209656207541348
874115504351794570520219208662862157046501295951312793744072467620
419226655674453334447296817148735449387338480166542826423783384831
756543833361744087321879219971430971939075615289979919334816845664
869894315760143802862633533136185723793167236606367549438005252967
139974035099407121933737585712045559496028444564046130603362226362
162934122457615116541938791684813280962469524445695462125087911893
539832219637899949870575517487718861051045258709120015502718111214
008330339459997728658704523419166730406855700471728611726335884968
271071745003538903363106665809112216112279535205973563154238786279
221174002792992766027230910087889644867197751064485285423676068067
832870271602149122089073835986791677907984654684765443288633275459
268997647136118219193637197094309189760958933074195091535789981594
562681740310911862136112387032663287459251238017221859237596420397
178011973301354548630311562876453973330103535199368908917165821184
472025394047093178330601239641672709312163693791933239184259773052
761479229302123013163652956137623330528454637744966783855724163055
532861053275520784389404424723308700149400756485394938970856366624
723511554968426370742241985340721884331711808624785109999817623225
805812020490727023675155996038558466728397347325959612710449694899
692807040872355613550188348609827334494211927951159638914217013371
362540595915840065763710336218594354090721495079719264247416878866
135096201313031939816564431842319103674142051255686332809855207709
323995574220458372892438309481108423300876415366308472416897637519
419399848086392769531790164372780297768880616249084193376410364509
612604065127369473343213647516686745418754235332490452514001261991
025504942206089908653489121851977852080353829793516473616363948528
497562849714885627036425437615253034856791421813834154676563036293
594327156888851139645341755011355523422660951773817818038938644309
083053992738653198839237082514434976695795125406640558213249534760
824464237959520467403716910402286506016440118821281688727839234273
692926062064096409195961459043145172341616179151070617767174151129
700974362635716917980979131076075544400727482316585363917076912591
900555112850732808167705134749074145011950248108427677735773081036
084500375556502686582708949066409611462996904292269838084349681389
149247988622487167128124089262797006509374129142801201881922065421
593897363381932259127071303848942162931911004907149225362821862035
617644685446995943076419072713387818263384790269051413488524088341
597040931667176458485165390460010963472932317024526860807864918007
702454260533859200916633150792778732483259016044217156687494057915
189677115913189275017804451824993743874329932914355437468094683402
608346425268170735136026784411711754768030257828432741271295550926
710857402304746960026445711893018058112189257572500241791066473020
112946937549533383927107678381585580887567061329996499158939499040
874977823550392105136301646716340862269365394034567695186527752685
603128680881568916991604601367935600028878486501738703611861366168
233700637624901718703548391653008880657523737679906815547888893864
623380433678814473862636975144463533151364503365250987795413093994

146760112222850127827345575515956198448726728886216911391278644418
265010715934333181605528809809313757602195448423668918140487612969
835740368011755189133005722699475919228724396947107244977040473296
751338485372898919851448791269339956272762863015717827057355238450
193665288694250301571288649098993055897745148064974007108137602067
660610028335398320724359456720594945121684402530561416115047237679
687125269315631930981608232979504258981667480087815264867736414493
569584287953879511112090041388243506999888209156555403289250228805
141696787929926626862224670525490667495362501326970031824510114073
519298152709116828763161525453362313242268045222889614970917397113
535255440123608618815454147085320467229946939071488188603326828261
722826964785169840975561328091090492994205890209975868027011829714
381130616650165606940509417447084136593172946036832314886783783401
584666526277938110347185652734290112646968995135220438138835925408
450875742934048304805257026367468199997111392499430823809481473192
576011528538247357208314910527160816992228141867532991179552447748
792024698247835770179058176843376667776890217764906219369958965467
659969428721801097813692136744622097478300409271819051376356123254
861272145222616805180293256818310931413966592453103442368843397067
352872663830004541951464423032623019071897598561247023586500542075
982524898199075031653803249502601693723058314817314752430435942498
914879189062802634091227267353344853777985327688970476167261585288
351406035252708851992921713307057857638749393745559400967615375217
782801162690377265289896203441261598810632168253206443816406129171
172120095567473839167222962355574612439015599054488322626441625687
126870485003449211415757614315487883822624493825719072052822435654
030668643394952786639197826196621288902931708091506933547609363069
503877964838065009708771258420744211499716985561589989747876513750
578536272453652178066289777507327157034985477471678902956663958351
111997725430882108300838719703001636037548232031811034519634199719
570801626375425606969661834362972690706622306143131863618116113316
841849516129647994635408155166288645312201056179623810144384620141
325246851026413793411662166660443555433967260839002933424985605923
047725430160485968987816153242523488947992749956804057508785961584
656399688277050582480803752624440992284265581071965313962147422223
415350770031361866522902424242733975223220119730089596891049854054
474276975638059626226908788476436765519375681951996304422809024719
659779814112299761130996689484065470304306161542840528984605556105
277431670945479765425699944325615151270411776840247262990518468739
384403174909227786713746504877565400352618233613582209691595165310
030299470261213798326995515479430045282504041161789922994791117641
217399269377416582020283502426115579535771019286950264605435924118
006680782334174983342235251194039578690357868099795735556646348184
109235356638053216250587339612730165179209152696307741603539343614
876508656958944166875931028197227084213006069890327681248136434088
291450693535007842690028338969289003676630651962125691137082514952
641307300205723426006143479478418466207633742474019652349063930296
622337730820640228704088095403944892602375593027578381867271119555
903626438180369441026989560997022402685189290570563411576345663453
530917836449127065514652145274516095709269601981935148250423083093
324020856938232573732465561978380507982367839148964413212119032538
371930512612143512054346721380249172084457240675607838911836144206
172196093241887871539065311934562423143050595975813896800145932726
803699031531485898178421841408627035413234057140637242334416230520
114600537243354544085804784915273835605370083298419441940878577289
428942989055641118489012798817424271309417325022464998977618499584
448243196333877136064170050575881120626018903546125859345154561817
568409731473384201495189375815899601208752575627603329500301183188

095642910867929936491408742632266721386849152241299032914629320268
237349095662579032064280453385167557256633596432829836906797154489
491441442844573661312147165257729283228387225191227818503331845753
752311813889104687301120253329343303322817674447909206656325018838
874991783124527795687803251857087877108213218175422991370299903463
408243198220018181430169501586756477231845517351601935397411806816
255498633469297427936383683122862090150084763296027154205540923472
197748755577372771253584379299733675504135390096260754601770478320
092090000437030477206239693112361996923069451921228075128062610903
396080855119939362576645605845474892984566105164377632302047629334
883313664553345734804735715674449977347178219815739262943566148533
256352573800753734245856962732264432925391218548350084718726153761
193599211755449468751722095340217149673323008543031277343008442170
392235658052374699781195238474449333837385774851142746225220393467
572123278506610526913279773063462887372622241958467166720221516808
291000526702236415126522740776004619794966850442414929033037526153
247556530093153145577415607854888437204157140600876512807613311400
021517609289824898629450626479863972781208733447929847854531512329
334051406847255746928486263150354770925719144201422185887802572791
283311779822123368077931168758654777139994623954398600178217140445
115877933764582521759199108819238300516633102828372361341272140722
462379539129338836418793155329932894879874861538613915230746891741
006626186077722679134871363221475165685084419917806948619546019340
893708192321419263827753375919457032645023630434756871734529583995
536709739473113745139433281977911222269397254591249383798231266070
963822259670190083814532862904610606586856320978015085422334848110
590617385229862052817896049500732570427222020393613638247958310354
325985507262140340985962778601721689559875030328828176804094685209
388640336365236494428576533381097953342025875230660994737779174834
099640562083733043167671087592982666684354670095997048589537484151
152214502249945441528386578029285301765856291013881441726693837902
070500341910121386791346354652287481407153382029019192351467212683
827510001739480517922357591031062941178267158381863781954648843122
973630207590729496131322642355108491026499847418870181274039872030
679358312315482878780386867207634549849519911344509912442473105052
272527668320660348538056734851263693194665299251629026264658941634
139609150972187236402755002697010883868324941421257120488696456582
963616098653685988378839028020706070296399620892916924201175646292
127178414438660944484153071327538274180512475604700845614196078604
954485925581307161527176818710961041702864624451063869927990313298
023938322923078600246111212562537492992069623605549739779337090550
915061599580746264769307061465473365729538801084659307737092643932
709617335897987551332985173533580576198203756071739649512102605682
421535394322065787806543336816683791839254310296299786255831381508
429023460414642850633182078026674085750429654935395449486518527564
708814351323195973497899171415169373256883389331628338964518488703
226398930556894518391912430829325156540236753850043094552275229862
193634999307995606896844661874598947488234136640851885321936731143
758946356570214222303717414812012726282910573318578392273347952606
800413122404444690695700343265791095617342284655138302877708170928
004370327526445576200902948987017264718228932761788234679959538966
801140286687052633670600630426129946084949956382755990602647776521
970253758306411814612875438760985782899634221059502253415043982609
618760983521652316543316977214412517700380390215981379748913202929
277554387117033911632248075246572497296231247650935179435674838114
315286413330290891237771466124690448645511649267993463415562118822
817564230240516948954442816831414049043805788605901073700671829849
936504074947027855738627203271084260273269569006412015558094691371

012984255290544957645064575600374031494587908210547355911363990672
780648145919170643387069714773665247784433863025569838810258987930
950197131284070891871969674939400265719405722159295868834578669810
318183594938102719311615251530174090403194517238322459633052678626
421000745736336797264614352971498884605529190782295721345692646383
479217594057805130367348879544947334464560679667691278267990494200
362880699002603522166525266488097224672121294616782282247427178341
053585849093818084382076967122622155649252446410116006638391181830
873085635422672150172188913491114434074231672018580154409683941721
845529247030666331743969920320999137230793920870633268149502702418
363237393557565948355864342758527153036475346746011816231218086111
379932483545148228986306253693327937473726404693126737565340199730
090761426212286501158568944820803714283612048583161747503907712876
046503361236135224312142049114096204585829225543574900902717114310
056202779664273282036840883514218997367661285154174170155055966929
543355338498868702324902061064458071692286334339185539443465974183
103315453291025913036064622666879779455734904546748823275317375995
937232273103710445211331153382893042477397241957274401165418484315
564894048921358055708557627558495534889191385643791638342408939602
209788019587504761416457873384344319808735157516674968200379153796
102973494432109476073270046363343661259071179260382965776504898339
968200528464234206854494699303871249646642485811604420004666933985
741685551729836982926358491044717933844683250433844717587252699366
862337570798586379951176474378774221029593262173881717992112564960
766549050364753011284605971998642239727843391967774038958231917557
325994193790085492825980660767894985484333355330520442978146864226
215463907056678047938913177651922049935761663882196322357224138758
048818728755477834305533714162429159181440724910183373607258613130
585839379636913731605046386537876161997656835278960391654122119712
316370646384350875058804657553196720080481063208311821537956138009
835355952609363700064531708064420288837726690826800942475061577365
306953699946473444264179908807236585691623899636517578076237318613
662803000677595254569830359350209310340106654882387605906309667152
580319027018056510774179659964177889506640602788471706807792755570
351022237147306795006509607538053426398202615407127213785603227432
886168024173389459790505032137974846614903095301740230095495752617
958896983609703142914084804583842017705933308727898829210653986085
497841770226800199431723125607279669350937846167380814534710813293
763452196474416319331178690649982482372761620561502444394472323379
106960839688560326743659447613243668623910583435263725870265527272
354681097361367537998854340224782973219586473847079849851417285386
752779230658409174320605010991022389298189386457216041689492340208
559404805979888719907538994483624575918179587264785482436871784280
511816570103599948961675645814411774359994155741564054198094077706
078181787327808839235166527298117294704518248948869402539784970404
012578501708525229480032644855398293395410250493410544461435613045
371236961682202427087546803225777224676453869069173584632909965978
927085724136068529472284189988811197694925777567347314920454188249
935386075448538327349316024944583018400520110059712112248818992601
409033905843014105055980718844415476335609338929558270335638391892
072441156624136346793755416738908930918686080312637892309129166075
500989808404308771738687684930623853335091504106000383060163948853
687921061238941057439403460624016371854842521771675451639760025505
022643961152599429430869349869074629783759970161295030843803660660
058922658529305637886695846678487572600253291839307185472610120143
531812300826282453907565263848166284306712414091535323017373577722
317054545331857330398636116290928079651400076258029586832521130356
252134998540067832905798100262663767805172062475401635370252168218

7355287204019963596188736069347306728409608128864989228165452185240
83282791281849386363527220300859827544599898999583511157436878788
812704855717381485740307803629420485942064433415790169383959681533
5852775087815743971924322779883170605463400533096961159954373203 94
1299551771974092493728193869104247191680745755805413172816833655 37
9652759510402582937660069379484763050243686693087498612911515579 02
9908914751147143616550977810891689159389043228611630980896160154 36
5423970713173398762556138393349278906057471453816915692648820151 02
6214721832503409165624542935311732839683741755506978877246043985 56
2610853373740287709972880476114915778576510475290891138178065469 22
2072171325415946797805559574054495325587792843232475048202572961 07
2119305427203445431119018432651599832951192425499568866292451206 15
5443548518778433760228457318552553020385780679964233347394328325 50
7976814317493529036535523570833622729540297603622459678702246796 10
8729006536915811032977241127196871846317120153108720228291216785 13
6832868288998410063083059699732951240183437928080758668778898496 04
3772754027589522929366853922675139928237161596447373298270175090 83
7568027446669159111149779944667113569108892437919930942472130807 30
8198426259314434796579085670082562885883611446330706901910636068 59
9518538704171062385680432411229940699769765217189489934971880450 38
6432175982864331340232731735034487552793783486413104199496595525 77
0669045671843550215620189672793973426821625686085922488113166647 34
1429891013875712757048305145943663609924661062720112440987239997 10
4207565439150686310201357598460146730265119903429865063967600069 66
8795828289243397825905874856782626926330346837233212406615776029 51
5653722610682290383661336834150499989593428093201865424703607359 07
6560816219599759343820172461807695817834722127150399123937330805 98
1643494623136717495999946304211763818147830191021334473569265625 88
0571016446879845566172037587428149099843393046523931220003564424 86
5028020022103872381508554360806108595381742785324654979231101510 18
1266741384662946267400340729092430677564917857934277516529509846 00
0986282198651935014863141311338234081864181019598887229593585603 43
7224236723960151462789656548533533174007419842426013606673552984 07
5440257317771495402192754876256636632097951348389232398473093428 27
2993909549226286852582803762571040814041407092153377982471219334 64
8252071838785374450072385259360567659576220402194519247929124130 75
8546485918127845559512533948537732743954653252016862250537285001 30
4537240004647444790745978251029444790475972689949375374692808933 11
5543550514205161236368344100734984299470708653487282616826119499 54
4459888850359607914367111963913932091120095413351288550899249339 28
5379476656164159254527588534790680348593042101431778577117245113 74
1846243215533722405654121494232234673410032864092372275714731703 80
9305846661141052866534929215704384371938758254891849894465897489 21
1239804355925364919086589066917390808867500913233054266548207715 73
6402520816248301658987303608659808379615767364117736273466976156 66
5392134829345642399129280785997987881520429221519091416907854973 61
8725169309999132270006749972335155765795143664747023748766149644 04
9261348608299760978362604927823173889497922468520977599508049882 69
7239249576598722306469511876779916054956726996908515258226529522 73
8588543930217342747557439187441137663399412859483123438484881279 45
7601367100667616597439589632554530670816842945121140912212009108 66
6989989150010205569248485237225542131071661391982827657429818829 17
5183372084175238696768280591023151992531280144537722164743682595 08
6088863644346720804079957456104290101965608808393098287160616049 12
1636045869086222897375645574135743071591089367242331664477332829 68
2418831492171649497251401194936905670952961527043291961756410101 85
1405960839542210112530043203244772904509568672868692837978994453 34
7325407832005428354880451308874136319369558168287460790465669459 00

4074428841873812325676996716469267998695588605208063729838321123862
246812028817043481558140629498830626345933449988840386526057374223
073857866400023774153128859094551257535393369408694442939407522182
847100107766580951275670201477540829825943655390077790618030037104
030191092185293284782411655189092287029012404160045214931709357753
608163420856552332014438453889583422068417138823995322706356387242
611330217236088753196924820117906522275084815465360634684320835251
951083321606843173343658468059130157408787018229598765822830020425
283556693204501981988171458261171598400011723232484622368023378498
390571958420833941883302500041002600378834221148367305474960967792
429704499983799479044054349710896265896766913002849909603860630462
400933379790920357551625516664005711218771723903001503960954051845
816999386430449804010399161286593474495582760668348248909337338629
266989646970531741560892296662428914381927372356726030305011034159
701503907594115991561792511656228924417675577202639271089785260599
471315335700459048301245358602245707716058212332185295875822030519
372890200177343206194287342147523788608300702997965315568103011287
992589391833877964700675202703688772405840664369190902874387633882
097145801017495101346458402812780113168139897806509007407674642209
638998045332620765149608259774522758423904134502684618616814579533
717594622683030636661453659920280300843252851498178817712725738675
355028513383679230567432436869620272756904956947214224246798843604
119226316915567882764842223911962740367145989741445431800616886293
376356239752548161092018062894420650865088651743884451744029361570
891066530518191344083524173853908952947331269090022881476173592405
472755741008722118602480706552734785464670810033252880494872818846
476645138719484647002739836639678691108722490689445254499301361359
823021009664966265824979074179330260447961646789612176304735470941
059054767787436276981114648195946765395331326021604518805685201238
185383599352509055867304816958939312668888710724516375807869185298
046443759849390149864088672912156151469350544680039277537716280028
444617088728346133201602784693514171037189813592856544404738893533
643422529953563067148643575822661507084722421213957490588123647260
807956653918210780697591919627299613768250527190168013501825936503
904314892374222182997294359105047665101984354967715836390356050902
744094547376200086625518953798973986955249442094528368937291622558
644588185723220050973402112420242701338138097506387073687862241346
162660761418658903675756728049468513929492469474976704482862785037
993942832787912203329713975438436447227794195248300533133083265941
268165481431836724185190716453711839456188537671861146345100987635
561039688240346932274316386856389366920782628786664631623058656232
080344670332241489658442908620119179775183607898117847087626296153
194003478154640506345659858453959336783920471778161961151978159915
334832397561116212221045289683087138354599880658578013548593749042
639560201729578681154940798878998952785944953129124758248171371088
590969140706193303618003033238919132168402423711785594147938175122
615373552928204846201190878355782410767958987282648380188363060516
258745813238071702127070311601599319566532110559086844637230112193
935288299328435695606597198930148418941924696514195047131003620913
846840871436786883238124871873805822177969166872677052869491723296
929757412937157650310486149826499639425425153552278926558176593281
228135199049986283389176950986498709388528652246416241498009133604
809416167206933424250017253335902412245206966274283806079157097461
019323432744228427903009219716781979659790595491272105538644724076
008310005880818190724478703436574542794750466602116861532820793670
422831576774109787065652889958092151207900910624889386465174860336
630488088385858365475419590635290169607959866719791951547267539998
847762188847851073660556792237455158009612734637037295470996414489

5440357070050979597124970791498940575042016503073921008375739413281657808571980285113042479613451450427776366605487057390164979963880067492763356999017421424708604276336870153889542554486051966156011457074311012678360618976337654085955844167396998989171468666484090241900149311117346206295823077870579486764638556758721099513646830997771609454655720168122853937767374099428303951557494531692065820371452045827753578337982711615753554754759598809028882500151326903062183753558815228008049946219926313951475900767150444120102864042232346757214652225543337464540769554429637336518329440811481655311231788856853449362565109233822325088751970064021785626240504392039311551274241981278661180457520379031352263821500210772130502406624183002862477655911114130894749764170442763287774736669514152962797298473622901963622315436319141841799696896280359277506155139875578536726635378140081715318318793314798031663073541838249854774614347582122035033039491346267250843739731331346134971878652215484795211328065589745100203248729792239256892753749027042598668561490405375284504662614426425991229576994984595616886612934274152168604536950858277105538068424706396860778132899310628551128799543943367009919152088861445556457442279253309475103278639998286086647782469776693646959682093306304832523027286162884091854021075857505952335491745175635055894316749129117086207384696004878978263910905625739599487492495979011090191614659080207484962639358279265058364767670838301196885550505186157980972185298078202754679077735284594885548642095718480957735502641837966220105606201767241016475961823177144419881009610279477760819625624682085457499375939175577255504390164427090909940333682021018189188094314468725119448839760726106428947637705083816809284728704530854716107278666310122036688758290646249653294421504042608960795596028483768115833106563981887154102229185636375546808614676806062617591254751326265896334165706265117268170549301094065863026692204229894830237644323671421946364900320018910295105537319742039399303809428707866552914769288138564587814966477980823314826640216655484671340624018597141217542440978712771828778534134383738258099537774555668639030970006149284077302470091720199524620624539198585923713450742399985171822495428895132343503182334483788319295955334879010992117189922536529429365336258259539094655296358137449729369974651187533853748709421770808174570174225160409043884612157412445285020237983903699911389696734778804970420888639312843891579868614995357206376948214920930628125131228080499666262553224283839917352025667452522990840994632586468341113042084589880324142878241486082621057749475330037706021516825542168588825520528913719387724986908202539336294047304750500997044070946935891969900533478304463581196491048316381606807432397475188737745048539320801189209217620032541285900192128507800878010809612186997215672787878360378342850502233591047386100379003368195821534795312420332192379118697973810932850103678278806081274528288310391831570441480372715269111958913827350206266167881367389300589434987192627542175867837758616929115641969549780500502717441482142253177166456489762559437582676430949126055292857753556527311492956087911082159619401757024920446262077946805377695544163790238444286270760013358829372290589561353109924487923773970012638010390621036297100540090233266205286855129238893654007663966398392572450824496898926245992439438450876557890902818985683433451099619638409116437670596054841952535056878723052067916205797398008695875004656119615450401629767770096547018528433497764467949602803722422923120925815523806451507317582639784897165956107629449587379456851984590609369134973227024282382935848347620097279573203973908244049026524593739625729544911772960859885910979831916191923715574314777305613275865030186712879223339994092210626790946258508116928279669238172379855127629935238608831771468972

5715591047969113529008536773754899551247186890395744660004847954014
4478028822320824256701845114754583859463997760412123218294512072077
9081768233151618455421959874974455890151199270623605189634073539348
7564095315603407069359582576081667695587492098240480666527562455
1923220032514529417553964682719023521957637963789759417505215867 54
2019285365596662014504563385527835712567120512822751960929539092 45
7841885540127128286062203184030416252304573349919862033683941854 91
9174431281417027468348053312363654640800952279867998081678614387 67
0229917642042193577030302889240633823371188316668923472566531471 54
2619736924881856774442368306762858080355776670852493943272675918 27
1305802820474498683109238845183965692912826914135802285087920263 81
5575944588577839176304153824777450273630047967685689590574290775 32
8240318388741953284725057582369989841855736097379347926210756480 01
2760660056848562895296270365033407284992457076821660600430851921 44
9939946159274018679067024925091880842549753825000573641152765627 88
2068316837456136116466105945296098764787617810657325699441300696 04
0441275748480590815431525219147593078104140991804367706639566726 34
4353562456193564075135654672716790941187948848086809230938328738 22
5004285130861556467382313448911976530330590349488507090436408551 17
0896815968188535559683677670828759269590012226813063407905218083 14
1753428838750131652921876633335743143710101634779665888361886988 34
9983289811640070259655020476119906884593064118801374335280937951 09
8305220586575519025533132473935518421572608823101683181764097241 79
6642821705539235733182359972105591467580990960157125453625914657 35
8345108084976967080717266943718308124966413928529324205814835226 27
4469375858569571687034502237757777835981585809545667455462729794 73
0797569320508562640503528302556722866775444828408620998138979303 70
9253264259640385349961705463860837230314894810013719990131970126 28
0108496986832790805995250749377542359999787837446577837123753575 78
5267771063814356136789079473724792426181599280191554236358409224 10
9269519498675837014008069125945944555925467047320392800470111600 15
5954170202879503592920177039686011345630444923339774354557165457 27
1495173534529046221587766932103972565405338682391305809660021212 32
4831138704907261634988177752506339880373528304413788504053529141 95
7344862622148033294954488890854507924805426837419406691885551675 72
1662211092089973185266767829352661329042761712071734330023452589 19
5373557475092687431529363428449640771785673995843814894210462851 81
0384964342773551924438540110560036409186090659830023373684591499 58
4014449901293693725893952889834811556455810610307945458047566465 04
8935767852752111834079945987442056755069846162829748169274340319 57
4481121269219891324355031983705466444249609636337484865587143693 42
4000842683069466472686782160543076055555157113010549963694214120 14
5286056715492814503563608857935420449883125571959955870785833653 30
5512108392849984112684790571297220246550138708205244749272341919 50
3603039394603476770851534725433807691354302103233118270994105254 37
3163717918996081338444036735092091110631733765874016000186973042 52
9842024488852703317251469249747539319823503525226176160943848097 05
3512457087513186892673700507442774120709790407346312262052900063 91
8929099033196435333533778277303380095643555202372811893414222131 63
8412466218762656292621316537474409523054540169190591210298332548 73
8443149969008179176662444562571055001406036680013324958090641028 34
8771641935647143040955057638038573852209871616795910485222080708 39
5443816652953500877460681136027308248856626139286677283717030323 78
3233467164064186510599162481767070634565314676407449265899040024 92
9433820347665482730341522657904235101310929756783048481363297639 76
3227467327299098939274496385721441290848706448070466163067183262 69
7301895505226175036704437656063860897547480919952639440353604654 39
9823773561902465026931295891402221059887340881632225625868617445 32

44115894930355509977482474151507373411947573337355652762741851916
24514620104987826294536895088446232117635800351184183592675986429
12813985384144690969171907995167404678189358750274435035511807996
77624987062847592913272616884488849391411287453205724835916236067
16163446388013123511612633214157350735873789089864603319101047904
51088053276243634380195320531364134355459364704198439973273192818
02576008576753025832169671464683435884003914203707242670826017616
53390298986715267562219813060352959484464640403968856006481766109
33056079065038768968446731694854343891368935379334168987646104050
60241093598555326643129997593902636796969251376936926159102285117
16541488921572622359773667701463984585521047074763889722549022179
57834738536300819334378988748156802769759994951212544157027137649
27751507877994910956148957426227398794033221282575324578456195527
49717773748923110717580044852610875339714409334236416700073947476
33876638025429586355293691304769868990613336245908473538052999169
52374065057657039349015473906556528925808194469628401221392455111
87603807405660994105231164360192568472205328510725875883708878783
40746177976015317304950058293812475052503021700236109573670907840
23511622259723813014440798479181330321054432443110271979100591373
09348377586361399771943672006559245699381470276191846661211804296
71866528310695776609243771615983151245936172801039016365520466193
09252512536313965592011778214276801942804765865348581587470311992
77097913350491107205165243246231253831574487512954156035527502636
96145461093441685357810273138996999546867343518103443255259516930
02330050592572799781590482022358905260479203482507504217173455466
25312528525947844068421333789797245599838145290249134127724342450
71795118644615062815283928650237212926436840813268942316931518881
95385268180047631030778038992441215187198527222544989996778751458
79160239707666044133305940017725196828827618138902540142141154032
22188075221493138730394389483863488048858725731296054402267540119
44344271239556902379071500713861064198583888795688054863351728084
46447248132772385956105213013091015106264293276316247222859852571
26371594629994013226550518804465739898003775442184629979193304006
51761618798602655576076891495679624033020920353382300641985751212
05490876812649998662968202160083935774256923900145096765518968300
14985803950066246338616802255066870759127483197530001455406015411
80784113494829465680601707418219448269420291459193117972944253523
13933101121868831678002326637691951938989304956559636443049982636
33222213173795863375272901598026762438208490894364283124967962162
00163529966893044625845804167249071090414279544743227764055860644
79937748159790612922202930619833526180042611796654967580548290181
68947372216120265716263043635302282129337245083943534370967854324
80583128895052786635657171288036985284548714950798562466555779317
50790289986859714364593077973507014015227544766934198239263898929
53534319021801693875850287786879702046168231973519928076697586512
60646823839169601486711150096038459388200510615266462562727086373
94715025771810723089620436264155255715212790458354295905910120518
08605199833582952027644524251235773515361451329122133578341196718
07585660635000297664587218996568468098354225556797699786152958316
06574320737210996844061946085275213997207746028379618294067782659
09969835866089743865103656242556250842534354563015121971111673283
65507058324791727356514276941098498635706661880252843453611977423
90001604109135980658253251047724384737585884666978580299992241973
65041155119628160047321576507005166289845399637919414471961275368
96384841840783539219495160760752984767084138274604403017707579996
67675686125361051400391716817256780501389783718658379689761501720
84806022721512087636711163552935619613040209273964185286936047265
39668755630400875358568683131412868609228255512422695956799303502

901136677093640349903748458741989108901894570519857812478440357578
671393197085549893809997065410579169020959889749873844727261372435
180656164993163897351119703310393978040928094799734337726502122497
143409823782365196589288627922327799039418205068569837671166237721
514412022766337949173743734228317972794011937453904905797144617183
602567322140552194621862859145438965623402489453798119559649687070
730218607805131930946785284844420212843249930715413564423079386272
526585208492269484399394853053527270682335863948608170577407516973
852120210628959417716079254130699134608143824632866935231265907343
036680953859560845016739322909654282885409786377872218259072743419
564661165959340871344812029957960400057641468485638419208402583268
552123795496248911586260094009876541358587061925113653719481486085
710770837602097237465955311440307339494234844861525237226662531720
908162269400021122759183425529828168997196876101438519190128988074
242805283555533252325715485875614477520119468011155094405429657310
193621595918758219482207475615307833480300854994330873982134027070
003112858879279673962736609207126971151381953771554641063374555849
196316917552555991892240897978331242735453841796027845958986470600
952841808664677110946415984315000959749470220759587493696093489235
131553087952267531922887519326905699269590112428008437808975802237
272111281427715809215786190314382328222158431197363867972722768495
813632814701827652361699870383165480571461752017797801074900520098
622253867134378987984881044503027177386068210671823485666103281598
409185714908477074125773721529623628951428193949344923100247552946
876881422826882633819921070500314822969078127068502360967952456156
317625378443908683885454717662505486886145388589401906509184101588
520883369612987731672752691975623864276095136945838426185218338957
051864132632025229313913483808212879643388136298455390423731285738
559006232871979159091218171033492088287365725960067510345169173034
840664773124728536498697032255058028512103145813971655005402199912
584671247522962330479476204174835733965129338846259860546690206872
743107989200936008629962585264955692634224434907588987120754055723
917788898374740062831103598936397536831431637915535084459949981517
903571916645957972640535635796228522323209995625290729565968625666
476118217436889365526584209735880483823563270384629174104274634032
764044247217919633892333004523529200287573013563038211172897133169
436372206175085815202908497246369055667628921842626517819765195385
246430364256203212959160895853359815407065024525216570882264388969
553200330712386017719942697998742711966035304852835818441460854913
450644431713086666567324744794322805473391376062758189042836565865
989546648856170985902335718936471102219155448014160810863286526572
025037347796586980289697568759695665517157299781741491255194508337
794974466980602686431815234229243167763265101550537467770920415647
646247512496723011428848395397060605607246531422732188958383550139
850224798061638234945366128994040935591780926586823606419849478639
492739255146759621855648434028716399842416564079224320499215193527
094274925509720983764055479950763696237008961585314208285497806440
657569461074012428301167709175408804880626657504874497001006448172
817033718571687699052050431269126758942354642426321926816126071352
559377998468768487666374637370848309130230318759755192524399178262
640279966163670943691125300862843502978866714838773557009540850951
092542672370871628508720499100146666069343525439681324227750520412
084311778362086425913741403701893905849130885307671803377759779815
450406004508428169265395494242417343968257979429633223312132181078
012932197936027503888526310458725788790499330172493716992903363545
290749651464056091275487528815748748504665647881571336432427015771
206508826472570911545293554151064551035094710720178800179246413597
213842994871045977552798427450697742656414883406830080923054646260

389483267223960406246465945742025221084288128266889675278980243468
265578962656266379745368910929346892092484109702357131627533023890
207698673143258427609481838812458575873401409706867189561813112229
470021978448241581544509819889152942428906934660562607956566439433
934279983588235711675751163292556214994709518352233124581091315865
577359175847036941484881503863264098660524416089944214237244673185
768540526164596080359046160459477472320562878910886992342019575526
455549151862881037098556289818492084536281760741755782246663879072
604677554834517443481890496880563765032503535926271374514988624054
829562473693334496233090273911370810584071237265540394440666165466
698127940161174380314425458361131387001800510642225047421267495399
707590971527103141655363347732005496354914667194992566437711135196
471224844213383344829914740191692970978273304820965702365744166216
404138597571994011075505481383567509753677020862148022476905759312
845796120520106065272215959974558956392927416101354477714866027222
280787891903104933048606423488894126536509198046804726679290797077
022905020257671384436845815634829002226658722453892420758678126480
742598431037231448350739349163285687087989353089830355384383950079
707801508993166772121535578504768244806681206008160925249005506560
688200539439752940429777997773909548121823388281599786284393448937
918615556384584794259298949054384503736976473332256852424021601959
044724016399444925891545222094738197285735608141629442012082969234
213675190569210533735923778160066566876364146366579043573782443665
166851049351957765790791933549005424537483625826410516843786445029
591835898396440568593688826368637105027687838531277770566599560300
157235823714715472190567142601433687966468436491439499957272215058
042899218100989507993449442141990214468925539286769175103982467582
463734380523973508302806871873446064187629932031217040704830461291
642631982086285078695172901899700143428968262441780781916264417195
044507576668563242456745817551859136043599958574598825035538466741
328204525370108050902351148404819530588806416040823552351294128140
484544881037085694262600027163923010086037214242484291264957195698
732190524275633949016224645462593474567030056728375084672498252798
367349561770898365165478278894261861051832769208503603621780033915
249337148444550141579116250506890910713827580224465050986098688327
677971183257939187162167678835622419886753868393157565897768652016
394528273886744065178756600689498217355748074443255775906927363784
818051335709627018971520589097081198620052276793499133040582845672
035856647241058894553087522364618384396402596011254852876608866284
830763601287006667228026700036149423026125416550458296169931638437
058296750705322926910916111967493612737291631205863688479052509527
515438062221932730028959924079390744385736690935294925868944010734
221780243707715281612425319088227173271543382374014745436218433325
680295994077130149833237045109699654532027453350702177037070611381
910516358803074747708198082653195105524034346189080408772855887102
619129910922959508822518192058518874419863486251882456654507803295
363481026348433030837241811365562783019391016183517403223845097914
687521623877144239223232455763641079747001255398324712260190530488
666649333228486329053808683517096435404402568667911688444342169402
517726916720054236587952458648729193194196398370591083465955654573
745542747225256387204919648468045612163467555780018391145805072029
104091774619788296504861235552872697552000423820381832076425536409
632089339245449675981523092151894730501978535100152995735305428112
836484236594743595960959569801620275302395941995334462820826497923
607942188680411060241587415085751945806156888083430185412537819545
969741423677874187066721584275223193527707018877628030323374028626
604207305052378520354210425772442559140427008749076435248269386810
716376693073037272374171754245852247735758270295996649854102311510

1387432370479915917879029944855016825586545153881254857425414604291
120123228556148895327177172401226279944108268691399729874382556815
811126238732626104264249191467303139324078996737861432904142084411
467415351674268973370321906902874770602008842219032125165528911717
385674777311366153734391751962707954921617093780054340457873789689
579406584026092226706663974706474619148511446524438074152055212868
620200671723263684717967231549153359492453428928874859317664269936
209673407497745053070568431410133263287775921305762314700874373845
076260305875774978732420714066513617994956945601081928431937367328
417989351895954235197022893472976977104975358649956731850469509873
966280139531524733606745957465367342250965950109876696237341406068
393503481983827182118684406017615765602547011395373572678345269455
960791770944971727234694733436780387223775747916868955520413621538
014282548673776952837038414279340044001305689783556227986071360705
966044685334333224708199605172761522010806671065237950190709749374
182163352993865189170077889883476360923618805269060064082807971343
497894275959288720261078515554112397373046097592317883468807253835
105908002186604490289791189672593675521396472906857973507367571821
697403667069896134578745061097120350147956537516416615312164732544
167892775072459436281923825623362981037565328913282392322272506794
170946913185669962267630084749332949237724820251680550666316199034
578005029621609509712783134975497484970807501469286177972053920877
355735426344654046917507878362978303712221263449953760458706825466
567329277539722860367819247326075875375036396055575515247044790468
927900741744080991621535352379551593168250391708411573389472148337
705878936954153482731607170302320249290945466553120552501263225317
414273729408935823231304140359670910492571838352019535775111030301
893738736672956883475900280466901502804156922062794107897682809626
966101213748811955631196777980468124930647837405316274760628345868
471645943824367753427608269557653234427605339971298708052089898561
442065934722157535135731353150964325686326997601181676887233090478
473878261088300271876502608262925233194476909940416700264006555597
136991814669791135677806574510572964711963826438069896023388113530
721498585368708628587178926879629780160894163936301209416355230277
734299630152634357751250413519834118736205833454731185380745726843
372069052208562550105061009379428561404741845592117933927270859725
115288005694028053614114492402092462087948741836224854453701673479
359020150069908947111950095477169960645156934098095760872306116798
593054474249458559276375426550790985078262274524142805641961957947
016181410188593967029288408817507132694912645147924587138834722095
701254537628711546135844710131132320149549094464014760003023763285
717139536547149001355586963306925811264047920053172809211791287009
678813893732959490687691623091782228643533340593396791602428932748
444663155945748561132045178306464916622418132462957675091859029883
332306551450236294043474054925561176422160938847117341895740719985
035273669869338669851702573938066023027910628085352549353166194585
293885401347619818297901927026997553976270972133207752142888313638
279403779548104363968462169524948229844322968969208533555308531740
953971002744873252835275736247945801278044550361060645585780357362
625255636064773490568638324600588264572996728670647068819718804899
591820953876986724126105812313371883281538730532406351716048837318
634831944878552453402131059605432697873627899027362358152686677286
484137632175406689989734882611860180029360022362615884959038938183
834781502164731089138369537380868316436990879808593012837352876220
600536227587287679465791680576358143240925305502388654829492572512
760977104308414241327149223014555024915380116515701072599196608891
033445877802018420198687255798348589279411579165489841807965598165
292440028600089283308995984612515413473641247553705658072496073372

```
8968639565510344975858300171880139293408159346577407491687314019 90
3828427712262333244605887567398385935007695131185563168457383865 55
1229294080306842203625672459181138606350480155226167063564964286 73
2345965669379924358729329116688498393642069797039019159319455970 36
1259262706370837171360797222924483897365994926322185943095293445 51
7054009459274870328435199388140267085915289496359507636380732347 05
3462309324415095756918504808919571739191653100024157142935668690 97
0675538485026108040743406457426342832522110207103450374538340721 71
9272860930797090878640274037560341962032609518023331944660470439 34
8005406358691029418314381980766263692292015196267454788905487300 85
3342208815974032892535678247804572344855566388429936517859381542 87
1473470540776250407980710868325712720965952470280931298490597903 06
1967508599442179885069831610963804318575734932089702792144339391
3428290098389029276009981034971675340055350266575485135820698171 89
4317365218737272703866524342059269683995858771658075362930491745 82
1027533012670236222733052137092747575492754032248665363239284288 78
8071811943447754439431574633737742190514462638401483845223060132 63
6502788451471704790583180583489408569494249944151554838633423772 04
0699601933580313753449760284449951541090113815606641323294310535 40
3566334982500900534136214959747529802823984619728367006210584613 97
7815827467657982601784729657646395894187742496331695884228391191 59
0565640228193496801758163840139429208142088204546902994637652059 98
8197831754480127119965562213173244271608021931664460718450670245 16
0461201179763827239211348339438798962905840179686360943255300650 62
8889732392513616327023907523959826534893994680658059487686462751 41
0940649931153419873217299143125910097771886869455712444493582861 38
1259764637551134279845737620234356256898312253042059021490709996 74
1603215467075388165029658639915531513429055133316532483088503470 19
5490556744841010321876589956758394558238268883108141862835473194 75
5252747115754025434804661774803878685983787156949134278530829472 88
7354120431902320905952954328610661326976076266926611352114662527 69
8412773408524191382685828095507583757875182944195381669964759338 05
8109744197040887688325257374385618259110897501964314793725720708 09
4058960553970982858004455630785998611082978451980399882309416250 98
6802635162822807560816507048348964416836183659463297691973326645 04
4332702652964407332608359487129720563680362999226922055550219361 31
3094392925616898258938095311543481328951849165428725462863519781 03
0237300349037991793076886122045326513181013816898791956784667866 14
4331058014381259912791415887667170290675999071222928172745278544 31
9176377518648855446905521418294754607553734560608556346642039617 6670
9575287454949012046603519646368735772929742823475054967865445927 36
0627608989246978241890713669103009266781913030559195116956931783 19
7540796240338421104664465240458186863932609646353033471129213154 36
9571442206723727019032161283163660683535359140279885260953147441 97
6705764010907530604721386570676654997265613995625901850853030455 5
9284076141246522180996543530716318507488647893313715980401910580 10
2425541713566189612060110697612033871217695362774814702404628795 94
4796568929166656151629117736946184946177683166359452851171641400 87
9610965586719421163816545789355943747416596019104026506996537608 91
0884905409088076624223642445253132852178568721171075072872580243 702
7495835664624563513977205964778734767213709697787437222222854441 50
5162581475900160010349873421642873794152091728087438528706874529 96
7585063462361565683806568468586659128839299398749192804509759935 76
2191530345339640241628163756457337985969012829274187962762503806 40
3057998223939350958921995278510291646380478329362091927804077150 41
8730068917857381782537931265322695642984805750570433859369934339 34
5604496232593724330435766714771166426205667161937737577018204936 1
```

498531194933972205256017243806783980806301980330387140108237370202
178023999196594847420810041624631399989387267812969837139653343697
806394667642860024152824873785639412927935383607707608300750085468
836684968344738138000194809994639279397254809261755712323072287991
472949962782931812117999531191393368291421708003917108392039962532
465712426711480762187323888663027326326051022748558768582482996273
785630375020526294883169608949012916313726348858049819248754553524
887326239439367504501654768934020682145856556617105077511943738058
941342569604131394758191970668230263242340902445054095883410876889
880583600190580085619949103323884501331396041945468828359061480627
917027425505698363038681904080769860746508844236776170546226084543
972221940355420265203310445526900704688277458214569663667446999428
841734711480702347951794307783615275740011675423810497822651699679
327017909913253596925641310212031767596026367955108692598919360515
293161163999379060522162182625734473633420263075057264252255255006
962503721338053244844653771497170577712173855502140364109117838972
141797635473176486099732370208765716623286273640666665252244484423
704743126027019405293253920388124561167839226071478001195718475045
561161525038448220826918667452950019454795547426703119533884633675
047534119243051728905583063960642730093178990339713439315840591610
085863938221382022717081924757782100150391638486660170908139091359
533406190146269042409568052624010705647776618407365199659831201591
486191247910453282084900376962357904520492814748144846581726874291
601112567800981177269122209070237851486611243714459198466857630947
375120942433520044650532973912516708301825542971302260674660980052
603919627557983895090692431900473756401836074549348591017944755771
628631505548876102867291818675864766444786596278272940399932099053
354969138475743042203580266820050183524856515170342099431107260374
750821643495885414321045735574198011829400651638459077831309473009
929939541827181805997731995522537656235226168798804828472031495890
625696894424278407617197478521211170867529446036705335557033361952
699406435233081909574370804656078412350061934156495100497331738362
040042273443789157896534958611591918135897244495607070322322782801
579148558088663267009240423420313927164689630137056221855039903462
294679176978329701524127573584580130897595258639855021546367705090
070332907975583265256973309925199423352342674324526287834348780393
209901478697413172811254964459042777973691266877075337938105295979
219560094519647245214566447811294088842597399502289315673204890035
956531488181813352488448698694800612536474277550005042040462103442
555588086621442523793246458613086709152614396887816336773496951240
900773916726414094124216456185364620858380521948089887377463428514
000439792350670242467066706973079235532297655668457047629025632259
783196183339724915469525525143514347973072505898539350344143010372
769330828701075552612213237948915324248501547845700977456836739570
646245323167834271605995132533503844764618045298891008925761423645
689209371216233577791901016985274650743313394038605583835605125299
114647525104974008392381880409524665697821906750077451273404137469
485989032430384217737576080925883639849428524916612397421340306639
968399457315314592189561854297369416392912745866021491932560629083
547889453870589099102877684263449092427581953822771544550755082820
784883600938857504071338064314464351066357725420010721512821487380
127832552194772666696435196399513880976645324332723557684264153138
097822980463213153774625235404290603580170561927446884847759849178
086745549832965851953461620010812542242676672446591985003737246700
094531418328513802244338672642501595675921141474962451849120472064
675609405933598879179790590013674885386450686962065614990834968225
785681134556483957272889037685647348587027874709244378417015400337
456982748232469221776707380599850755861668713787186830968096765583

702166111407722078522106402682698887307289605168437835203674022500
330129422087997280733552033208498251879476366174290552837591285641
649741372819365114332541321596654479750889663784405834138448245573
385931223675087756917458077706663761431383703360433458674650157199
994935709263149831353947601099746000005376581449457332674586106330
493215029738393935442737099386415060481731338107605825309439419878
575323657233230479592495750580234655485760174078300407648946768258
640951038804374399269553415069260955209966849636219609754199025668
927300183039884115526354875841019186258565195894991754367013775262
962123556260890759814472451063453168437420392696341251225494255324
466039223854149218025474882876575364066956656854499049481491528050
382535285564672004425228943146357459705605413211134776183459143500
680409751657893967117854851652292067783571075255139531452823122246
077432648578021471069671258249054953252002980118743298038560982084
749878105162425738536217494683456079440138680427259737217799597613
484926836925904945447285453485554767285509262696633351543953191992
620902229011791650354053205354155732396286587788501001210509448364
369355456565298702510427337311270368643270185393046229863901849849
779085270466381007410606419934676834879214901823792623153577676004
525398309488349953129731567381103576713809506026898810326066044317
643550299533074351217835363742263073089411890983341196386055152202
496844393942530531427282384519772446016604039958354727913725799487
690563099638920470629376050565218205421345799773554893538368033652
820760812514894362469001742501483694891032497863941911460054023062
161855699086102467315486460988020278107347373777049188343981529606
960617171547709155804539147213794488630702751062591007953739995139
486174490302297892181523395588210157649640209459194090016061165874
111119807343351033819020401202992932403144329877164721941875892567
634592321990922901557239337288860713694061776164509345671522804998
274750927830263599319816399168556260449567993519483203844703965009
899853801126586133337947755213032889161978793373276844741328316215
382135050223298247951985584253064406273275670404854215795371733438
301365792271697658152542272531902691721583563479065501433932221184
545774337318654527185594102106229488252043473087838122229831687235
778373373445155994823292339269578929447998242009493926642707394323
274713200971603475707441072850307963039075727028380502095916175507
000501662779242931812450523867239968585193177959038840467556513325
575821494315351795949879017593995401869958616206697153893332575600
180920779578501271585250132437026798381532168315105880273745589398
225165079655680674055293358041645869285231652892506244103450725475
174669695766475991206556847983177883862444164695419191341166383135
950942698407897055494658072945983183865462177521063114551043363353
661775730503740634292089127750223620941830938203794204943018325648
815146217473149531227249665194033266694654817482534172522912474970
511461616005428188054035419894723457255907901419851829814814599027
140514326607102622963491948125934214533789872458309760486188866958
513039072917339207722568960983652091279311482179747506446559761340
387575922402239603473570849183378112149892582953233544437853183336
101743721712707556161914380532726602449486684750343807525183922459
239571783023471419022350350380794112727748728950400230804740722455
600124762073159921250457886191338649903391268514724330910608559436
605624848676475330103363659857748963154641346528176374126158110078
173670124954479654116012249009394603499330999402392749438135040080
294489916879393667610867550354692386570894147039461647784558930701
189503601696843007815886653291356205568916972515781275439554162151
952105300131336221871938145719744354678463679883187413333055129221
001574730478222747354362338060820098336645368558689256331574692024
315168312340672977314170507985368300211465536273578120633869732468

790648595875816722867648874376546504490483805651802297321117954056
543794461504202156340664560806217490834492965788882953793035047260
862168232015498493360658850369596066660761363412546677142657669669
382533335382968667780185547637587413756149740851683629386444454372
825282127596573341880697804038637562432135935333827691436403187220
519444882699899867804379050317265195333242884761680065162980299075
023247883443424065678288128860769637406496025630715656767505203705
308391361662839216418450982844674305122844423983346413015302739217
504201192661457275082813903767336357639254460529716017653757738989
469139770671718421230296681287204971364532552899308640123305232260
540816113878892531948786172327631527480204769970108414222399792911
010289569196329428033277083535253890351431858828631937429265422212
927628526004225453979981005955366783999892392918091503877736140092
815308194078606647484238225957635285178492414744484368434252066821
565966919769728805736670362156355124994436122668933058039480454627
750978873567271612375825713839139410872431955195160533765155558925
355182697971475606689317353105319152122983591730230267583927416493
142243938974431008744911962224480737158524947552758284137168733885
684513971935717435137665104895239029017063092712264684066927834839
988628595942396793645641735587184019901875725463460676207062627896
232203480537183635179469108776385199110783793669022647896142652819
899498944095836469265144316565821247179207892033514059366782849401
779896529798150547543580317758562255869061012302340109361295535357
635843299746307928840816702336678897012814595884742819942874986614
377118597010460786118312228567494691319004482564302826200724878752
539439790125220351799067010886444417333294270385047776295948438149
909989126253488800224701126853866059437662227036508267221232223515
725503765038540952531017575868733835311996936024166003965092807274
380713754704848445688684891968721231098199098730747953205414010395
973619696752305244216405619005267427599397913726868627853543055179
125750471576601849237182239348493919692953944998838019657266250365
175749403311969597941172125762371383165114796260557917844027818812
255383408963982704978907257430443033411791081090599505417122077373
797477503036812582030995844119486799985794011171393324239526270361
992706377241703397811133323827156827734201479054726165434019322671
442180196105337050263337427319045518794871348524986266822221111131
891445576221042289834739034998125977811080901318642567088910367429
803049136365142132498203398923826233114577540076316382512686668485
031584111114115588642679072134604092217050745982472070245524314035
201195653139245833100914253634958789790743937136597095527255566660
070242028391007459462466313074544675113059937775120411928055649729
151234915055327810229864306084053764409891744431076998717276003815
163442860652183069210091799279417093193629424774580681733533597017
598932860314853201568769791565212418967004030959172775708160301869
459714079982443633328754319221407359969526268303856595845208956554
977810132745394340864386526954140660420525015130837808664572997492
569037617093002816162250318700303273714486051251640723900700882382
390681147399580435559035124122182329827436677789038538586239181478
158835325813789319130516472390159051500732925827462042891439266753
495215494292409278561294142586937295146893142705444227000932416109
334447817626188849164416925608135907757944738620601592404012063374
998954252911634095244853142318247458686792135102696628751705210646
078470497466615451881415103516734981578308880506290252472784913288
358585719687703163300975539049048845664489746888824842250422741076
906915478241586198518421957909459139269455393497074170826012991361
372933199089961244761127027704388927170172348861763196368502467208
266987608481975265151178468397433083172604878540303329427864436091
148976287974130203367539268931859458016183291794401009583249805875

1219528492148963929423810955982270927599730329947375056874498728129
3470260662447761583466617049162697571797587242929114187975074878217
1533419974526805573225600314170463422031897578207730237386246978504
1650979758445271645852204355139759287529508954652280662696943449901
4880200418118642039774220404270266955446032990925495943520252796488
7345800543458468494529753539158379291130570376177366337579523977108
7393379547332118548790619268542240083953610368778991522104521065020
0380051808347709316151054412972685089966422824644897764232319476756
0243809769463100168877605725679893692800865024874467608245459575013
2838100012297473056539991376611276067858345129580303840502530416631
1739822211379220747439396630034070496076432882198733987733380285979
3609821535465591007243170971570609710596868869066479067951508101151
9705136357516361120759637386375738584999837864530579030443943012950
4109743783727473215821070222676757039661418608774439690962477761429
8268125725518207382106291429179289825779700233074929888582353993193
5639425268061948642067083324510468170670405542134187651641921977618
8680295892187243673912979296702170026047079540759988069653829470682
6294750079920917805450108721318367103021403412399398867410140472913
1764442090801109198035244321133365358285271828432623672503700241162
5948225597448808317606737015485628911866544655045630521377904505732
8185120197966540943027004974463254612122414112832936643794032198447
9276656109271707635594012205535902446730707378406810891604840226316
1316537882261306234764932281552291952234092518396017149557062533930
1919067459718990077654358539453730570858767339377522556188759086266
5572607148136026843048094633781089487053325334693152865228184850150
3993800336638789733893411288434577352332199976257543876194782906084
1049231708697927266856501771784535760144406871716869009528068034318
9335630427097277870650886333372970310015932022324797004101814946764
6133696884479094536114901744629897493230037580253191769550524162500
6552842611437307622654208268221345946753703366218421816566443477572
0963001368851514596794720360121393994632611454144468275264866158675
6681602323921704745125701348659061643086005885568792084783360624630
4195846417470830336606533005342062042836863188883242668160351752140
0746402690075876108947946351508449595617000050182770896682426327475
5293913446482687560096276222425072962721883787469558539680869873126
8923748126981350128702594852987093853722712995055630159371662858218
8659160520740380572603304513197216792914771867563052935572768823902
9226621973058048731140117530813389217011861801173372514366353087565
7342089417048072193359887479736426861987941028542129529410436548061
6664665609535068126798608377234268522061587747745045440859724123572
8136293950248772132291814474603528240905004010177366269864170218167
0351818974670512042796054366627945217494148564864034233759590549060
1360969309168410629163679468923218912091274019517061803872122843070
8796077311372544130605417298995054837882772704666386419113779886974
0632776799960810327556562870906770161485811851671525572067310925943
2660248555938871841243042256167746151083897288342422589148505084729
0517606189977758300706565252084740820884217337391076739811488077333
2201658891141005158542238840636728656725089712885038452940316291883
7144537878661305400091705011115471854363558310332072119813348586311
5342360192029371830435261690844019550048145065893768715238471241858
2070564413530487420515164512086660809688655364730701451711555839964
5835678003490949039327445144117799163796310338254450612672966298207
6892834738348275637617138396079392508938287519388908374248283953342
2566401300588877665791783597740406707501771087193911457846852542500
1211040115678138129566257255109176460365967780579961308628200115012
5167924849476048498842037333939543468668590235457523095407381530411

950082322121220319485821924434755293375301739351818146692053006768
354983260897626736030117845855487527030732200353241223910296706648
167195592545322134782497402500270274805998168376214118183387608348
792580981381516616041464208075202053745495801305135529753878317955
606609534527502899952843052599586314977990312599592685238675997576
441360225760470651198719323526264910813019359159967624775420054683
329136080933231845309103266426957027363688615868419863558986966216
126369423462698706529516460365908897763094369532892149718025971263
107919863423433683379854278159243753610595231367670588425147268669
259622556233388254449153389514807803531600962362723626293210938381
213442925961689776071296109653285812638563528406918707269095087199
084875880597680615434384983307862237329950538595446528725800917384
821639524761866919428440920203252858635973427365210841779240653376
994870919040065362840026579910278085881694281124198678032126766119
082120687694390256898318550295073581362833258813293497875611996570
647703233546013593301573718699852759527884015523130446664670070445
701673747302944778425837971339579810234192743011416633563104720020
623034667200434736336209186074063793874108377983726591022622662816
683681746145088810586794020916962370702672270785567024669662152359
232489065565411423216530012306683158137095011751649747417707710478
671144231270308259649702806523097265225953502609376032046418104012
825702971529099663017972874967160781873864346043656031260091308019
905591344969824305981044812142232391988323305174876177610603802422
486933607826983487990531871201936565718811379882854700036618033764
616418080056211065665713544573803557216270206698706659611630266928
133512859723422734740435504503018436615705975860259189729717629378
207585152143663005844137543530152873638639375592024940199122961478
312053390204021524957162375177139207404812163206956168783744067514
276611819357040926225442812572446747935669902240001616279356999773
793622293288995109667188147254724474423324508328161135885062617781
347522637741668930679618906981738542620711683452020866172255402151
315203014261536352924176248872401938432847031453356855321163460324
119109690049806616365370483004401011812918656108974698069557691913
585155593833791589067981878573696873349166531702934832744826234967
893671344072677226840840390785044733709169016194834174928447685766
055823894997626657072609591728102611237088220424048967417981761059
712165243418987697325405183576390689675246430494598140399601983368
186282170586073372014699356972850002318515413576999413002897984546
387206092599165675004257455772138558216066238188184326085590864884
230577290245837548332719465922189060822557719303102442845088023804
118245874545954059911879389866524346777606716241118610010104090734
913030671369690732159434848129745453214650616117015870792378267767
522436635663919196042731264289021440187347592854707425670344996917
675233598478138865565705899983318601234615036475938171158603856970
478924993590444127976041889820913034833021495306782619690302406082
509918409496241117147501936682547196684447339851530387850051106979
020064973535455938575707883336768892461107934627144419802729030596
670946542694668000365572482505385372657003464552984375485605766436
548468959198703255509085909374209804862413099267432372863561176180
971636873658852587542928998684070667409131052333116913901759061177
908405550410409730126750877167686004326404717317308748944579536828
065166828841880976387687751677225401507003933693798837135823136755
015852875240337553986870947839756157946308525991462120723856095229
220102419422643655009643728162123659295649212034193108558048057925
207056097073311003610872573365512443639741726877515322065422563933
914249879192202992430401513532618304251639821759987994032717706306
529601965946603316609193225213978517420275604532822558009110705570
160851977806571014316301211880912558064986030948049146676107720745

502634506956158153283070441964462644401978953041673808332923846545
153725133316840331525638965894711008798811829532364680046133945376
701491217704282194282850506622188463050880409783570115326654916255
249526385096275796749475771603492347359628017615562674390623303470
044538119409528697550938580679702661145684844846308991494315152488
178024416974099890603908439441850253047357246937305616185379340688
294601425402211423373139724085744945862377619322551855689110364637
684070516003606140643570811847797770405995641200104601413000890788
089377595295745047655039053183599146854554075702541944535881728230
217184087340156065947706698217296638659132127716519251666291242160
026082404556418182824207155593041412847921238148595144839965671505
764602713610407345488817072271800832968140120396622375240917625930
501196457437538992864618902187410750169415517307487965557943412213
401861945711114145143476791105087385843795432281462392510321525178
061022002985061171511522924684406233670927089226453241078178623493
002036486152993070136846970506636958762130387191849673626286128773
135307721316622088806901178452231994936532272443177479044080521012
426828275776057685207211032353363551733222846585385579072592997658
497868869011934529274642275255585048782417415962774392726950863003
739197232432809642332098580674697455115723670387955932445758453122
600239564518634241370010986150265195096501276333828196736597643559
400008983705072192268375625342457964215208141538140352834232745745
282188653998454027744122254522942265414501024628838852570646391990
412738586374723771970181661501559306552580402449092982449535013277
809532564263426263432798995168669229691978769462307638213428791853
401663826586138981055665067401063420828539478180020453642269679016
347991779681426970196243141837017933239208206963911456865343575178
937024664269595259600654326091604270903277334124879376220897854569
431847419434510620397466555147205617061019461222686681596904838394
290430992831677354144429372219257639217779234223773397148488181910
169982018278621131812845363263984386215655096776531988541783185586
741582390043053451170737372867280188233545785306959967788067417964
300979384132540544812380831903058654085153227854231354283742353758
721688236322055070652706495251356963675212104632064184322393563779
520998965454720016968351074307701939884197870709476949874864200855
470745726706021712740956692665431438337769024013142789997756778724
179571357223060632305238563514763132557264467597733776986284175094
033938121692142673563864623543402066943063507513940274428897658237
003075649301065734518698212694757885041191961361046093431644074153
374395772435885256502313780947543195773054518785207209863861162273
043345329587547485512733832797221918503504787189788424928710689618
210899417158677612083801516188865145400128585971646301364496516055
149382279283444908376743281989211429104310611108686921655279207982
016493293745813421507655118784892767382548793511783235161120855178
410837621175281500785185477818607679428641484323323319109726685003
293412891404585820443155761077878576257742383848993157237395983811
060781246778635855689965273688452409425287045964359207605793466843
241453255963562487488211791208183970347846417549291422486198816983
583681912923190240717129570007687463345058546053618974136529114874
878826670127663175041210072031578108892437664036773091175530904092
817119613621053923413873342259376787685262571351224341348244924378
633188527682753743104903544552421435713693138564029040006569936253
681779918785587184473077058252974350574866412708465442603847274453
318352984893683898897808289018623074484047084490852849403039434295
546385274085475901526881600037477252281178116115742371398195074067
417534318414635054434243835745192486115808800396066834394019089873
781924404002198322845207651547921136925507478741658366024221854621
946430775152186135581519887540463551761409590543738570052501358093

904859672162021606984416156907877168406812741437661409111813963108
559915166148107954451532495992136682549963271237356256841474541431
231206048819565493502823579799437677757183566590069020417993369091
728241049062313271244249416014285160962807790636038335841419927091
542276842581774992219930572580325951237712012994424840701227968679
444734180679258265793495891476788837891544093263165670060889473109
325610561988030860548652087745645452474853531540501116812428474957
904372159415539436702907125778653689754228949640116185024437131408
434630354358064772127114350431362548438660924361550031965108550050
907158337041502556889102458400418193352025212449845716767925568221
802326639623157096458907968699494137343262118149547389785624882172
010558278984533333136548489450888785675190404459592653195215811924
489811352902213455038356598271936949317656578186760047007731691816
455034194766774380181590333099302566595666806010250926530115150160
622468616389719309021432141704002191457726858407865209722168098063
403409743086907298822309751951983100298612670489081778827687757961
178330223290184600190716087476842315224050774360779330699714208266
188407940469258063274247275041565742546714723309962603957286538590
552888005911222577468976219282389040655521371974945223212768338451
551457684661127766882078834758858600648865735763165275569994119108
346614456186706812749463392864273661511558912240868597078927007503
394219875836535973430041069559226761035533059931059176312793511629
714079606501253293619401366506307941570050211188043581494533791320
858732548430463659635754453470236895764075477044155714931768679302
277753119882184837301911786289306940111589359951274829912053068129
551929824938272147795541012944377345175642511716526216166999185835
646874649357924097724551332802362936675709907697852514226641191193
585434501374096079376005086552834228065726478022042061307951699639
533247616622891546684694096914515252563415426976727096542892366968
403192535094906738450902029724318091231270182551951383460029378821
762077676614701791056987095327706600308416181059206587486560051483
952184824992543254856132508585197436417072553803196406928071563221
322173345451876615575526390318339396568420029460701119129003740322
343976701062591895494110816617775621921623121137090313971995109300
060103007153333452901597889358798595884784180052537960087983471109
265787548954190753861066220189959789540593437873947063842367677502
183369288728660527834544522024058057113629600759746651977711112630
969469503423846510531433667095110760686290587829078822087133464200
364390366332980988500851337814963166188577108066312558044324207698
647512216582361620834681484140831525275396050627066696526301902593
848744034126606357473771924725215165229394066955245342248129991832
656594423759120559882004728642042067074222872759074097139821572379
632145499167296730802864864484068322190336849026989929710200920411
865787025178591835750552703347688775638585462739607509976147400572
192898182966726201203114582608158553826922251013825611025429443067
962436400899781005440067802134276707642549925936710102284674866225
941175294051667555818626941750593997115376753996650983301615736970
922700557309695955649272385182575787553417888752639788555964624474
923264074821628542338036335949374899526806016754142197883509023731
057687503281918259052125331849318306100702202678580527851563024324
195553938565711330622522462341479711447932578993736265220222845799
230346011771041574409412468197295002911413986761550352599819480735
239154528581118222981815649447927563553322350629049623760218885772
940818935145056394333893533977655456340866555282658136000816463334
525449484505955566705163107708872505206626022560573629214451674721
138860169162930054205124206415555272401945587359509752625533729093
899907528808524234255268789623261274535575780817153091024596353553
948147627719691641984261118275886258905804366154875620681637434080

047600164419843802937341468286777176160414285412606425641580937439
615912924271363136731776124458993385917773061133298846655245748283
492523119458706998912381060083820227026598562576333919033672940001
264290699668384238179436127469866593189619045322223293603166329601
494368718280932743049152389322562551911147300753148128807206426842
226981136185520154479808760107287609450269249454061487952597780986
360069669101377814123490020686919838926415376722551768749515204014
883031204113020227806459645894805871377996768873493918703798214391
672064590869651897225691698509990310203096657363301877215791078781
426445725617411584185877699635635291576611517161175561912146377213
801136522636271278503521453330658424042719346570805618056428626998
831932727372468508080637253458677431435647134329913610845871145670
176806024156398745240385833637939835633934944595030714427694213921
659335151610028068260018621710802758990916345462460226985867174542
310977297306019504557502121786672926863366512580710505001747213369
207269807192779063301291903342918220130117130206861246373615361763
948055611226790359599834582073680303215605083664016691546402608726
050665198490757496212963311920460864247010599662752040679908521111
887393952622217171839456743584366250259407567787455206883177021018
226392892933319946189556214339398753777418233490776385599354008719
353215578106343492646921668019730695877934717222544807911811119639
263927648001235177257192747883057839711366906064552543319190322289
193609549718432491009045406627292385027405633838548590188268149435
843645843980262611116171765842816097957652506787611795116323992635
170032619534150929750055340488664096583519165979491935086348821926
159814484369243754556511220419480552682307533247697408847277874354
523592721088040526258353686891993198446098971608446742636735916800
552847863406269127172173171756061677171463475561619807884390311358
477716426051047474576636143832085499367219745739979786652277503539
806189140888838590932137437527203362302578779561047292856386085130
915778464960008736333920310489778169199045483721157693261472210169
373395677086569137611108691532783540556894860507108229542480918055
808095852840667352878148038653802146646757143896547580860434512955
393551309586932110862993111060583993942496576010657495240264494636
552444240730599036528490896664804046794551760568902763171719187687
277257489033656717785638232165305692129115050326412815732707501138
355197893094089107488034261090882741413711940912309437167869613636
072477657104623486150406068547046457718789166038214014347509730536
910311084407969550456237753811982755215952136501877563397073543958
012047196601951288251054503317305162163410905181922605255463122643
553225929574757288200162627080823604244459035813619904596044916754
037553727206181988999514777161493276079799935405323179303743527584
995426817187213747300025933561536192111129261638916218469566956203
356497059650933237168855187842033304180750306650556062517416050526
233166409192522382588709552189028812957505217116559791713082534046
084330797746547688166691968144768973284839179217767764271032151745
244795073588076328905416731603181192610241700381775659611825654152
518096798760261634301727837032796173292508134704765485765605901072
767352138545982768929873329758329944685365659919272027231284196896
663259359477266722350011371950264673084492628609598526205224095218
223592003147069826977926672250423655929192492054350354442390940880
576201050465309269773134941085727997638011304927973986558419898876
583320159334396104687507963520178472987317304408427266965840609618
054654656319302505014959888401925085596318809233280147303879129195
795825101292043776533474108918075807247127240276166629686262222231
660704487522921471507146159607733512382721669154552729130788761367
040033447710520770059428997271177365919242991321208097064896312558
843911944264248345550207274615226720992564456528346798994906603417

413684761557731344073469800379804214122671320372465321073217357376
049193206275564676654903913028689978015277912027247510532924595527
398642066245295718008680915733555397019651293100483231470413504942
935965118265724049820443139756703147053709850613146155991545967908
038206332711270539764389461306833524669156764448058479053185626495
783935454683629709750886407255782366929906508127064320767425390436
885713810940745855659674181134810267297801287659770581626728457561
593281266064575328698356754169437335171869154494298039532809562532
967176474278419217105496385341428321986214861895267914830400454230
243724462494276958818851304787948051509240221847272687432602896204
859856831807437521486290992133913992180695380744363471106230210202
399080166428312197903931088989028681277449139877816012369634935383
790483373886164973988642408564038860012173560371256630282419328441
393715260355025536500684691379951253301570881693176198059576609611
328145362486381437470813760997339279340713871021835604437946559576
321085972638056113173862779961866282810658000636060566965160500275
463200064283833990047068610621589780135918708023883757689557911171
327021871913861244609228509466217880091235664671425284581316883233
438617026345453106357326861714173268252109927195832490783232192898
048512298230337937859269722693195895033354126067763684519202799011
294351329008525896049061348176846124481858634526732494412395033702
428955285667455763776548313039154490724174469903334896301032696085
126405368878222121462152194238250907818894026435367515689104488568
329212836025815748009984583205487653084082425613508935972296031858
838858038358185664412153867087676012483101046308447443218801447909
336747468844785641481744590912453981032300885920637101563587565164
309596112739640158617697813140879507371428831776043668981962641347
500697194056519504550516774239794019689986252789008523744580732886
706974177396357254506098542345678985204250732286070942026394841828
669256606166544407204577568393932312265407812478316668801803018422
504520053551868684850558453085253849754261205794305353308074775082
649608852944557278500344139558937933505118403022529870616291505964
225999600585648957233653179869136696994427866578912525684626041479
811086568557173907213307893310852590333311371858772870349262790271
567329168662716674908199318258316683282850015757078016119316922193
151475493775515980465409283999109493742010371708560860588178544900
570410413604043513764246899815268060925540112346532950434918053747
735616666710469298309567892031648206393173922124840551203478063211
131681337322406321645541558823784609194273808850283831236226654974
430055809989829957425843235376429863146563505528356047709075747328
263677043963097923465629794934456641396085146437130393213677421247
090445272154068792154263064259726029201899465529811514261260490076
386714173023572727683904155972345066669386646058829201247114417883
178223482153389187605836327618181943322769555311258190484751746290
562001346899644071619823205434718461102051131555093022682510749199
014960817856260510885903658474515037638491513400032951639910621924
055728300810352176197961683221398169240835763955621165712611219290
508716325525855864961663825419359148218187619592329205699550637645
818268557522115278701180299433546741536276207749785405413330313633
542368241010846476374906388527984149007646469764854009479635895497
546144813763697059163569983681198752505479306935320757076678014844
770142471624190816682249007420711186488154772891718653596776539579
933503342728214605416964960098470697958559264304287036366471307131
478233061157641991322242064609989883076268583605552740990478467610
760424178421506285175573529996478625529542836742987066457943375801
014074021161861448432976574426342852870477855630830963143527878304
194501970294657577773281674685808745393160393725331589928057943463
140873586086177882633492774615118491165513068184671367734882334108

513640394793920887688633633946138235834479408156961091429387734713
8934237736191096460564244474779082076049660271356168954106444483213
659808293890972961891211834291490616389638610693752089534688398334
446718982124347807238740745769755450743684674713502485881839966556
819634452881194183317263682505061186490039412552057457120360355780
251419043526718372192138482990580322469584243231589844325103965443
535053543229216747040778614684859762557446153511880031430569954927
847167454497269761283933251838197222328360707522781292813010656941
262948730634268837338181742170608647548276394242391402753218042951
903411635170469807423351556057857562450999253201787499636640473477
038985587306507603870997731843128109897898820854355955094325390237
189521682023344245572575307879263398550901645594237339662522335164
875058955694217297244895998825089232112034795894154654603037878617
591571661398869326873749684730549653293782147564810579380828530053
244708050656929422340010959348294614539078890661626402150130735330
033192074563726377077099939992288621224324880206263485088853036010
723436890136064275814252839878594917997961121963797576519245218670
960880921371119775000878159304307293448839309575741592413752859777
972918934538505080383198677459002518657917237080857416429715380788
406071306868036198241971577476389507253468404569192759531937223702
229015580065607604738547359904477996748749969769427137668695533195
125337764098587096683863263926164945608684140374568420719405950701
743035469182150900466493998551741389385197573121568261622862231881
096729747606013028331193716114087472706762558567775119956667486151
964912970193318084994109618139296492789360902125354433273750642606
242994120327362558244174983450947309453436615907284163193683075719
798068231535737155571816122156787936425013887117023275555779302266
785803199930810830576307652332050740013939095807901637717629259283
764874790177274125678190555562180504876746991140839977919376542320
623374717324703369763357925891515260315614033321272849194418437150
696552087542450598956787961303311646283996346460422090106105779458
151